Principles of Electronic Instrumentation

A de Sa, M Sc, Ph D, M Inst P

School of Physics
University of Newcastle upon Tyne

A HALSTED PRESS BOOK

JOHN WILEY & SONS
New York

First published 1981
by Edward Arnold (Publishers) Ltd, London
Published in the U.S.A. by Halsted Press, Division of
John Wiley & Sons, Inc., New York

Library of Congress Cataloging in Publication Data

De Sa, A
Principles of electronic instrumentation.

"A Halsted Press book."
Includes bibliographies and index.
1. Electronic instruments. I. Title.
TK7878.4.D38 1980 621.3815′4 80-28240
ISBN 0-470-27135-3

Printed in Great Britain

Preface

This book is essentially a collection of topics in electronics and physics which are the basis of the modern electronic instrumentation used in physical, chemical, biological and medical research. The stress is therefore on the physical principles and their practical implementation rather than on the mathematical and technical details which can generally be found in the references at the end of each chapter. However, essential mathematical 'tools' such as the use of Laplace and Fourier transforms are included for ready reference at the beginning of the relevant chapters.

In the fast-moving electronic instrumentation industry, most instruments have a market lifetime of only a few years. Descriptions of specific circuits have therefore been deliberately avoided and the 'electronic building blocks' approach adopted so that any instrument can be analysed on functional basis. The book should therefore be useful to all experimentalists using commercially available electronic instruments or innovating their own instruments for specific applications.

Newcastle upon Tyne
1979

A. de Sa

Contents

1

Modern semiconductor devices

1.1 Introduction

The purpose of this chapter is to outline the physical mechanisms and characteristics of a wide variety of semiconductor devices, currently used as elements of discrete and integrated circuitry in modern electronic instrumentation. Basic electronic processes in semiconductors are discussed at a mathematical level that is adequate for a quantitative assessment of device performance. Rigorous mathematical and constructional details are available in the texts on solid-state and device physics listed at the end of the chapter.

It is hoped that the physical basis established in this way will be broad enough to cope with most future innovations that might eventually supersede the present range of devices.

1.2 Physical basis of semiconductor device operation

The operation of all semiconductor devices is dictated by the behaviour of carriers in semiconductors under thermal equilibrium conditions and under the influence of external fields. The **chemical bond model** of semiconductors is adequate for presenting a qualitative picture of their physical properties, while the **energy band model** leads to a quantitative estimate of electrical, mathematical and physical concepts. These concepts are now listed in a simplified form.

Elementary energy band concept of semiconductors and insulators

When isolated atoms are brought together to form a solid crystal, there is an interaction, which implies that discrete energy levels will broaden into energy bands. Both semiconductors and insulators may be represented by an elementary band diagram. In Fig. 1.1, all the quantum energy states in the lower band, i.e. the valence band, are completely filled and those in the next higher band, i.e. the conduction band, are completely empty. Electrical conduction can be achieved by exciting electrons across the energy gap into the conduction band. The empty energy levels (**holes**) left in the valence band can also contribute towards conduction, because electrons can be accelerated and moved into them. However, a 'hole' should not be considered merely as an 'absent electron', but as a wavepacket which can have momentum and energy, which can collide with and which can exchange energy with other holes and electrons.

The energy required to excite electrons from the valence to the conduction band is given by the value of the energy gap,

$$E_g = E_c - E_v \tag{1.1}$$

The difference in conductivity between insulators and semiconductors may be explained in

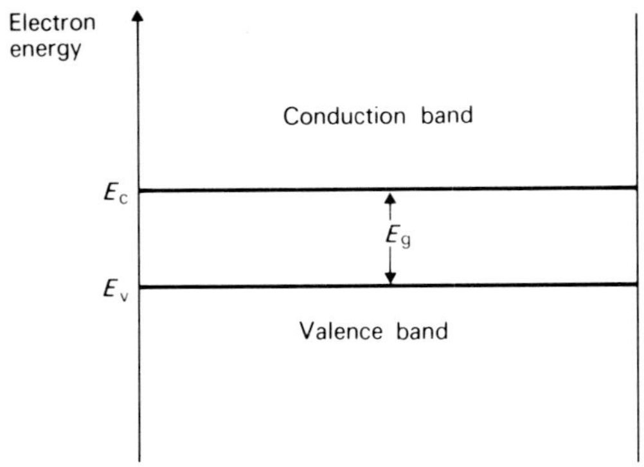

Fig. 1.1 Elementary energy band diagram of a semiconductor and of an insulator

terms of the difference in their energy gaps. For example, at 300 K and atmospheric pressure, insulators have an energy gap of several electron volts whereas $E_g = 0.66$ eV for germanium and $E_g = 1.12$ eV for silicon, provided that these semiconductors are in a condition of high purity.

Intrinsic and extrinsic semiconductors

For all practical purposes, intrinsic semiconductors may be defined as those with zero chemical impurities, so that the density of conduction electrons is equal to the density of holes. However, extrinsic semiconductors made of tetravalent (group IV) materials such as germanium have pentavalent (group V elements) or trivalent (group III) chemical impurities added to them. The pentavalent impurities, or **donors**, contribute to the number of conduction electrons, and such semiconductors are called **n-type**. The trivalent impurities, or **acceptors**, contribute holes and such semiconductors are called **p-type**. Those charge carriers in excess are called **majority carriers** and the others are called **minority carriers**. Thus in n-type materials the majority carriers are electrons and in p-type materials they are holes. Similarly, in n-type materials the minority carriers are holes and in p-type materials they are electrons.

At room temperature (300 K), in both p- and n-type extrinsic semiconductors, the concentration of carriers due to impurities is higher than that due to thermally activated electrons. However, there are cases in which the impurity concentration is such that a semiconductor behaves like an intrinsic material at room temperature and like an extrinsic material at 50–100 K below room temperature. If the material is intrinsic at room temperature and p-type at a lower temperature, then it is called **π-type**. Similarly, if the material is intrinsic at room temperature and n-type at a lower temperature then it is called **ν-type**.

Most Si and Ge materials used in the manufacture of diodes and transistors have a resistivity between 0.1 and 10 ohm cm^{-1}, which corresponds to impurity concentrations between 10^{14} and 10^{16} donors or acceptors per cubic centimetre. The term **doping** is often used to describe the impurity concentration, and heavy doping (about 10^{17} impurities per cubic centimetre) is resorted to in materials used for the fabrication of high-power and high-frequency devices. When the doping is in the range 10^{17} to 10^{20} impurities per cubic centimetre, the semiconductor behaves more like a metal and is called a **degenerate semiconductor**. Heavily doped materials are sometimes denoted by the symbols n^+ and p^+.

An electron wave in a crystal

A simplified quantitative picture of the conduction processes in intrinsic and extrinsic semiconductors can be obtained by considering an electron wave with a given energy in a crystal. Since the electric potential near the nucleus of each atom varies by a large amount over a short distance, the propagation constant of the wave will change rapidly. The Bloch theorem states that if the potential energy $E(\mathbf{r})$ is periodic with the periodicity of the crystal lattice or the arrangement of atoms within the crystal, then the solutions $\phi_{\mathbf{k}}(\mathbf{r})$ of the Schrödinger equations,

$$\left[\frac{\hbar^2}{2m}\nabla^2 + \{E_{\mathbf{k}} - E(\mathbf{r})\}\right]\phi_{\mathbf{k}}(\mathbf{r}) = 0 \tag{1.2}$$

where h is Planck's constant and $\hbar = h/2\pi$, will be of the type

$$\phi_{\mathbf{k}}(\mathbf{r}) = \mathrm{e}^{\mathrm{j}\mathbf{k}\cdot\mathbf{r}}\,U_{\mathbf{k}}(\mathbf{r}) \tag{1.3}$$

which is known as the **Bloch wave function**. The crystal momentum vector **k** with its cartesian coordinates k_x, k_y, k_z defines the concept of momentum space and denotes the motion of the electrons in the crystal. The allowed values of $E_{\mathbf{k}}$, which define the energy bands, are obtained by determining the Bloch functions; a somewhat complicated process. However, if we consider a one-dimensional Bloch function $\phi(x, t)$, then the energy–momentum relationship that satisfies equation 1.3 can be illustrated diagrammatically as in Fig. 1.2. In the classical free electron case, all values of energy $k_x{}^2/2m$, where m is the mass of the electron, are allowed. Hence E_{k_x} is a single-valued function of k_x (broken line parabola). The quantum mechanical solutions for a periodic lattice (thick line segments) indicate the allowed energy bands, the discontinuities being the forbidden energy bands or **energy gaps**, E_g. The momentum space may be divided into regions called Brillouin zones, Fig. 1.2, which depend upon the shape of

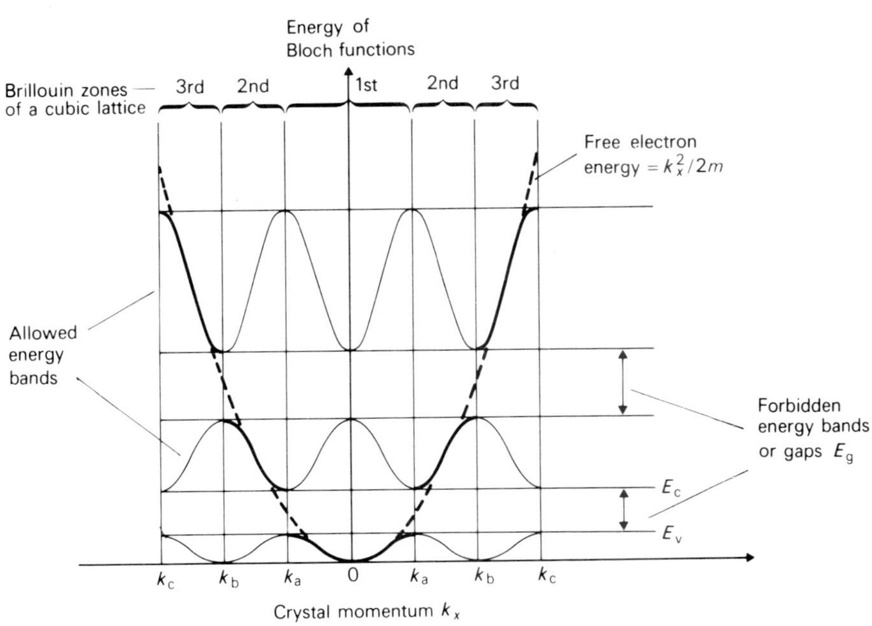

Fig. 1.2 Allowed and forbidden energy bands from one-dimensional Bloch functions

the reciprocal lattice of a particular crystalline structure. Knowledge of one Brillouin zone is sufficient to determine the energies like E_v and E_c near the band edges, whereas all the zones are needed in order to determine the total number of energy states within the entire band.

Similar results are obtained when the energy bands of crystals are plotted in three-dimensional space. Theoretical techniques such as the pseudopotential method[1, 2] are used to study the energy band structure of semiconducting materials. For example, materials such as GaAs (Fig. 1.3(c)), in which the conduction band minimum is directly above the valence band maxima, are called **direct gap** materials, while Si (Fig. 1.3 (b)) and Ge (Fig. 1.3 (a)) are called **indirect gap** materials.

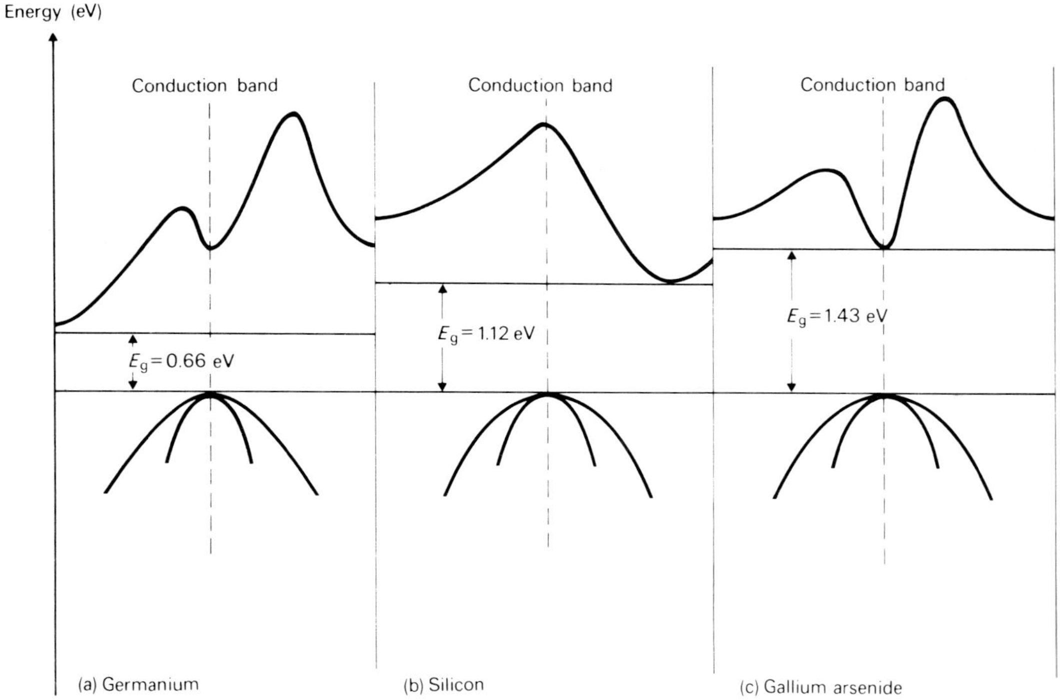

Fig. 1.3 Some aspects of energy band structure of (a) germanium, (b) silicon and (c) gallium arsenide

Lastly, when an electric field is applied to a crystal, an electron is accelerated like a free electron but with a modified mass known as the **effective mass** $m_e{}^*$ given by

$$m_e{}^* = \frac{\hbar^2}{\delta^2 E/\delta k^2} \tag{1.4}$$

The most direct measurements of effective masses are provided by cyclotron resonance experiments, in which the semiconductor sample is placed in a microwave cavity so that the current carriers are accelerated by a microwave field which is perpendicular to an externally applied DC magnetic field. The electrons or holes spiral at an angular frequency ω_r given by

$$\omega_r = q\mathbf{B}/m_e{}^* \tag{1.5}$$

where q is the charge on an electron and $\mathbf{B}$ is the magnetic field in webers per square metre.

When the angular frequency of the microwaves ω is equal to ω_r, the energy absorbed by the semiconductor increases greatly.

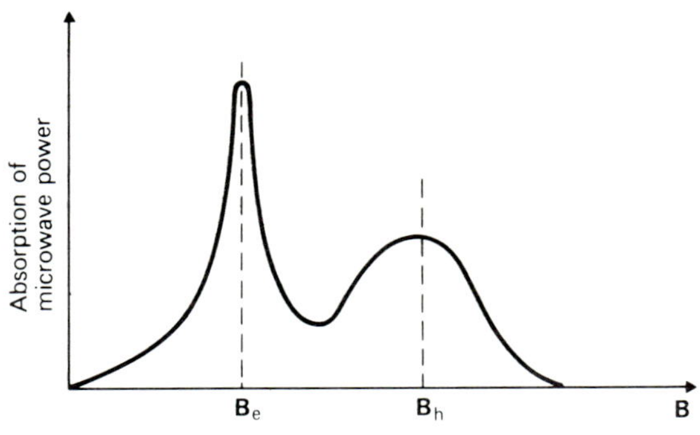

Fig. 1.4 Cyclotron resonance peaks for electrons and holes

In practice, the absorption peaks are located by scanning **B**, which is generated by an electromagnet, and keeping constant an accurately measured microwave frequency. The microwave power is monitored, Fig. 1.4, and the values of $\mathbf{B}_e$ for electrons and $\mathbf{B}_h$ for holes are determined. Hence from equation 1.5, when $\omega_r = \omega$,

$$m_e^* = \left(\frac{q}{m}\right)\left(\frac{m}{\omega}\right)\mathbf{B}_e = 1.75 \times 10^{11}\left(\frac{m}{\omega}\right)\mathbf{B}_e \tag{1.6}$$

$$m_h^* = \left(\frac{q}{m}\right)\left(\frac{m}{\omega}\right)\mathbf{B}_h = 1.75 \times 10^{11}\left(\frac{m}{\omega}\right)\mathbf{B}_h \tag{1.7}$$

The effective mass is a function of crystalline orientation and the ratio of the effective mass to the rest mass is often quoted as a range of values rather than a single number. For example, for Si, m_e^*/m lies between 0.19 and 0.43; for Ga, m_e^*/m lies between 0.08 and 0.36.

Density of quantum states in allowed energy bands

The energy near the bottom of the conduction band of an intrinsic semiconductor, Fig. 1.1, may be written as

$$E = E_c + \frac{k_r^2}{2m_e^*} \tag{1.8}$$

where k_r is the magnitude of the radial momentum vector and m_e^* is the effective mass for conduction electrons. Similarly, near the top of the valence band,

$$E = E_v - \frac{k_r^2}{2m_h^*} \tag{1.9}$$

where m_h^* is the effective mass for holes in the valence band.

Let $M(k_r)$ be the number of allowed energy states per unit momentum per unit volume of the crystal. Since each energy level, i.e. two quantum states, is associated with an incremental volume $\Delta x \Delta y \Delta z$ and momentum volume $\Delta k_x \Delta k_y \Delta k_z$, we have

$$M(k_r) = \frac{2}{(\Delta x \Delta y \Delta z)(\Delta k_x \Delta k_y \Delta k_z)} \tag{1.10}$$

By Heisenberg's uncertainty principle, $\Delta k_x \Delta x = h$, where h is Planck's constant. Hence

$$M(k_r) = 2/h^3 \tag{1.11}$$

The volume in momentum space between spheres of radii k_r and $(k_r + \Delta k_r)$ can be written as

$$\Delta V = 4\pi k_r^{\,2} \Delta k_r \tag{1.12}$$

The density of allowed energy states in the range ΔE corresponding to the volume ΔV will be ΔM, given by

$$\Delta M = \Delta V\, M(k_r) = 8\pi k_r^{\,2} \Delta k_r / h^3 \tag{1.13}$$

Hence the density of allowed electron energy states per unit energy in the conduction band will be

$$M(E) = \frac{M}{E} = \frac{8\pi k_r^{\,2}}{h^3}\left(\frac{\Delta k_r}{\Delta E}\right) \tag{1.14}$$

Differentiating equation 1.8 with respect to energy, the derivative can be expressed as

$$\left(\frac{\Delta k_r}{\Delta E}\right) = \frac{m_e^*}{k_r} \tag{1.15}$$

and $$k_r = (E - E_c)^{1/2}(2m_e^*)^{1/2} \tag{1.16}$$

giving

$$M(E) = \frac{4\pi}{h^3}(2m_e^*)^{3/2}(E - E_c)^{1/2} \tag{1.17}$$

Similarly, the allowed electron states in the valence band will be

$$M(E) = \frac{4\pi}{h^3}(2m_h^*)^{3/2}(E_v - E)^{1/2} \tag{1.18}$$

where m_h^* is the effective mass of holes.

Equations 1.17 and 1.18 are good approximations for small values of $(E - E_c)$ and $(E_v - E)$ near the edges of the bands.

In the case of free electrons, using the Maxwell–Boltzmann distribution, the density $N(E_2)$ of free electrons per unit energy at an energy E_2 is related to the density $N(E_1)$ at energy E_1 by the expression

$$\frac{N(E_2)}{N(E_1)} = e^{-(E_2 - E_1)/kT} \tag{1.19}$$

where k is the Boltzmann constant $= 1.38 \times 10^{-23}$ J K^{-1}.

In a crystal, the density $N(E)$ of electrons is limited to a maximum value $M(E)$, and this relation is given by the Fermi–Dirac probability function

$$F(E) = \frac{N(E)}{M(E)} = \frac{1}{1 + e^{(E - E_f)/kT}} \tag{1.20}$$

where E_f is the Fermi energy level. The value of $F(E)$ varies between 0 and 1. At $E = E_f$, $F(E) = \frac{1}{2}$ and is independent of temperature; the value of E_f is chosen as a reference energy level.

The plots of $F(E)$ against E for various temperatures, Fig. 1.5, illustrate how abrupt the transition between filled and empty energy levels becomes at low temperatures.

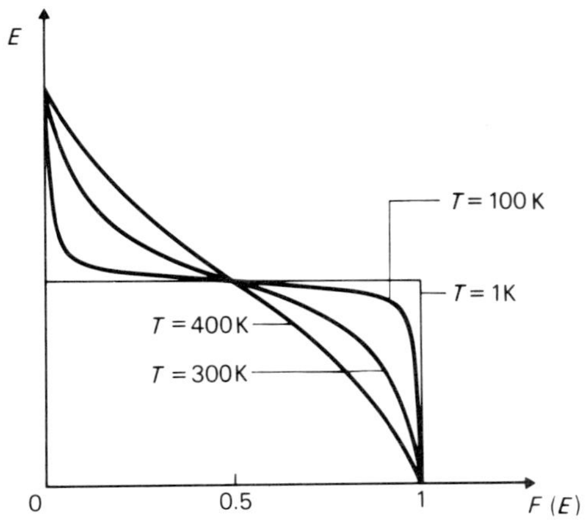

Fig. 1.5 Plots of $F(E)$ against E for various values of temperature T

The total density of conduction electrons is obtained by integrating $N(E)$ over the desired part of the conduction band, i.e.

$$n = \int_{E_c}^{E_1} N(E)\,dE = \int_{E_c}^{E_1} M(E)F(E)\,dE \tag{1.21}$$

Sometimes n is written as

$$n = M_c F(E_c) \tag{1.22}$$

where M_c is a constant defined as the effective density of allowed energy states in the conduction band, with dimensions of energy states per cubic metre. The quantity M_c can easily be evaluated from equations 1.17 and 1.20 by rewriting equation 1.21 as

$$n = \frac{4\pi}{h^3}(2m_e^*)^{3/2} \int_{E_c}^{E_1} F(E)(E - E_c)^{1/2}\,dE$$

which can be simplified by approximating equation 1.20 to

$$F(E) \approx e^{-(E - E_f)/kT}$$

provided $(E_c - E_f) > 3kT$, so that

$$n = \frac{4\pi}{h^3}(2m_e^*)^{3/2} \int_{E_c}^{E_1} e^{-(E - E_f)/kT}(E - E_c)^{1/2}\,dE$$

$$= \frac{4\pi}{h^3}(2m_e^*)^{3/2} e^{-(E_c - E_f)/kT} \int_{E_c}^{E_1} (E - E_c)^{1/2} e^{-(E - E_c)/kT}\,dE$$

giving

$$n = M_c e^{-(E_c - E_f)/kT} \tag{1.23}$$

where

$$M_c = \frac{4\pi}{h^3}(2m_e^*)^{3/2} \int_{E_c}^{E_1} (E - E_c)^{1/2} e^{-(E-E_c)/kT} dE$$

Evaluation of the integral and simplification gives

$$M_c = \frac{2}{h^3}(2\pi m_e^* kT)^{3/2} \tag{1.24}$$

Similarly, the total density of conduction holes can be written as

$$p = M_v[1 - F(E_v)] \tag{1.25}$$

In the valence band, if $(E_f - E_v) > 3kT$ then

$$[1 - F(E)] \approx e^{-(E_f - E)/kT}$$

and $$M_v = \frac{4\pi}{h^3}(2m_h^*)^{3/2} \int_{E_2}^{E_v} (E_v - E)^{1/2} e^{-(E_v - E)/kT} dE$$

Evaluation of the integral and simplification gives

$$M_v = \frac{2}{h^3}(2\pi m_h^* kT)^{3/2} \tag{1.26}$$

and $$p = M_v e^{-(E_f - E_v)/kT} \tag{1.27}$$

From equations 1.23 and 1.27,

$$n.p = M_c M_v e^{-(E_c - E_v)/kT}$$

or $$n.p = M_c M_v e^{-E_g/kT} \tag{1.28}$$

Equation 1.28 shows that under thermal equilibrium conditions, the product $(n.p)$ is a property of the semiconductor which depends on the energy gap E_g and on the density of allowed energy states but is independent of the position of the Fermi level E_f.

The Fermi level in intrinsic and extrinsic semiconductors

In intrinsic semiconductors, $n = p$ by definition, so that from equations 1.23 and 1.27,

$$M_c/M_v = e^{(E_v + E_c - 2E_f)/kT}$$

Therefore

$$E_f = \left(\frac{E_v + E_c}{2}\right) - \frac{kT}{2}\ln\left(\frac{M_c}{M_v}\right) \tag{1.29}$$

If $M_c = M_v$, the Fermi level is exactly in the centre of the forbidden energy gap, Fig. 1.6(a). In practice, for Si and Ge, $M_c \approx M_v$, so that E_f is within kT of the centre of the forbidden band.

In n-type materials, each donor impurity atom produces one allowed energy state at the impurity level, Fig. 1.6(b). If M_d is the number of allowed energy states per cubic centimetre at the donor level, and N_d is the total electron density in the donor level at E_d, then from equation 1.20,

$$\frac{N_d}{M_d} = F(E_d) = \frac{1}{1 + e^{(E_d - E_f)/kT}} \tag{1.30}$$

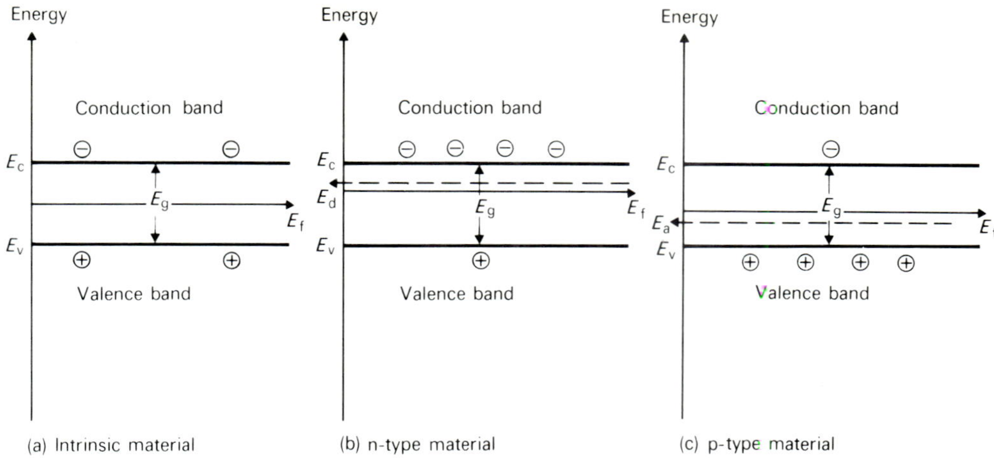

Fig. 1.6 Energy band diagrams

The density $(M_d - N_d)$ of holes at the donor level represents electrons which have been removed by excitation into the conduction band. The location of the Fermi level is determined by equating the density of electrons in the conduction band, n_n, to the sum of the hole density in the valence band, p_n, and the hole density in the donor level, i.e.

$$n_n = p_n + (M_d - N_d) \tag{1.31}$$

Provided that $(E_c - E_f) > 3kT$, equation 1.23 gives the value of n_n. Similarly, equation 1.27 gives the value of p_n provided that $(E_f - E_v) > 3kT$, and equation 1.30 may then be written as

$$(M_d - N_d) = M_d\left(\frac{1}{1+\mathrm{e}^{-(E_d - E_f)/kT}}\right) \tag{1.30a}$$

Substituting into equation 1.31 from equations 1.23, 1.27 and 1.30*a* gives

$$M_c\mathrm{e}^{-(E_c - E_f)/kT} = M_v\mathrm{e}^{-(E_f - E_v)/kT} + M_d\left(\frac{1}{1+\mathrm{e}^{-(E_d - E_f)/kT}}\right) \tag{1.32}$$

Equation 1.32 may be solved for E_f provided that M_c, M_v and E_g are known.

Similarly, in p-type materials (Fig. 1.6(c)), the thermal equilibrium values of electrons and holes, n_p and p_p respectively, are given by equations 1.23 and 1.27 so that

$$p_p = n_p + N_a \tag{1.33}$$

where N_a is the total hole density in the acceptor level at E_a, M_a is the number of allowed energy states per cubic centimetre in the acceptor level, and

$$\frac{N_a}{M_a} = F(E_a) = \frac{1}{1+\mathrm{e}^{(E_a - E_f)/kT}}$$

or
$$N_a = \frac{M_a}{1+\mathrm{e}^{-(E_f - E_a)/kT}} \tag{1.34}$$

Substituting into equation 1.33 from equations 1.23, 1.27 and 1.34 gives

$$M_v\mathrm{e}^{-(E_f - E_v)/kT} = M_c\mathrm{e}^{-(E_c - E_f)/kT} + \frac{M_a}{1+\mathrm{e}^{-(E_f - E_a)/kT}} \tag{1.35}$$

which can be solved for E_f provided that M_c, M_v and E_g are known.

In the case of degenerate semiconductors (see Section 1.2.2), E_a and E_d no longer represent single energy levels, but must be treated as energy bands since the impurity atoms are sufficiently close to interact with one another. The Fermi level moves towards the edge of the forbidden band and the assumptions $(E_a - E_f) > 3kT$ and $(E_f - E_d) > 3kT$, implicit in the derivations of equations 1.29, 1.32 and 1.35, are no longer valid. At the largest impurity densities ($\approx 10^{20}$ impurities per cubic centimetre), the width of the impurity level band increases so much that it overlaps the conductor or valence band, and the Fermi level lies within the impurity band.

Lastly, it must be noted that the intrinsic carrier density $n = p = n_i$ increases very rapidly with temperature. Since M_a and M_d are constant, a semiconductor that is extrinsic at room temperature may become intrinsic if its temperature is raised excessively. This temperature-dependent behaviour is of great importance in the design of electronic devices because it often leads to a degradation of the electrical characteristics of the device, or even to a total functional failure. In devices meant for operation at high temperatures, it is often advantageous to use materials with larger values of E_g and with higher impurity concentrations, thus ensuring a predominantly extrinsic behaviour.

Alignment of the Fermi level in junctions between dissimilar materials

Two types of junctions are commonly encountered in modern semiconductor devices:

i) metal–semiconductor junctions, Fig. 1.7(a);
ii) junctions between two pieces of the same semiconducting material with different types and concentrations of impurities, Fig. 1.7(b).

In general, when two dissimilar materials in thermal equilibrium are in physical contact, their Fermi levels must be aligned across the junction. The rigorous proof of this statement can be found in specialised texts[3] but the physical mechanism can be visualised as follows.

Let $N_1(E)$ be the electron density per unit energy in material (1), and let $M_1(E)$ be the density of allowed quantum states per unit energy in material (1). Let the corresponding values for material (2) be $N_2(E)$ and $M_2(E)$. The Fermi functions of the two materials will be

$$F_1(E) = N_1(E)/M_1(E) \tag{1.36}$$

$$F_2(E) = N_2(E)/M_2(E) \tag{1.37}$$

Let E_1 be an absolute value of energy that is identical in the two materials in physical contact. At this energy level, if there are electrons available in material (1) and empty allowed energy states in material (2), then if both the materials are at the same constant temperature, electron flow from material (1) to material (2) should not involve any expenditure of energy. Moreover, the electron current in a given direction is proportional to the availability of electrons in the 'source' material multiplied by the availability of empty energy states in the 'sink' material. Hence the current due to electron flow from material (1) to material (2) will be given by

$$I_{12} \propto N_1(E_1)[M_2(E_1) - N_2(E_1)] \tag{1.38}$$

Similarly, the current due to electron flow from material (2) to material (1) will be given by

$$I_{21} \propto N_2(E_1)[M_1(E_1) - N_1(E_1)] \tag{1.39}$$

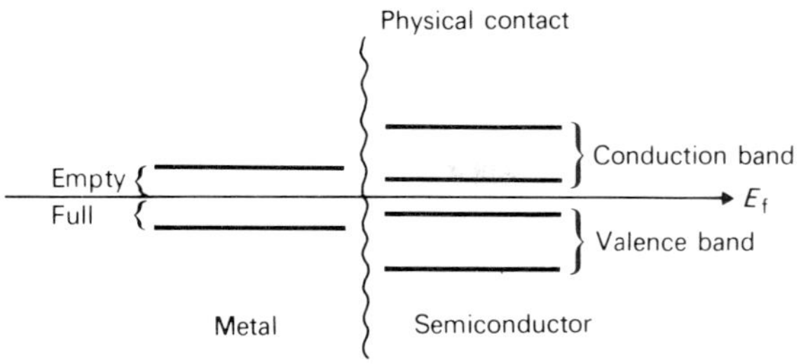

(a)

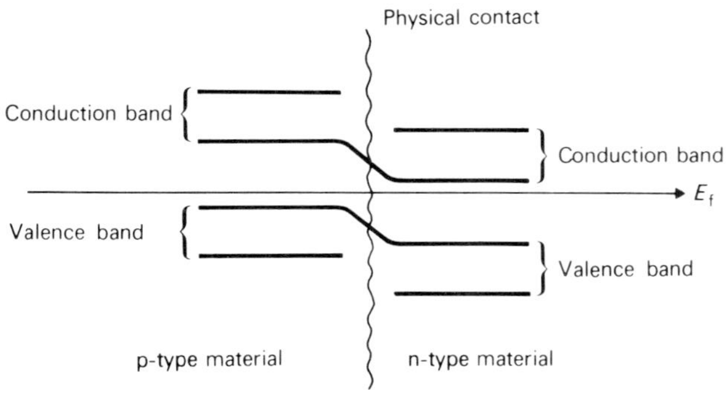

(b)

Fig. 1.7 (a) Metal–semiconductor junction; (b) semiconductor–semiconductor junction

But the net electron current across the junction must be zero. Hence,

$$I_{12} = I_{21} \tag{1.40}$$

Substituting,

$$N_1(E_1)[M_2(E_1) - N_2(E_1)] = N_2(E_1)[M_1(E_1) - N_1(E_1)]$$

Therefore

$$F_1(E_1)[1 - F_2(E_1)] = F_2(E_1)[1 - F_1(E_1)] \tag{1.41}$$

Equation 1.41 is satisfied only if

$$F_1(E_1) = F_2(E_1)$$

or

$$E_{f_1} = E_{f_2} \tag{1.42}$$

where E_{f_1} and E_{f_2} are the Fermi levels in materials (1) and (2) respectively. This means that the Fermi level across the junction between two materials in thermal equilibrium must be continuous. A general discussion of a p-n junction under zero, forward and reverse bias is given in Section 1.3 (p. 14).

Conductivity

The average time τ between collisions that reduce the momentum of a charge carrier to zero is related to the drift velocity v, due to an applied electric field $\boldsymbol{\varepsilon}$, by the relation

$$m^*v = q\boldsymbol{\varepsilon}\tau \tag{1.43}$$

and

$$v = \mu\boldsymbol{\varepsilon} \tag{1.44}$$

where $\mu = \boldsymbol{\varepsilon}\tau/m^*$ is the drift mobility, m^* being the effective mass of and q being the charge on the carrier.

The conduction current density $\mathbf{j}$ in a material is given by

$$\mathbf{j} = nqv = (\mu nq)\boldsymbol{\varepsilon} \tag{1.45}$$

so that the conductivity σ is given by

$$\sigma = \mu nq \tag{1.46}$$

where n is the number of carriers per unit volume. For intrinsic semiconductors,

$$\sigma_i = qn_i(\mu_n + \mu_p) \tag{1.47}$$

For n-type extrinsic material,

$$\sigma_n = q(n_n\mu_n + p_n\mu_p) \approx n_nq\mu_n \tag{1.48}$$

For p-type extrinsic material,

$$\sigma_p = q(p_p\mu_p + n_p\mu_n) \approx p_pq\mu_p \tag{1.49}$$

Ohms' law written as $\mathbf{j} = \sigma\boldsymbol{\varepsilon}$ is valid only if σ is independent of $\boldsymbol{\varepsilon}$, and this condition requires the carrier population to be constant and mobilities to be independent of the electric field. Also n_i increases rapidly with temperature, and intrinsic semiconductors exhibit a negative temperature coefficient over a limited temperature range.

Diffusion

For most semiconductors, the transport processes of electrons and holes are considered to be independent of each other. Hence the total conduction current is simply the sum of the electron and hole current components. The diffusion current in a material can be expressed in terms of the carrier concentration gradient in the material as

$$J_n = -qD_n\frac{dn}{dx} \quad \text{A cm}^{-2} \quad \text{for electrons} \tag{1.50}$$

and

$$J_p = -qD_p\frac{dn}{dx} \quad \text{A cm}^{-2} \quad \text{for holes} \tag{1.51}$$

where D_n and D_p are the diffusion constants and dn/dx, dp/dx the carrier concentration gradients.

The diffusion constants can be expressed in terms of the drift mobilities for non-degenerate semiconductors according to the Einstein relationships

$$D_p = (kT/q)\mu_p \tag{1.52}$$

and

$$D_n = (kT/q)\mu_n \tag{1.53}$$

If n_p is the minority carrier density in p-type material, p_n is the minority carrier density in n-type material, and n_{p0}, p_{n0} denote the thermal equilibrium values of p- and n-type materials respectively, then for most device applications, the excess minority carrier density $(n_p - n_{p0}) > n_{p0}$ but is small compared to the equilibrium majority carrier density p_{p0}. In the one-dimensional case, the effect of diffusion on the minority carrier density can be represented by

$$\frac{dn_p}{dt} = D_n \frac{d^2 n_p}{dx^2} \tag{1.54}$$

and the effect of recombination by

$$\frac{dn_p}{dt} = -\frac{n_p - n_{p0}}{\tau_n} \tag{1.55}$$

Under small signal conditions these two effects do not interact, and for èlectrons injected into the p-type material,

$$\frac{dn_p}{dt} = D_n \frac{d^2 n_p}{dx^2} - \frac{n_p - n_{p0}}{\tau_n} \tag{1.56}$$

For holes injected into the n-type material,

$$\frac{dp_n}{dt} = D_p \frac{d^2 p_n}{dx^2} - \frac{p_n - p_{n0}}{\tau_p} \tag{1.57}$$

The solutions of equations 1.56 and 1.57 give the variation of minority carrier density as functions of time and distance.

The average diffusion distance L_p travelled by a minority hole before disappearing as a result recombination is called the **diffusion length**, and is given by the expression

$$L_p = \sqrt{D_p \tau_p} \tag{1.58}$$

Similarly, the minority electron diffusion length is given by

$$L_n = \sqrt{D_n \tau_n} \tag{1.59}$$

The continuity equation

The minority carrier concentration in a semiconductor is also affected by the presence of an electric field. Thus the drift current density of electrons in a p-type semiconductor, across a surface normal to the flow, is given by

$$J_n = n'_p \mu_n \varepsilon_x \tag{1.60}$$

where $n'_p = (n_p - n_{p0})$.

But $$\frac{dn'_p}{dt} = -\frac{dJ_n}{dx} = -\mu_n \varepsilon_x \frac{dn'_p}{dx} \tag{1.61}$$

from equation 1.60. Thus the excess carrier concentration at a point is decreased by the presence of an electric field. Similarly, for holes in an n-type material,

$$\frac{dp'_n}{dt} = -\mu_p \varepsilon_x \frac{dp'_n}{dx} \tag{1.62}$$

If this effect is combined linearly with the effects of diffusion and recombination, equations 1.56 and 1.57, then the result is the equation of continuity for excess minority carriers. Thus

$$\frac{dp'_n}{dt} = D_p \frac{d^2 p'_n}{dx^2} - \mu_p \varepsilon_x \frac{dp'_n}{dx} - \frac{p'_n}{\tau_p} \quad \text{for holes} \tag{1.63}$$

and

$$\frac{dn'_p}{dt} = D_n \frac{d^2 n'_p}{dx^2} - \mu_n \varepsilon_x \frac{dn'_p}{\tau_n} - \frac{n'_p}{\tau_n} \quad \text{for electrons} \tag{1.64}$$

Thus the rate of change of excess minority carrier density is governed by the net diffusive flow, the effect of an applied electric field and the recombination rate.

Ohmic contacts

An ideal ohmic contact is one that has a linear and symmetrical current–voltage relationship. It is characterised by having no potential barrier, and an infinite surface recombination velocity. In practice, ideal ohmic contacts can only be approximated. A metal–semiconductor contact is approximately ohmic provided that the semiconductor is heavily doped.

A direct contact between a metal and a semiconductor does not generally give an ohmic contact, especially when the resistivity of the semiconductor is high. For example, in the case of Ge and Si, in order to form an ohmic contact, an Au–Sb alloy must first be evaporated onto the n-type semiconductors. The contacts are then alloyed, at the corresponding eutectic temperature, into the semiconductors under an inert gas such as argon or nitrogen.

1.3 Single junction devices

The p-n junction in general

A qualitative picture of the physical mechanism of p-n junction is easily visualised in terms of its energy level diagram under the following conditions:

i) zero bias (Fig. 1.8(a));
ii) reverse bias (Fig. 1.8(b));
iii) forward bias (Fig. 1.8(c)).

The region between x_1 and x_2, which contains the metallurgical junction, is known as the **depletion region** or **depletion layer**, and the regions between 0 and x_1 and beyond x_2 are called the **neutral regions.**

Zero bias

As the Fermi levels of the n and p types of materials in contact must be aligned (Fig. 1.8(a)), the height of the potential barrier $E_{b0} = (E_n - E_p)$ = contact potential, where E_n and E_p are the quasi-Fermi levels, or 'imrefs', for electrons and holes respectively.

Under thermal equilibrium conditions, the minority electron current I_r to the left of the point $x = x_2$, i.e. flowing from the p-side to the n-side, is proportional to $e^{-(E_g - E_p)/kT} = e^{-E/kT}$, where E_g is the energy gap (see equation 1.23). The majority electron current I_f which flows to the right of the point $x = x_2$ consists of electrons which have climbed the potential barrier and is proportional to $e^{-E_{bf}/kT} e^{-(E_g - E_n)/kT} = e^{-E/kT}$.

Hence, $I_r = I_f$ and the net electron current is zero. Similar arguments are applicable to hole currents, as the barrier to the flow of holes from the p-region to the n-region is the same as that for electron flow in the opposite direction.

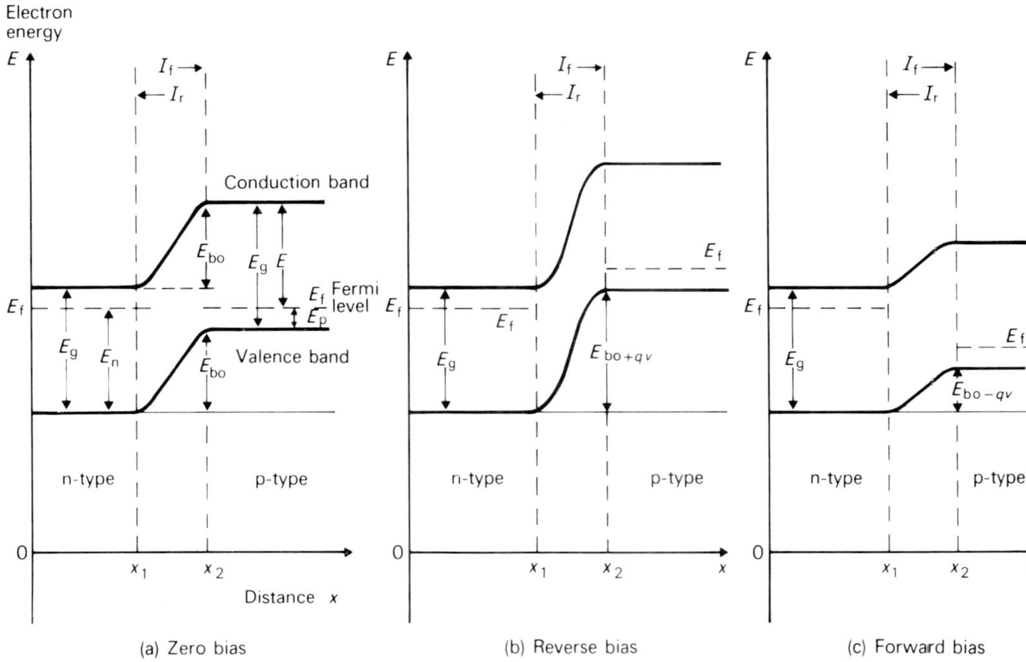

Fig. 1.8 Physical mechanisms of a p-n junction

Reverse bias

When a reverse bias voltage, v, is applied (Fig. 1.8(b)), the barrier height is increased to $(E_{bo} + qv)$, but this has no effect on the reverse current I_r. However, the forward electron current I_f will change by a factor $e^{qv/kT}$. Since the equilibrium value of $I_f = I_r$,

$$I_f = I_r e^{qv/kT} \tag{1.65}$$

The net electron current I is then given by

$$I = (I_r e^{qv/kT} - I_r)$$

i.e. $$I = I_r(e^{qv/kT} - 1) \tag{1.65a}$$

Using a similar argument for holes moving in the opposite direction, the ideal I–V characteristics (see Fig. 1.9) of the junction may be represented by

$$I = I_s(e^{qv/kT} - 1) \tag{1.66}$$

where

$$I_s = (I_{re} + I_{rh}) \tag{1.66a}$$

is the reverse saturation current with contributions I_{re} and I_{rh} from electrons and holes respectively. The value of I_s is usually less than a microamp in Si junctions at room temperature.

Equation 1.66 is known as the diode equation and involves the following assumptions:

1) the concentration of carriers in the depletion region is negligible;
2) the electron and hole currents are constant throughout the depletion layer;

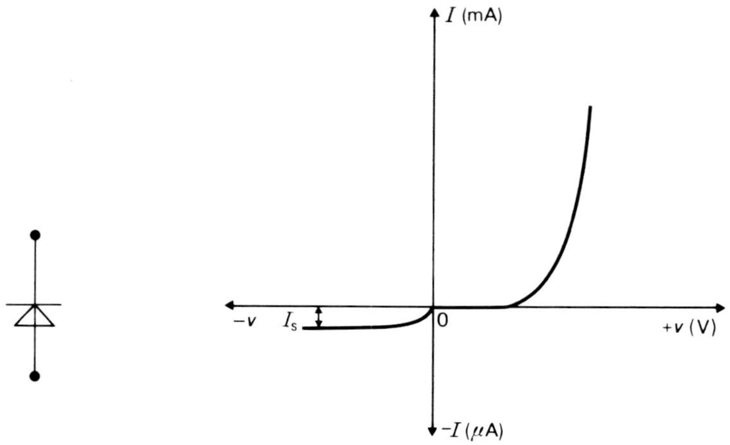

Fig. 1.9 Circuit symbol and *I–V* characteristic of a p-n junction diode

3) the majority carrier concentration is greater than the minority carrier concentration under all conditions (low-level injection);
4) the depletion layer has abrupt boundaries and the electric fields throughout the undepleted regions are nearly zero;
5) the minority current is independent of the bias magnitude and polarity but is determined by the rate of thermal generation in the p- and n-regions near the depletion layer.

In order to calculate the value of I_s in terms of the physical parameters of the junction, consider the n-neutral region between 0 and x_1. In the steady state, from equation 1.57, since $dp_n/dt = 0$,

$$D_p \frac{d^2 p_n}{dx^2} - \frac{p_n - p_{n0}}{\tau_p} = 0 \tag{1.67}$$

where p_n is the hole density in the n-neutral region; p_{n0}, the value of p_n under equilibrium conditions, is a constant.

The value of p_n is 0 at $x = 0$ and p_{n0} at $x \to \infty$. Figure 1.10 shows the minority carrier density profiles under strong reverse bias conditions. Equation 1.67 gives

$$p_n = p_{n0}(1 - e^{-x/L_p}) \tag{1.68}$$

where $L_p = \sqrt{D_p \tau_p}$.

Similarly,

$$n_p = n_{p0}(1 - e^{-x/L_n}) \tag{1.69}$$

where $L_n = \sqrt{D_n \tau_n}$.

The diffusive carrier flow is proportional to the concentration gradients $(dp_n/dx)_{x=0}$ and $(dn_p/dx)_{x=0}$. Under reverse bias conditions, the reverse saturation hole current I_{rh} will be given by

$$I_{rh} = -AqD_p \left(\frac{dp_n}{dx}\right)_{x=0}$$

where A is the junction area.

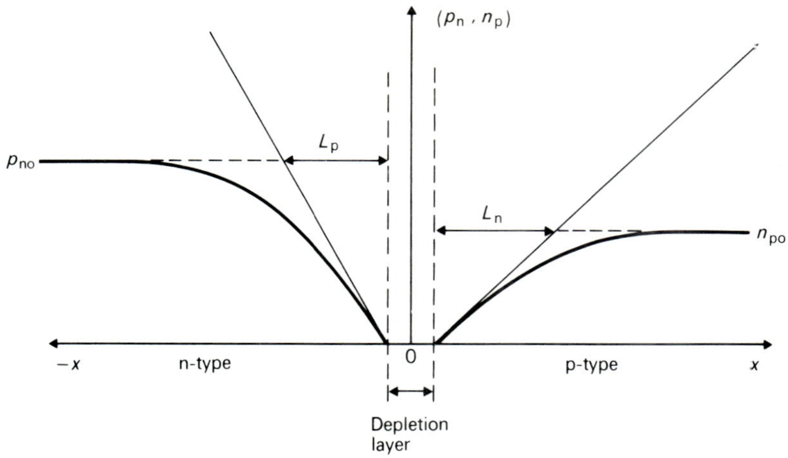

Fig. 1.10 Minority carrier densities in the p and n materials when the p-n junction is reverse biased

But from equation 1.68,

$$\left(\frac{\mathrm{d}p_n}{\mathrm{d}x}\right)_{x=0} = \frac{p_{n0}}{L_p}$$

So

$$I_{rh} = -\left(\frac{AqD_p p_{n0}}{L_p}\right) \quad (1.70)$$

Similarly,

$$I_{re} = -\left(\frac{AqD_n n_{p0}}{L_n}\right) \quad (1.71)$$

Hence $I_s = (I_{re} + I_{rh}) = -Aq\left(\frac{D_p p_{n0}}{L_p} + \frac{D_n n_{p0}}{L_n}\right)$ (1.72)

The quantity I_s is thus very sensitive to temperature, and in most modern junctions its value is made as low as possible.

The law described by equation 1.66 and Fig. 1.9 is for an ideal p-n junction, assuming low-level injection. At high-level injection, the majority and minority carrier concentrations approach each other and a more realistic form of the diode equation is

$$I = I_s(e^{qv/nkT} - 1) \quad (1.73)$$

where I_s may not be the same as before and $1 \leqslant n \leqslant 2$.

Forward bias

When a forward bias voltage is applied, Fig. 1.8(c), the Fermi level in the p-region is lowered below that in the n-region by an amount qV. The barrier height is decreased by an amount $(E_{b0} - qV)$ and the forward current is increased by a factor $e^{qV/kT}$, which is greater than unity for positive values of V. There is no effect on the reverse current and equation 1.66 applies for both reverse and forward bias conditions.

Depending on the device geometry, the doping profile and biasing conditions, the p-n junction can perform various circuit functions. In the following sections, several devices are described, based on modified current–voltage, capacitance–voltage and breakdown characteristics.

Rectifier diodes

These diodes perform functions of rectification and detection by virtue of their low forward resistance and high reverse resistance. The forward and reverse dynamic resistance can be calculated from equation 1.73, i.e.

$$r_{\text{forward}} = \frac{\mathrm{d}V^+}{\mathrm{d}I^+} = \frac{nkT}{I^+q} \tag{1.74}$$

where $|v^+| \geqslant 3kT/q$ and I^+ represent the forward applied voltage and forward current respectively, and

$$r_{\text{reverse}} = \frac{\mathrm{d}V^-}{\mathrm{d}I^-} = \frac{nkT}{I_s q} \mathrm{e}^{qV^-/nkT} \tag{1.75}$$

where $|V^-| \geqslant 3kT/q$ and I^- are the reverse applied voltage and reverse current respectively. The small-signal rectification ratio is given by

$$\frac{r_{\text{reverse}}}{r_{\text{forward}}} = \frac{I^+}{I_s} \mathrm{e}^{qV^-/nkT} \tag{1.76}$$

Thus conduction in the forward direction will be poor until the voltage **threshold** (≈ 0.5 V) is reached. Hence, these diodes should not be used for the detection of low-level signals, because the signal-to-noise ratio deteriorates if a DC bias is introduced.

The switching speed of these p-n junctions is proportional to the minority carrier lifetimes. Diodes used as 50 Hz rectifiers usually have speeds between 50 and 500 ns and can typically dissipate between 0.1 and 10 W. In higher frequency diodes, the lifetimes are reduced to maintain efficiency.

High-voltage rectifiers, from 50 V to 2 kV, have a low donor density and a wide depletion region, so that the electric field never reaches the breakdown value. For higher voltages, p-n junctions are connected in series.

Varistors

The non-ohmic behaviour of a p-n junction can be utilised to construct a voltage-**var**iable res**istor** or **varistor**.

The small-signal conductance of a junction can be calculated by considering a DC operating point $\mathrm{P}(V_0, I_0)$ on the characteristic given in Fig. 1.11 and a point $\mathrm{Q}(V, I)$ with a small-signal component such that $V = V_0 + \delta V$ and $I = I_0 + \delta I$.

The diode equation 1.73 can be written as

$$I = I_s(\mathrm{e}^{q(V_0 + \delta V)/nkT} - 1)$$

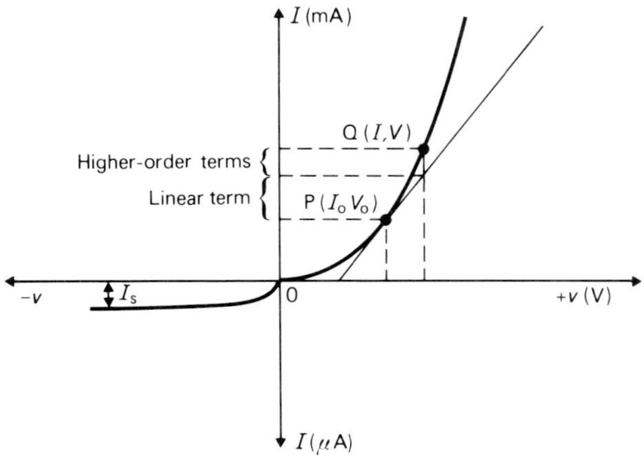

Fig. 1.11 Small signal conductance of a p-n junction

Therefore

$$I = I_s\left[e^{qV_0/nkT}\left\{1+\frac{q\delta V}{nkT}+\left(\frac{q\delta V}{nkT}\right)^2\frac{1}{2!}+\ldots\right\}-1\right]$$

Neglecting the higher-order terms,

$$I = I_0 + \frac{I_s q\delta V}{nkT}e^{qV_0/nkT}$$

Therefore

$$g = \frac{\delta I}{\delta V} = \frac{I_s q}{nkT}e^{qV_0/nkT}$$

$$= \frac{q}{nkT}(I_0 + I_s) \qquad (1.77)$$

With reverse bias, I_0 is negative and tends to I_s, so that $g \to 0$. For small forward bias, $I_0 > I_s$ so that $g \propto I_0$.

Similar characteristics can be obtained for metal–semiconductor junctions.

Devices dependent on junction capacitance

The capacitance at a junction can be attributed to two mechanisms:

a) the depletion layer capacity when the junction is reverse biased;
b) the diffusion capacity when the junction is forward biased.

Depletion layer capacity

For an abrupt junction of the type described in Section 1.3 (p. 14), the width of the depletion layer, as illustrated in Fig. 1.12, can be calculated from Poisson's equation[4] as

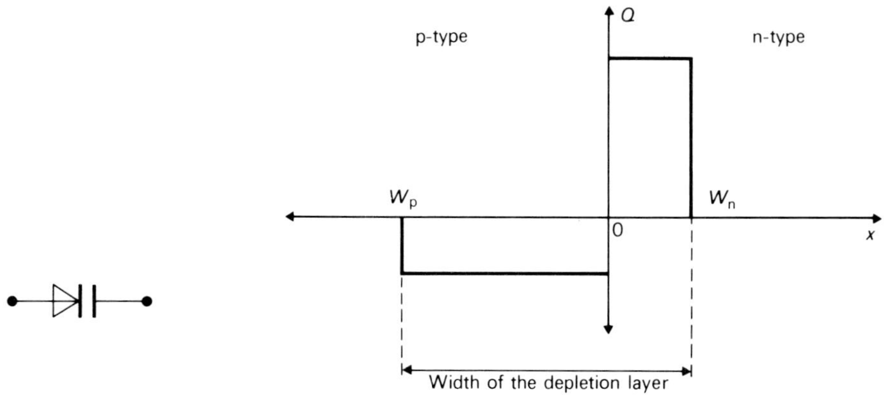

Fig. 1.12 Circuit symbol of a varactor diode and depletion layer capacitance of a reverse biased p-n junction

$$W_p = \left(\frac{2\varepsilon\varepsilon_0(V_d - V)}{qN_a\left(1 + \dfrac{N_a}{N_d}\right)}\right)^{1/2} \tag{1.78}$$

and $$W_n = \left(\frac{2\varepsilon\varepsilon_0(V_d - V)}{qN_d\left(1 + \dfrac{N_d}{N_a}\right)}\right)^{1/2} \tag{1.79}$$

where W_p is the width of depletion layer on the p-side; W_n is the width of depletion layer on the n-side; V_d is the potential required to align the Fermi levels (contact potential); V is the applied potential; N_a and N_d are respectively the acceptor and donor ion densities; q is the electronic charge; ε_0 and ε are respectively the absolute permittivity and permittivity.

The maximum value of the electric field at the metallurgical junction $x = 0$ will be given by

$$E_0 = -\frac{qN_d}{\varepsilon\varepsilon_0}W_n = -\frac{qN_a W_p}{\varepsilon\varepsilon_0} \tag{1.80}$$

where $N_d = \sigma_n/\mu_n q$ and $N_a = \sigma_p/\mu_p q$ (see equation 1.46).

The capacity associated with the reverse biased junction is

$$C_j = -\mathrm{d}Q/\mathrm{d}V \tag{1.81}$$

where Q is the total charge.

For the abrupt junction shown in Fig. 1.12, on the n-side we have

$$Q = AqN_d W_n = A\left(\frac{2\varepsilon\varepsilon_0 qN_d(V_d - V)}{1 + (N_d/N_a)}\right)^{1/2} \tag{1.82}$$

where A is the junction area.

If $N_a \gg N_d$, then

$$C_j = \frac{A}{2}\left(\frac{2\varepsilon\varepsilon_0 qN_d}{(V_d - V)}\right)^{1/2}$$

Substituting for N_d gives

$$C_j = A\left(\frac{\varepsilon\varepsilon_0\sigma_n}{2\mu_n(V_d - V)}\right)^{1/2} \tag{1.83}$$

For forward bias $V < V_d$ and the capacity does not become too large compared to the reverse bias value which varies as $(-V)^{-1/2}$.

Voltage-dependent junction devices have trade names like 'Varicap' and 'Varactor junction diode'. Varactor diodes are widely used in parametric amplification, harmonic generation, mixing, detection and voltage-variable tuning.

Diffusion (storage) capacity

The diffusion, or storage, capacity is the result of the rearrangement of the minority carrier density in a forward biased junction. It is associated with the charge stored in the p- and n-regions as a result of minority carrier injection. The diffusion capacity C_d can be calculated[4] as

$$C_d = \frac{q^2A}{kT}(n_{p0}\sqrt{D_n\tau_n} + p_{n0}\sqrt{D_p\tau_p})e^{qV/kT} \tag{1.84}$$

where the symbols have the same meaning as before.

For diodes with short minority carrier lifetimes, the transition from forward to reverse bias is nearly abrupt and the time taken for the reverse current to decay to a given level, i.e. the **recovery time**, is short. For example, in an OA10 diode, Fig. 1.13, when a forward current of 10 mA is removed and a reverse voltage of 7 V is applied through a 300 Ω resistor, the recovery time for the reverse current to decay to 0.5 mA is about 0.18 μs. Fast recovery diodes are used in switching circuitry.

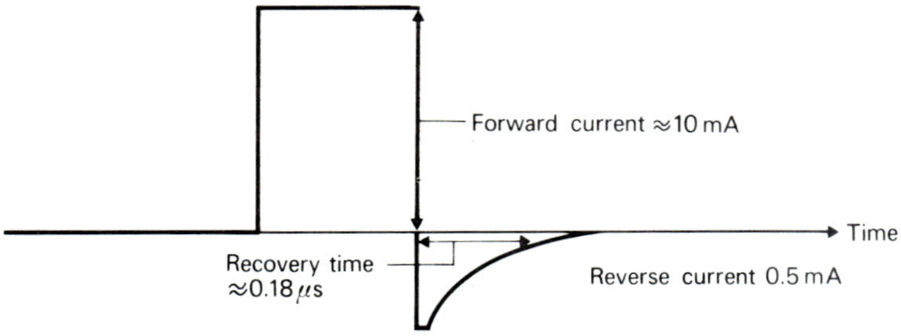

Fig. 1.13 Reverse current and recovery time of a OA10 diode (Mullard Ltd)

Diodes with long carrier lifetimes conduct in the reverse direction for a short period of time, after which the current decays rapidly to its initial value. This effect is made use of in step-recovery or snap-back diodes which conduct in the reverse direction for a short time, and then cut off the current. The waveform is rich in harmonics and hence step-recovery diodes are used for harmonic generation in frequency multipliers.

Zener and avalanche diodes

When a p-n junction is reverse biased to a sufficiently high voltage, electrical **breakdown** occurs, resulting in the flow of current which is nearly independent of the voltage (Fig. 1.14). With a high impurity concentration on both the p- and n-sides, and when breakdown occurs at less than 4 V, the mechanism is by direct internal field emission or tunnelling (see p. 23). This mechanism is known as **Zener breakdown**. When the breakdown voltage is greater than 8 V, the mechanism is by the process of secondary ionisation or **avalanche breakdown**. Between 4 and 8 V, breakdown results from a combination of avalanche and tunnelling.

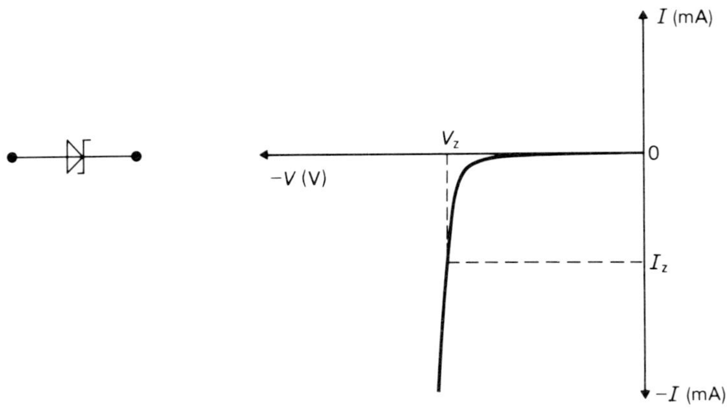

Fig. 1.14 Circuit symbol and *I*–*V* characteristic of a Zener diode

Ionising collisions produce hole-electron pairs, the rate of ionisation $\alpha(E)$ being defined as the average number of ionisations per cm. A single electron will produce N hole-electron pairs as it travels across the depletion layer of width W, where N is given by

$$N = \int_0^W \alpha(E)\,\mathrm{d}x \tag{1.85}$$

If $\alpha(E)$ is the same for both electrons and holes, the initial current I_0 will give rise to a current $I = I_0(1 + N + N^2 + \ldots)$ or

$$I = I_0/(1 - N) = MI_0 \tag{1.86}$$

where M is the multiplication factor and $N < 1$. Breakdown occurs when $M \to \infty$, i.e.

$$N = \int_0^W \alpha(E)\,\mathrm{d}x = 1$$

Junction diodes operated in the reverse direction at the breakdown voltage are called **Zener diodes**, although both the processes described above might coexist.

As seen from Fig. 1.14, after the breakdown the slope resistance is small and the voltage is almost constant at V_z, provided that $I > I_z$. Using the circuit shown in Fig. 1.15, a regulated voltage V_z can be obtained from an unregulated value V provided that

$$(V - V_z) = (I_z + I_L)R \tag{1.87}$$

When the breakdown mechanism is mainly by avalanche multiplication, the temperature coefficient of V_z is positive, whereas for the Zener mechanism it is negative. If a diode having a positive temperature coefficient is connected in series with one having a negative temperature

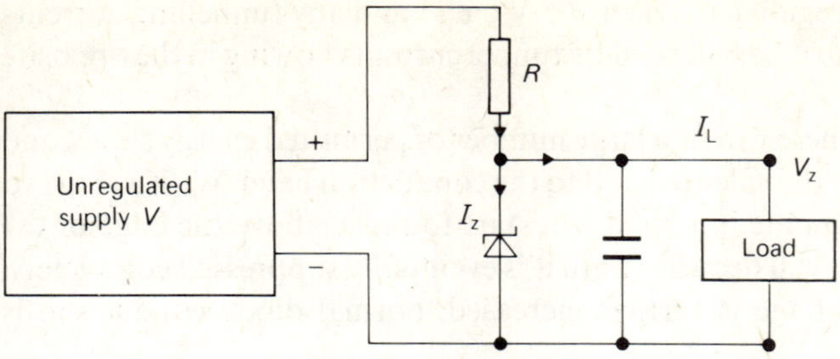

Fig. 1.15 Regulated voltage V_z from an unregulated supply V

coefficient, the combination will have a very low temperature coefficient ($\approx 0.002\%$). Such combinations are used as sources of reference voltages.

Tunnel diodes and backward diodes

Tunnelling is a quantum mechanical effect which occurs when majority carriers on one side of a p-n junction are able to overcome a reverse potential barrier at the junction, even if the height of the barrier is greater than the energy of the carriers. When the impurity concentration ($\approx 10^{20}$ levels cm^{-3}) of a p- or n-type material approaches, or is greater than, the effective density of states, then the Fermi level is located within the conduction or valence band and the semiconductor is defined as **degenerate**. In a tunnel diode, both the p- and n-type materials of the junction are degenerate, and majority carriers will be able to tunnel through the barrier provided that there is a vacant energy state at the same energy level on the other side of the barrier. Figure 1.16(a) shows the energy level diagram for zero bias. In this

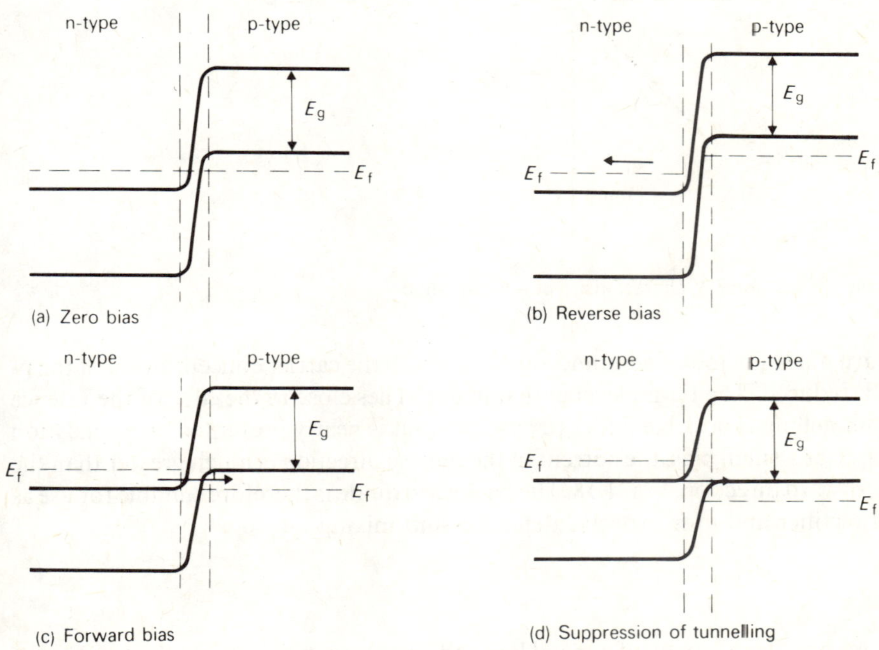

Fig. 1.16 Energy level diagram of a tunnel diode

case, the field across the depletion region can reach 10^8 V cm^{-1} and any tunnelling currents which may occur in one direction must be balanced by tunnel currents flowing in the opposite direction.

With a reverse bias, Fig. 1.16(b), there will be a large number of permitted energy states, and a large tunnel current will flow from the valence band to the conduction band. With a forward bias, Fig. 1.16(c), the edge of the conduction band will start to move above the edge of the valence band and the tunnel current will decrease until it is eventually suppressed completely, Fig. 1.16(d). When the forward voltage is further increased, normal diode conduction is resumed.

The I–V characteristics of a tunnel diode are shown in Fig. 1.17 and can be divided into three distinct parts. In the AB region, current flow is mainly due to tunnelling. Failure to sustain the tunnelling mechanism gives rise to the negative resistance region BC. From CD onwards, normal diode conduction is resumed. The points B and C are known as the **peak** and **valley** points, and tunnel diodes can switch between these points in picoseconds (10^{-12} s); hence their use in fast switching circuitry. When operated in the negative resistance region, tunnel diodes can be used in high-frequency oscillators and amplifiers.

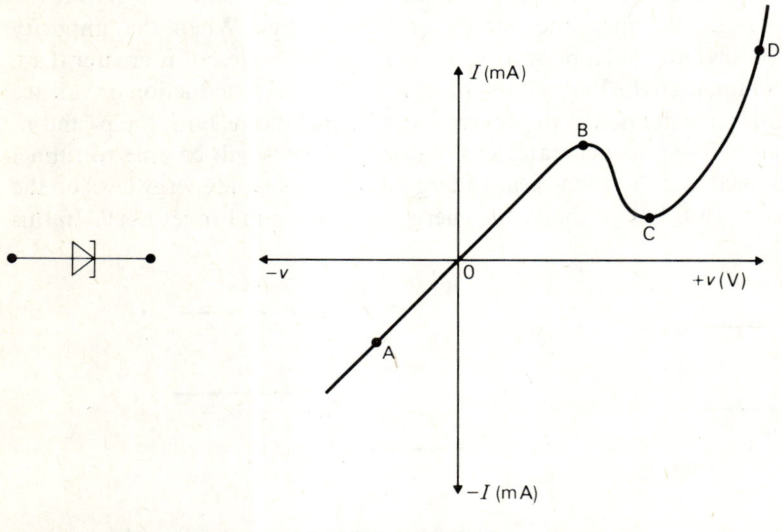

Fig. 1.17 Circuit symbol and I–V characteristic of a tunnel diode

The **backward diode** is a modified tunnel diode in which the carrier concentration of the p-type material is reduced. The Fermi level in that material lies close to the edge of the valence band, so that tunnelling is possible under reverse bias but is nearly prevented on application of forward bias. For a small bias, the current in the reverse direction is much greater than the current in the forward direction, Fig. 1.18. The backward diode is therefore suitable for use as a small-signal rectifier and in microwave detection and mixing.

Schottky diodes

Schottky diodes consist essentially of a metal layer on an n- or p-type semiconductor. For an n-type semiconductor, the metal is such that its work function, ϕ_m, is greater than ϕ_s, the

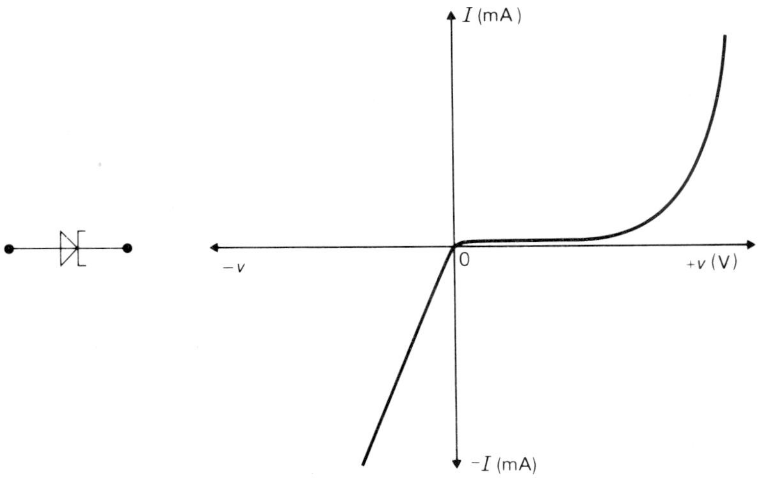

Fig. 1.18 Circuit symbol and I–V characteristic of a backward diode

work function of the semiconductor, but not large enough to cause inversion, i.e. a change in behaviour near the surface from n-type to p-type material. The result is a rectifying junction, similar to a p-n junction, in which the electrons diffuse out of the semiconductor leaving behind a depletion region which will make the semiconductor positive with respect to the metal, thus stopping the diffusive flow. The Fermi levels are then aligned.

With a forward bias, energetic electrons or **hot electrons** enter the bulk of the metal and the device is sometimes called a **hot carrier** diode. Because the current is carried by majority carriers, recombination is no longer important in removing the stored charge. The diffusion capacity is small and the device can switch off in about 10^{-11} s. Hence, Schottky diodes were initially used as detectors at RF and microwave frequencies.

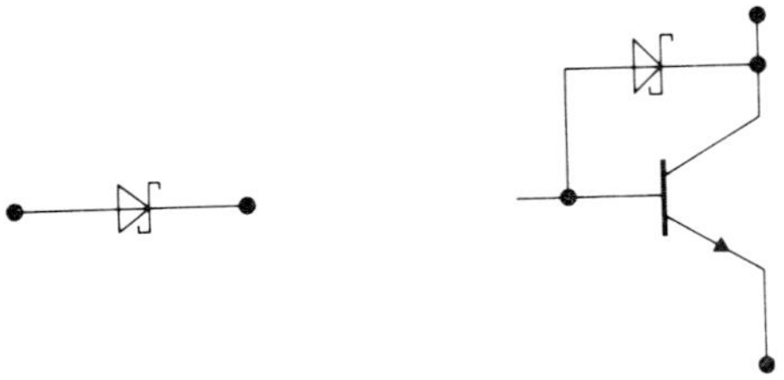

Fig. 1.19 Circuit symbol of a Schottky diode and its use in conjunction with transistors

The I–V characteristics of Schottky diodes are similar to those of Si p-n junction diodes, but the threshold of forward conduction is lower. Hence, they are used to protect the base-collector junction of transistors, Fig. 1.19, with the added advantage that the minority charge stored in the base of the transistor is reduced. Schottky diodes are built into integrated logic circuits for protection and for reduction of propagation delays.

pin diodes

The p- and n-regions are heavily doped (denoted by p^+ and n^+ in Fig. 1.20) and separated by

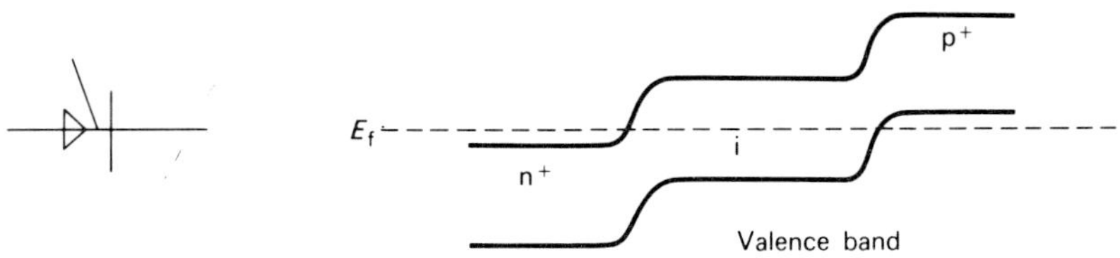

Fig. 1.20 Circuit symbol of a pin diode and its energy band diagram at zero bias

a high-resistivity p-layer (π-layer) or n-layer (ν-layer), whose impurity content is reduced to a very low level; hence the name pin, the i standing for **intrinsic**.

When a reverse bias is applied, the π layer is swept free of carriers and the depletion layer capacity remains virtually constant at a small value. The diode has a high impedance even at microwave frequencies. Under forward bias conditions, holes from the p-layer and electrons from the n-layer are injected into the i-region, making its high-frequency impedance low and mainly resistive. Hence, pin diodes are often used in microwave striplines, as variable impedances and as switches.

When the width of the i-region is large, the reverse voltage required to produce breakdown is high and in this case pin diodes make good high-voltage rectifiers. The pin structure is also used in step-recovery diodes. The width of the i-region is made small and as the stored charge lies in this region, it is quickly removed when the bias on the diode is changed from a forward to a reverse state. Thus the voltage across the diode at first stays low and then abruptly changes to a high reverse value.

1.4 Multijunction devices

It is wasteful to describe the ever-increasing range of modern multijunction devices, because each device is simply a variation of the same basic principles, designed to suit specific circuit applications. However, the devices can be broadly divided into the following types:

(a) junction transistors;
(b) metal oxide semiconductor (**MOS**) devices;
(c) devices used mainly for switching and power control.

Junction transistors

Modern junction transistors are usually made by the planar process described in detail in Chapter 2, and are either of the npn or pnp **bipolar** type, or of the **field-effect** type.

The circuit symbols and structure of bipolar transistors are shown in Fig. 1.21, where $^+$ and $^{++}$ represent heavily doped and very heavily doped materials respectively. The thin middle layer is the **base** (b), and the outer layers are the **emitter** (e) and **collector** (c) respectively. In the npn transistor, the base and collector are made positive with respect to the emitter. Electrons move from the emitter towards the base-collector junction, but as the latter is reverse biased, the electric field sweeps the electrons into the collector terminal. Figure 1.22 shows the curves of the collector current I_c against the collector-emitter voltage V_{ce} for various values of the base current I_b, in the common-emitter configuration. A change of a few

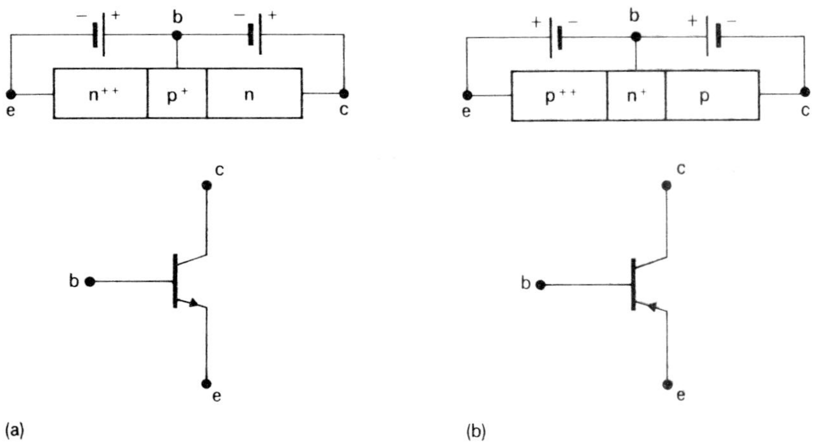

Fig. 1.21 Junction transistors: circuit symbol and structure of (a) npn transistor and (b) pnp transistor

microamps in the base current produces a much larger change in the collector current, the ratio

$$\frac{(I_{c_3} - I_{c_2})}{(I_{b_3} - I_{b_2})} = \frac{\Delta I_c}{\Delta I_b} \gg 1$$

being the current amplification factor. Also, for a given base current I_b, the collector current I_c saturates as soon as V_{ce} reaches a few hundred millivolts, so that the device acts as a source of constant current. The pnp transistor gives similar results, except that the polarities of the applied voltages are reversed.

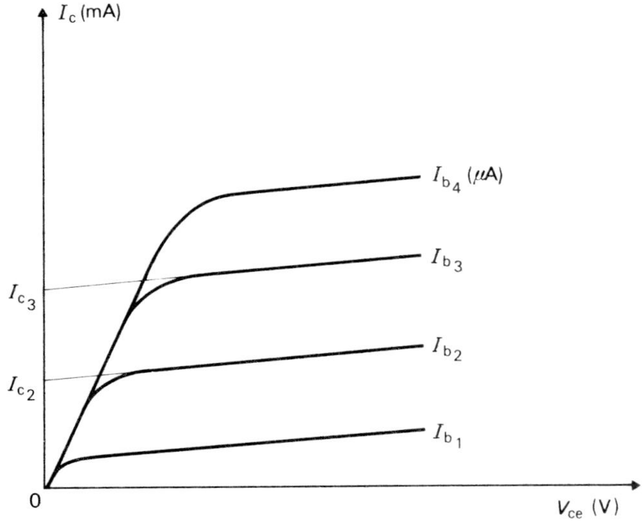

Fig. 1.22 I_c–V_{ce} characteristics of a junction transistor

Depending on the circuit applications, junction transistors may be used in the common-emitter, common-base or common-collector configurations, discussed in general terms in

Chapter 2. The switching action of the bipolar transistor and its uses in digital circuitry are discussed in Chapter 4. In the common-emitter case, the base current is related to the junction voltages by the equation

$$I_b = I_{se}(e^{qV_{be}/kT} - 1) - I_{sc}(e^{qV_{cb}/kT} - 1) \tag{1.88}$$

As V_{cb} is large and negative, the second term can be neglected, and the input current obeys the diode law. However, the reverse collector-base voltage must never exceed the specified value because the base-collector depletion region spreads through the base to the high-conductivity emitter region, resulting in a collector-emitter short-circuit. This phenomenon is known as **reach-through** and is often confused with junction breakdown.

Unlike valves, bipolar transistors are current-controlled devices, and their constructional details determine their actual characteristics, frequency response and power handling capabilities. Low-frequency transistors, switching transistors, microwave transistors and power transistors are some of the varieties currently available. Complementary (npn–pnp) pairs of transistors greatly simplify circuit design and are widely used in both analogue and digital instrumentation.

A **junction field-effect transistor** (**JFET**), unlike the bipolar junction transistor, is a voltage-controlled device with a high input resistance. It consists either of an n-type channel embedded in a p-type substrate, or of a p-type channel embedded in an n-type substrate. Such transistors are usually produced by planar epitaxial methods.

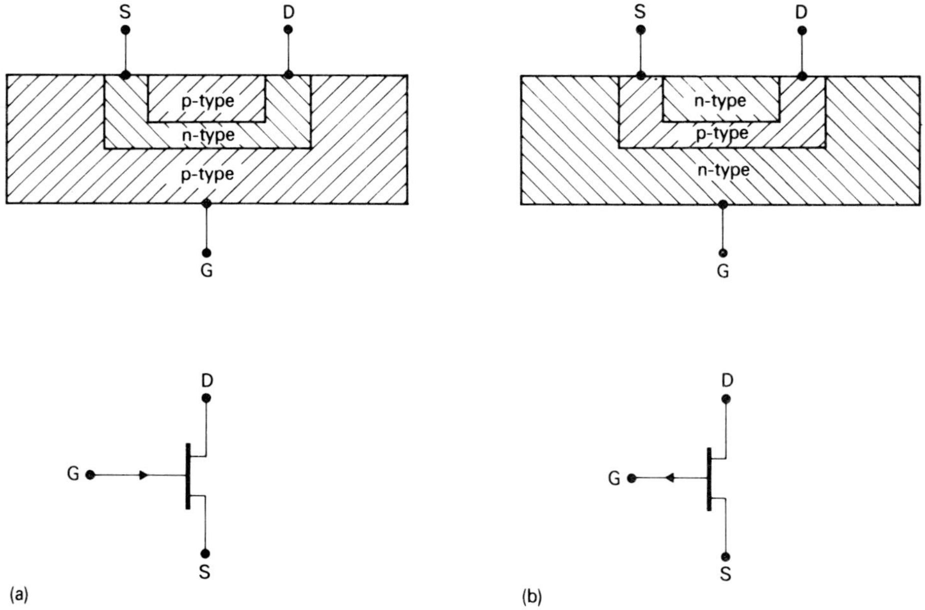

Fig. 1.23 Structure and circuit symbol of (a) n-channel JFET and (b) p-channel JFET

The two ohmic contacts on the channel are the **source** (S) and the **drain** (D). The third contact on the substrate is the **gate** (G).

In an n-channel FET, when the gate has a negative voltage with respect to the source, the depletion layer of the reverse biased p-n junction will extend into the weakly doped n-channel and reduce the cross-section available for the passage of current. If the gate-source voltage $|V_{GS}|$ is greater than $|V_p|$ then the channel current will be cut off entirely; V_p is called the gate **pinch-off** voltage. The gate is thus capable of controlling the channel current and, unlike the base-emitter junction of a bipolar transistor, the gate-source junction is reverse biased and will have a very high resistance.

If the gate and source are short-circuited, $V_{GS} = 0$ and the drain is made progressively positive with respect to the source, then for $|V_{DS}| > |V'_p|$, the drain current I_D will reach a saturation value I_{DSS}. The V_{DS}–I_D characteristics for various values of V_{GS} are shown in Fig. 1.24. The quantity V'_p is also called the pinch-off voltage, and for each value of V_{GS} the points A, B, C and D indicate the onset of saturation. Obviously, for a p-channel JFET, the polarities of the bias voltages must be reversed.

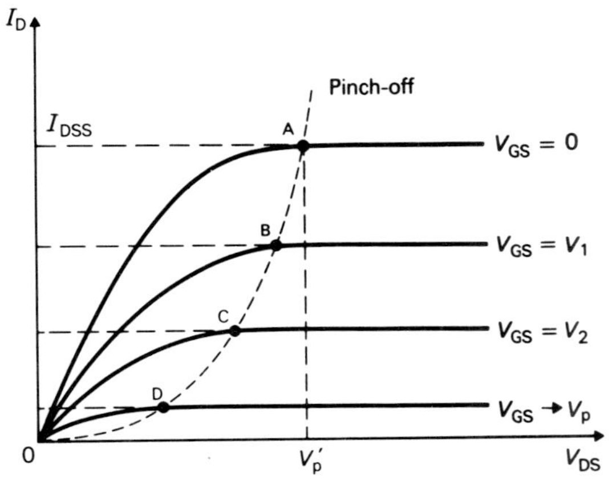

Fig. 1.24 I_D–V_{DS} characteristics of a JFET

The value of I_{DSS} is an important characteristic of a JFET and depends on the channel majority carrier mobility, the gate capacitance, the gate pinch-off voltage and the ratio W/L where W is the channel width and L is the channel length.

In the saturation region, the dependence of I_D on V_{GS} can be described by the expression

$$I_D = I_{DSS}\left(1 - \frac{V_{GS}}{V_p}\right)^n \tag{1.89}$$

where $n \approx 2$ and V_{DS} is a constant which is greater than or equal to V'_p.

Equation 1.89, with its delightful simplicity, is a fairly good approximation[5] of the transfer curve of most FETs and is quite an accurate approximation for FETs with narrow channels, irrespective of their manufacture process. Figure 1.25 shows the square law approximation to a FET transfer characteristic: this feature is very useful in the design of circuitry using FETs as active circuit elements. With $n = 2$, the transconductance g_m in the pinch-off region can be deduced from equation 1.89 as:

$$g_m = \frac{\Delta I_D}{\Delta V_{GS}} = -\frac{2}{V_p}(I_{DSS} I_D)^{1/2} \tag{1.90}$$

Thus for a high g_m the value of W/L must be high. On the other hand WL, the area of the

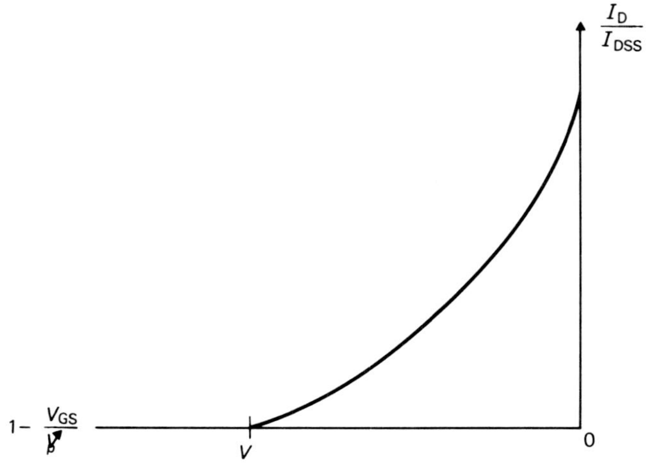

Fig. 1.25 Square law approximation to an FET transfer characteristic

channel, controls the gate capacity, which must be low. In practice, optimal design aims at the lowest possible value of L and a compatible value of W.

MOS devices

There are several types of metal oxide semiconductor field-effect transistor (**MOSFET**). The type known as the **insulated gate** FET (**IGFET**) are among the most common discrete MOS devices. A metallic gate electrode, insulated from the channel, is used to control the channel current. There are two operating modes: the enhancement mode and the depletion mode. In the **depletion** type, or normally ON MOSFET, a conducting channel exists with zero gate bias. As in the JFET, a reverse gate bias depletes the number of charge carriers of the channel and reduces its conductance. Unlike the JFET, a forward gate bias enhances the number of charge carriers and increases the channel conductance. In the **enhancement** type, or normally OFF MOSFET, the channel conductance is zero at zero gate bias, so that $I_{DSS} = 0$. The n-channel enhancement type and the p-channel depletion type are comparatively rare, because the processes involved in their manufacture are more complicated.

Figure 1.26 shows the circuit symbols for the four types of MOSFET; in each case 'b' is the substrate.

Figure 1.27 shows a p-channel enhancement type MOSFET. The heavily doped source and drain p-type regions are embedded in an n-type substrate. The gate G consists of a metal plate isolated by an oxide layer, so that $I_D = 0$ when $V_{GS} = 0$ irrespective of the polarity of V_{DS}. When V_{GS} is negative, electrons are repelled from the substrate under the gate and a depletion layer is formed between the source-drain regions. If the gate is sufficiently negative, free holes are attracted into the depletion layer and a p-type channel or **inversion layer** is formed at the oxide-substrate interface, through which drain current can flow. The gate voltage required to form the channel is called the **threshold voltage** V_T, and an increase in the negative gate voltage above the threshold induces more free holes in the channel and enhances the drain current. Figure 1.28 shows the I_D–V_{DS} characteristics for $-V_{GS} > V_T$.

Figure 1.29 shows an n-channel depletion type MOSFET. The source, drain and channel regions are all n-type, so that drain current flows even with zero bias. The substrate forms p-n

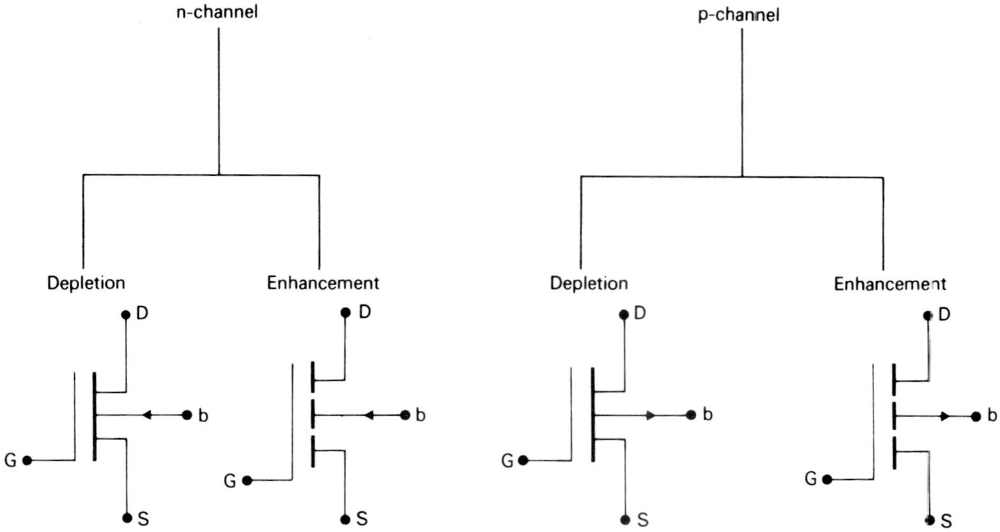

Fig. 1.26 Circuit symbols of four types of commonly-used MOSFETs

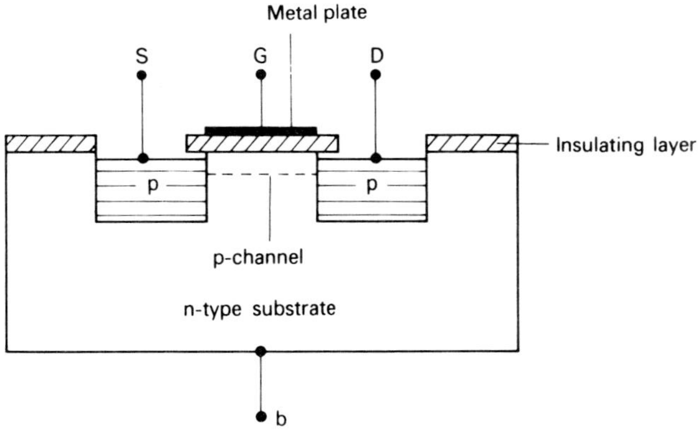

Fig. 1.27 p-channel enhancement type MOSFET

junctions with the heavily doped source and drain regions. When the gate is made positive with respect to the source, additional electrons are attracted into the channel from the n-type source region, thereby increasing the channel conductance. If V_{GS} is negative, free electrons are repelled from the channel and a carrier depletion layer is formed at the oxide-silicon interface. The channel is thus constricted as in a JFET.

If V_{GS} is sufficiently negative, this depletion region extends through the channel and the drain current is reduced to zero. This value of V_{GS} is the **pinch-off** voltage V_p shown in the I_D–V_{DS} characteristics of Fig. 1.30. In general, the I_D–V_{DS} characteristics of the MOSFET do not differ basically from those of the JFET. The drain current in the pinch-off region can be written as

$$I_D = \tfrac{1}{2}\beta(V_{GS} - V_p)^2 \qquad (1.91)$$

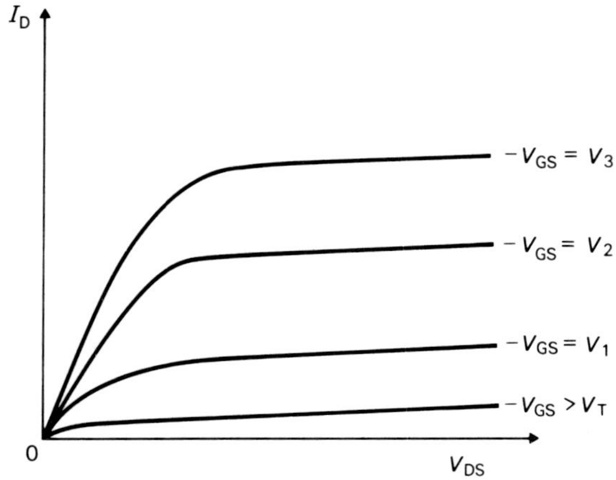

Fig. 1.28 I_D–V_{DS} characteristics of a p-channel MOSFET for $-V_{GS} > V_T$

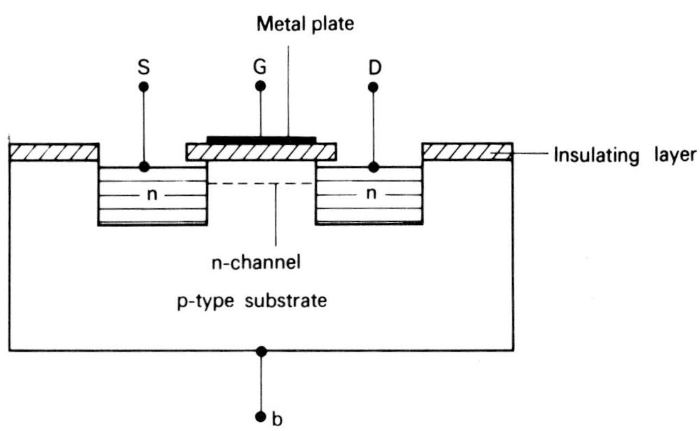

Fig. 1.29 *V*-channel depletion type MOSFET

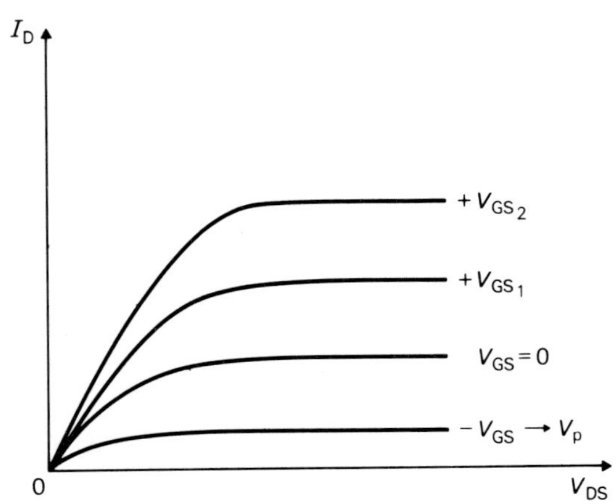

Fig. 1.30 I_D–V_{DS} characteristics of an n-channel depletion type MOSFET

where $\beta = \mu_s C_0 W/L$, μ_s is the surface mobility of the charge carriers in the channel, c_0 is the capacitance per unit area $= \varepsilon\varepsilon_0/d$ and d is the thickness of the oxide layer.

For normally ON MOSFETs, equation 1.91 may be written as

$$I_D = I_{DSS}\left(1 - \frac{V_{GS}}{V_p}\right)^2 \tag{1.91a}$$

where $I_{DSS} = \frac{1}{2}\beta V_p^{\,2}$.

Various types of MOSFETs, such as the ones with two gates and integrated circuits incorporating both a MOS device and a bipolar transistor, are currently available. However, one recent development is the **vertical MOS (VMOS)** structure which merits a brief description (Fig. 1.31).

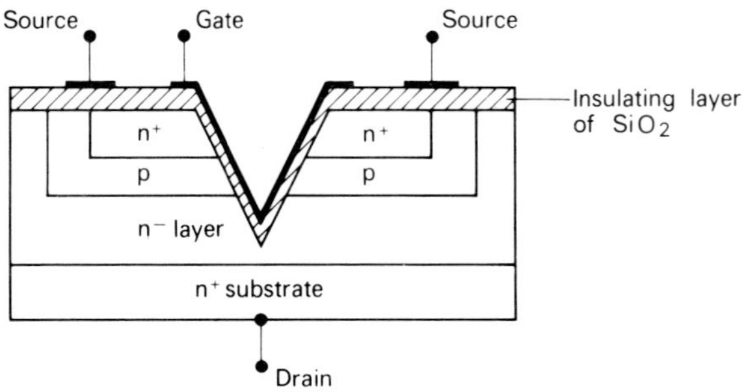

Fig. 1.31 Basic VMOS structure

An n-layer is grown on an n^+ substrate to form the drain region. The p-type body region and the n^+ source are then diffused. A precision V-groove is then etched into the chip surface. It is centred on the n^+ source region and penetrates the n^+-layer and the p-type body into the n-type epitaxial drain region. The gate oxide and contact metallisation are produced by the standard photolithographic techniques outlined in Chapter 2. During enhancement, two vertical current paths are created, one on each side of the groove, so that the current carrying capability is automatically doubled.

The physical VMOS structure has the following advantages over the 'lateral' MOS structures described earlier.

1) The channel width-to-length ratio is optimised for a low 'ON resistance'. This is important in switching applications and is compatible with a high transconductance.
2) As the drain region is also the substrate, no top surface drain contact is required and the chip surface area is reduced. Also, the heat conduction efficiency between the chip and the package is substantially improved.
3) Being short-channel majority carrier devices, VMOS technology does not exhibit delays due to carrier recombination. Fast switching speeds at high currents are therefore possible.
4) The effective input impedance is high and the drive power requirements are low. The devices can be driven directly by CMOS (complementary MOS) logic and TTL (transistor-transistor logic) signals.

Devices depending on VMOS technology with power ratings of 90 V at 2 A are currently available commercially. Future developments are aimed at VMOS power FETs with ratings such as 100 V at 10 A, 200 V at 7 A or even 400 V at 5 A.

Devices mainly used for switching and power control

The **unijunction transistor** (UJT), or double base diode, consists of a high-resistivity n-type bar with two ohmic contacts, base 1 (B_1) and base 2 (B_2), and a low-resistivity p-type insert, the emitter E, in the middle, just like a JFET (Fig. 1.32). However, the main constructional difference is that the emitter junction area of the UJT is much smaller than the gate junction area of the FET.

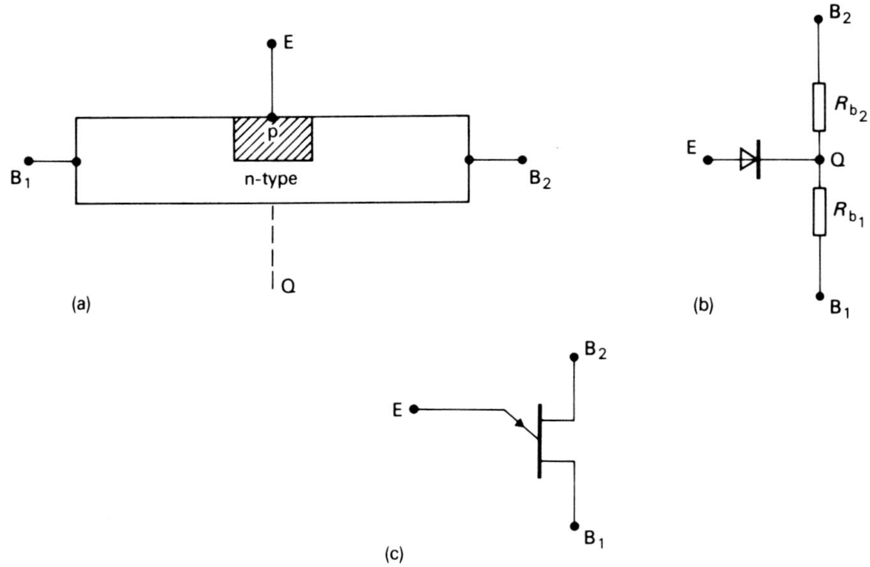

Fig. 1.32 (a) UJT structure; (b) circuit representation; (c) circuit symbol

The mechanism of the UJT is as follows. The base B_2 is made positive with respect to B_1 by applying a voltage V_{bb}. If the resistance between B_1 and B_2 is R_{bb}, that between B_1 and Q is R_{b_1} and that between B_2 and Q is R_{b_2}, then

$$\eta = \left(\frac{R_{b_1}}{R_{b_1}+R_{b_2}}\right) = \text{intrinsic stand-off ratio} \qquad (1.92)$$

The emitter E is made positive with respect to B_1 by applying a voltage V_{eb}. As long as $V_{eb} < \eta V_{bb}$, the junction is reverse biased. When $V_{eb} > \eta V_{bb}$ by an amount equal to the forward voltage drop of the p-n junction, holes are injected into the bar, thereby causing an increase in conductivity in the region between the emitter E and base B_1. Hence, the emitter current increases and V_{eb} decreases, so that the device exhibits a negative resistance between the **peak point** V_p and the **valley point** V_v. The region with current less than I_p is known as the **cut-off** region, and the region with a current larger than I_v is known as the **saturation** region. Complementary UJTs with p-type bases and n-type emitters are also available.

The commonest applications of the UJT are in the relaxation oscillator circuits used in pulse generators, sawtooth generators and timers.

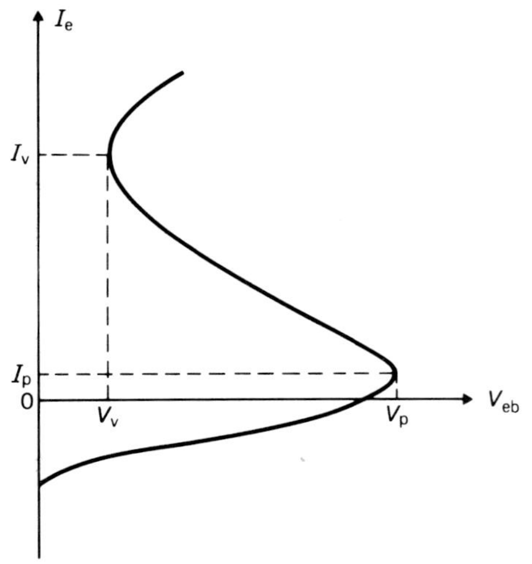

Fig. 1.33 UJT: *I–V* characteristic of the emitter

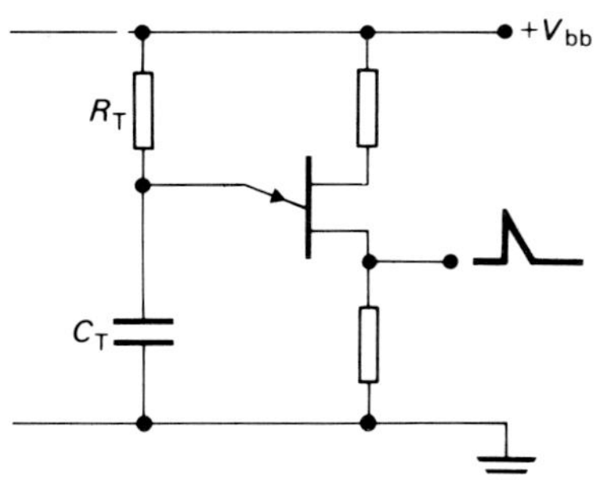

Fig. 1.34 An UJT oscillator

Figure 1.34 shows a simple oscillatory circuit in which the capacitor C_T charges through the resistor R_T until the value of V_{eb} reaches V_p. The resistance R_{b_1} then decreases rapidly and discharges C_T. With C_T discharged, R_{b_1} increases and the cycle is repeated. Thus the condition for oscillation is

$$\frac{V_{bb} - V_p}{I_p} > R_T > \frac{V_{bb} - V_v}{I_v}$$

The frequency of oscillation is given by

$$f \approx \frac{1}{R_T C_T \ln(1 - \eta)} \tag{1.93}$$

The upper frequency limit is set by the **turn-off** time of the UJT.

The **thyristor** or **silicon controlled rectifier** (**SCR**) is a pnpn device which may be represented by a complementary pair of transistors in a closed feedback loop (Fig. 1.35(a)). When the anode A is made positive with respect to the cathode K, no current flows until the device is triggered. The triggering mechanism can be seen by reference to Fig. 1.35(b). A small gate current I_g produces a collector current $\beta_1 I_g$ in transistor T_1. But this is the base current

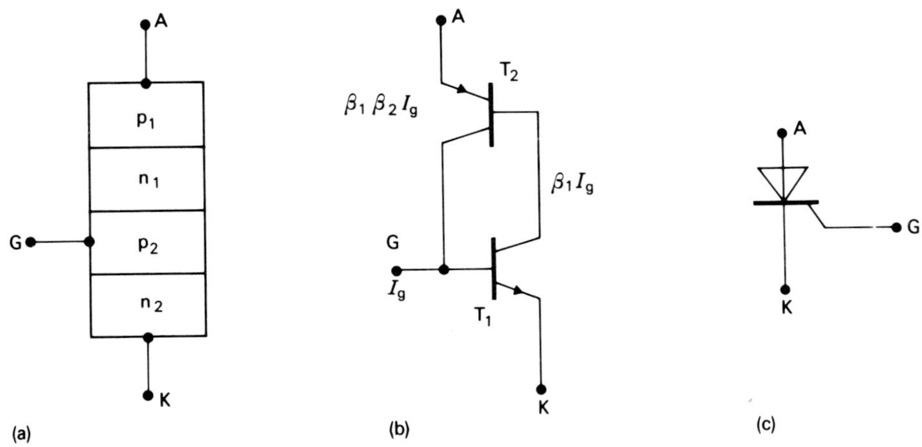

Fig. 1.35 (a) SCR structure; (b) circuit representation; (c) circuit symbol

of transistor T_2, which will give rise to a collector current of $\beta_1\beta_2 I_g$. Hence the condition for triggering is that

$$\beta_1\beta_2 > 1$$

Once triggered, the gate loses control of the anode-cathode current, which is limited by the external circuit. In order to turn the device off, the anode current must be reduced below a holding value. When used in DC circuits, 'turn-off' is achieved by reversing the anode-cathode voltage. In AC circuits, turn-off occurs automatically during the negative half of the cycle.

The anode current does not respond immediately to the application of a gate current and is characterised by a **turn-on** time

$$T_{on} = \sqrt{t_1 t_2} \tag{1.94}$$

where t_1 and t_2 depend on the widths of the depletion layers of the p-n junctions and the hole and electron diffusion constants. To reduce the turn-on time, the device must have narrow depletion layers. This requirement is not compatible with that of a high breakdown voltage. Hence, high-voltage thyristors have a long turn-on time.

To switch the device back to its blocked state, the excess carriers in all three forward biased junctions must be swept out by an electric field or must decay by recombination. If I_f is the forward current at the time of turn-off, I_h the holding current, and τ_p the minority carrier lifetime, the turn off time T_{off} is given by

$$T_{off} \approx \tau_p \ln (I_f/I_h) \tag{1.95}$$

in the absence of any back bias usually applied to speed up turn-off.

A wide variety of pnpn devices, differing mainly in their power handling capabilities and in the mode in which they switch ON and OFF, are currently available. For instance, a low-current, low-voltage SCR which is triggered by shining a light beam in the vicinity of the reverse biased centre junction is called a **light activated SCR** (**LASCR**). Circuit techniques for using the SCR as a power control element will be discussed in Chapter 6.

The **programmable unijunction transistor** (**PUT**) is also a pnpn device, used to perform the functions of a unijunction transistor (Fig. 1.36). The anode A corresponds to the emitter

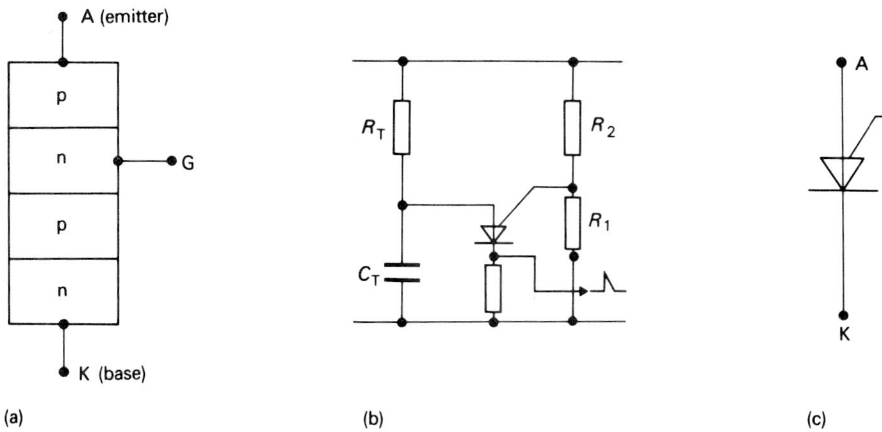

Fig. 1.36 (a) PUT structure; (b) an oscillator similar to the one shown in Fig. 1.34; (c) circuit symbol

of the UJT (Fig. 1.32), the cathode K to base 1 and the gate G to the point Q of the UJT. Figure 1.34 shows the oscillatory circuit of Fig. 1.36(b) with the UJT replaced by the PUT. The **stand-off ratio** η is now defined by the external resistances R_1 and R_2, i.e. $\eta = R_1/(R_1 + R_2)$. The main advantage of the PUT over the UJT is that its stand-off ratio, its peak and valley point currents, and its interbase resistance are all programmable.

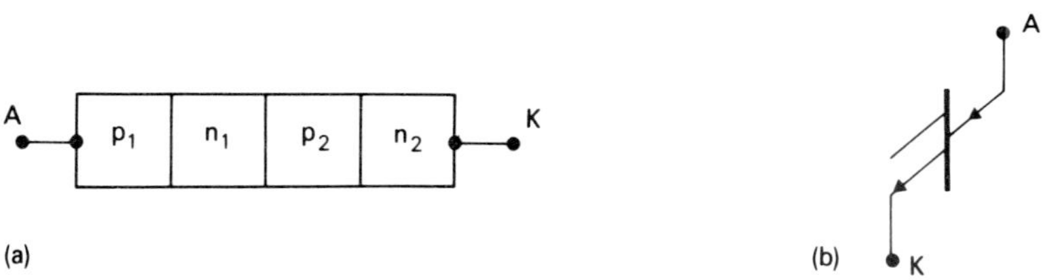

Fig. 1.37 (a) Shockley diode structure; (b) circuit symbol

Another interesting trigger device which may be considered as a pair of complementary transistors is the **Shockley diode**. However, the structure is such that conduction occurs only when the anode is a certain amount more positive than the cathode, and ceases when the anode current falls below a holding value. It is often used for triggering thyristors and its *I–V* characteristics are shown in Fig. 1.38.

There are several multijunction devices that can be triggered into conduction in either the forward or the reverse direction. The simplest example of such a device is the **bidirectional switch** or **DIAC** and the **gated bidirectional switch** or **TRIAC**. In the case of the DIAC, Fig. 1.39, when the terminal 1 is made sufficiently positive with respect to terminal 2, the section $p_1n_2p_2n_3$ switches from the blocked state to the conducting state. When terminal 2 is made sufficiently positive with respect to terminal 1, the section $p_2n_2p_1n_1$ will switch to the conducting state.

Similarly, in the case of the TRIAC, Fig. 1.40, when terminal 2 is made sufficiently positive with respect to terminal 1, the $p_1n_3p_2n_5$ region acts as an SCR which can be triggered into conduction by making the gate positive with respect to terminal 1. The junctions p_1n_1, p_1n_2,

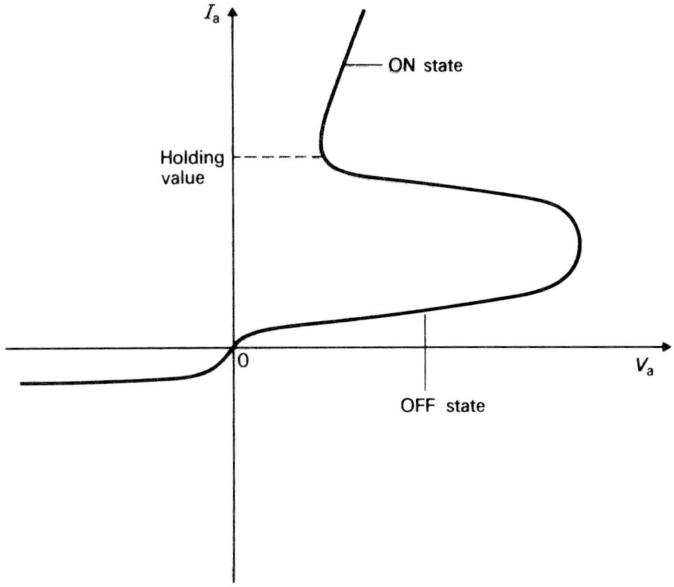

Fig. 1.38 I_a–V_a characteristic of a Shockley diode

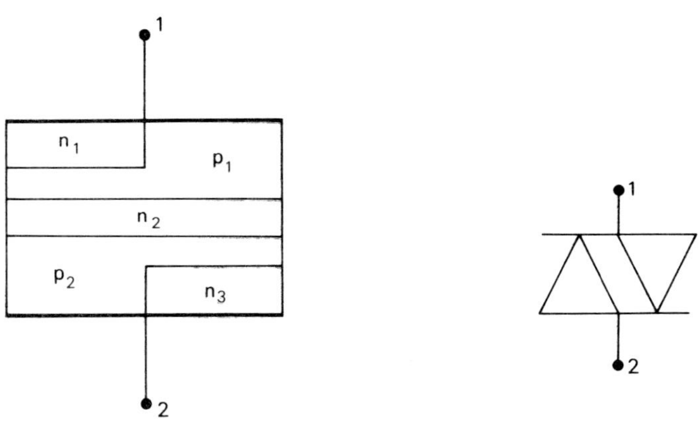

Fig. 1.39 Structure of a DIAC and its circuit symbol

and p_2n_4 are then reverse biased. When, however, the terminal 1 is positive with respect to terminal 2, the region $p_2n_3p_1n_1$ will switch into conduction when the gate is made negative with respect to terminal 1. In short, a TRIAC can be turned on by a negative gate current when the terminal 2 is negative with respect to terminal 1, and by a positive gate current when the terminal 2 is positive with respect to terminal 1. In practice, however, either gate polarity will turn the TRIAC on, but the above 'preferred' polarities require a smaller gate current.

DIACS and TRIACS are mainly used in full-wave AC power control and greatly simplify the trigger circuitry.

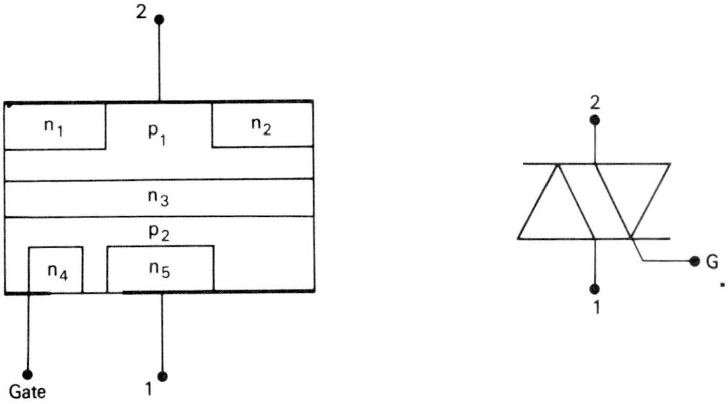

Fig. 1.40 Structure of a TRIAC and its circuit symbol

1.5 Optoelectronic devices

These can be broadly divided into (a) emitters and (b) detectors of radiation. Each of these types can be subdivided as shown below.

- Optoelectronic devices
  - Emitters
    - Laser type
    - Electroluminescent type
  - Detectors
    - Photo-conductors
    - Photo-junctions
    - Photo-voltaic

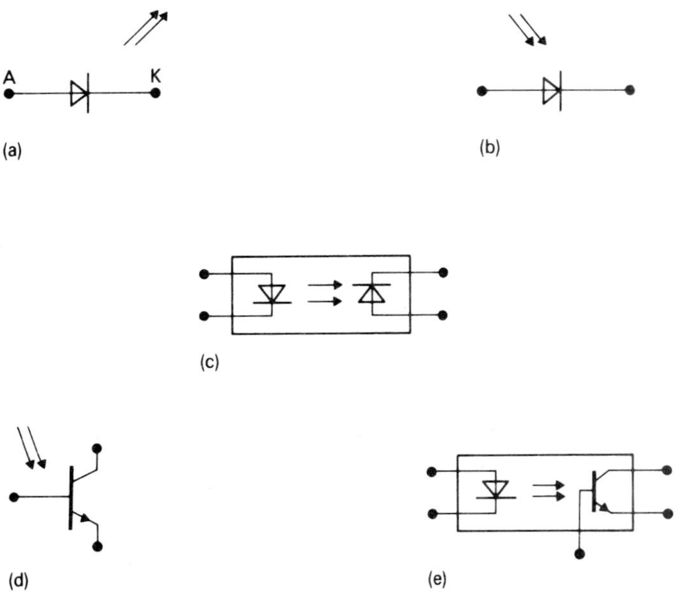

Fig. 1.41 Circuit symbols of (a) an LED; (b) a photodiode; (c) an LED-photodiode opto-coupler; (d) an npn phototransistor; (e) an LED-phototransistor opto-coupler

Emitters

Laser type emitters

Laser type emitters consist of p-n heterojunctions, i.e. junctions between materials of different band gaps, pumped by electron-hole injection. Composite materials such as Ga-Al-As are grown epitaxially onto GaAs to produce heterojunctions with a high injection efficiency to one type of carrier. They emit coherent radiation and when used in the pulsed mode can deliver up to several watts of peak output power.

Electroluminescent devices

Most of these devices use an electric current to enhance the number of electrons at higher energy levels, giving rise to incoherent radiation. There are several mechanisms for enhancing the population in the higher energy levels, but gallium arsenide infrared sources and gallium phosphide light emitting diodes (LEDs) rely on forward bias injection of charge carriers across a p-n junction doped with Cd, Zn etc. A fairly high electroluminescent efficiency is obtained and the practical applications are in opto-coupling, illumination and numerical displays.

Detectors

Photoconducting devices

These devices consist of thin-film semiconducting materials such as CdS and PbS with two ohmic contacts. When light of a suitable wavelength is incident on the surface, carriers are generated by band-to-band transitions (intrinsic), or transitions involving energy levels in the forbidden gap (extrinsic), and the resistivity of the material decreases. The initial resistivity is high and when a voltage is applied the current flowing in the external circuit, in the absence of illumination (dark current), is small. The wavelength of the incident radiation must be such that the energy of the incident photons exceeds either the band gap value or the energy separation of the band gap levels.

The performance of such devices is measured in terms of their quantum gain and response time. The **quantum gain** is the ratio of the number of charge carriers flowing in the external circuit per second to the number of photons absorbed in the same time. The **response time** defines the speed of photodetection. In general, photoconductors have a high quantum gain but a slow response time and make better devices when combined with p-n junctions. Special purpose photoconductors with a fast response time are also available.

Photojunctions

These are reverse biased p-n, pin, Schottky barrier, avalanche or heterojunctions in which the incident photons give rise to electron-hole pair production, thus changing the reverse saturation current. The type of junction mechanism used depends on the speed of response, the dark current and the wavelength range at which the device is to be used. Photojunctions have a faster response time and a smaller dark current than photoconductors, but their quantum gain is ≈ 0.8 to 0.9 except in the case of avalanche photodiodes, in which high reverse fields are used and electron multiplication takes place by impact ionisation.

Photojunctions are also used in phototransistors, in which the reverse biased base-collector junction is exposed to radiation, and current gain is obtained by transistor action.

Photovoltaic detectors

The reciprocal effect to electroluminescence is the photovoltaic effect, which is essentially the conversion of radiated energy into electrical energy. Apart from its use in light intensity meters for photography, the main application is in the manufacture of solar cells. It is common to use p-n junctions with large surface areas and long minority carrier lifetimes. Electron-hole pairs, produced by the incident radiation in the vicinity of the depletion layer, diffuse into the depletion layer and contribute to the reverse saturation current I_s so that the open circuit voltage V_o generated is approximately given by

$$V_o \approx \frac{kT}{q} \ln\left(\frac{I_L}{I_s}\right) \tag{1.96}$$

where I_L is the contribution of the photogenerated carriers. The maximum power is generated at a slightly lower voltage.

Solar energy conversion efficiency of about 15% is obtained by using materials like CdS and GaAs.

1.6 Bulk semiconductor mechanisms

Bulk-effect devices consist of a single semiconductor with two or more ohmic contacts. Their action depends on the bulk properties of the material rather than on p-n junctions or metal-semiconductor interfaces. Apart from photoconductors, the following commonly-used bulk devices merit description: (i) Hall effect devices, (ii) thermistors, (iii) Gunn effect devices.

Hall effect devices

If an electric field, applied to a bulk semiconductor, produces an average drift velocity in the x-direction, and if a magnetic field is applied in the y-direction, then the charge carriers will experience a Lorentz force deflecting them sideways as they travel through the semiconductor. For an isotropic semiconductor, the open-circuit Hall voltage V_H developed in the z-direction is given by

$$V_H = \frac{rI\mathbf{B}}{qn} \tag{1.97}$$

where n is the number of carriers per unit volume; I is the current through the sample; r is a

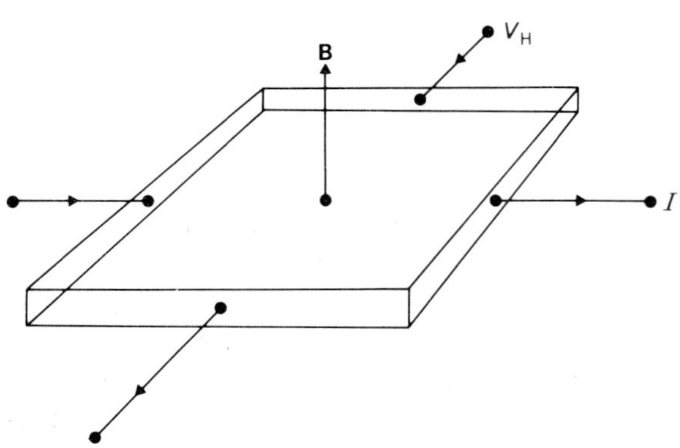

Fig. 1.42 Hall effect device

constant equal to unity for the simple case without considering phonon or impurity scattering; $\mathbf{B}$ = magnetic field; $R_H = r/qn$ is called the Hall constant and the Hall mobility $\mu_H = R_H\sigma$, where σ is the conductivity.

As the magnetic field will force charges of opposite sign in opposite directions, p-type and n-type extrinsic semiconductors may be distinguished by the sign of the Hall voltage generated in them. Also, knowing I and $\mathbf{B}$, the carrier concentration can be calculated from the measured value V_H.

Thermistors

These are temperature-sensitive resistors, usually made of semiconducting materials such as metal oxides, with two ohmic contacts. Their conductivity depends on carrier mobilities and concentration (see Chapter 1). However, the most temperature-sensitive parameter is the majority carrier concentration which varies as $e^{-B/T}$, where B is a constant depending on the band gap and impurity concentration.

The variation of resistance with temperature can be represented as

$$R = R_0 \exp B\left(\frac{1}{T} - \frac{1}{T_0}\right) \tag{1.98}$$

where R_0 is the resistance at the ambient temperature T_0, and the power dissipated is given by $W = C(T - T_0)$ where C is the dissipation constant.

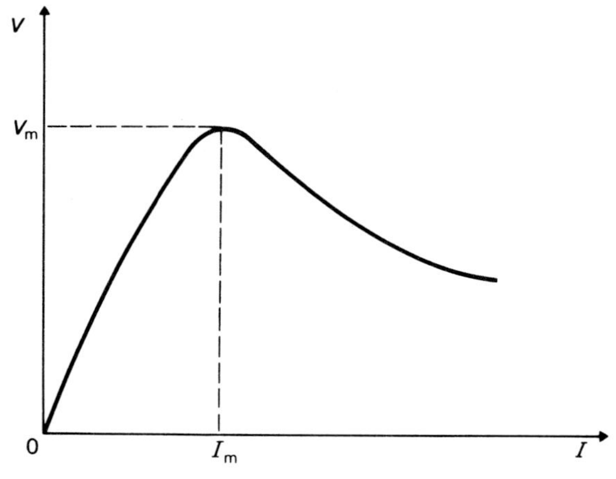

Fig. 1.43 *V–I* characteristic of a thermistor

The *V–I* characteristics are constructed by noting the steady-state value of the voltage across the thermistor for each value of the current through it. For small currents, the temperature of the thermistor does not change much and Ohm's law is obeyed. Beyond a certain value I_m, the temperature of the thermistor increases and the carrier concentration increases rapidly, offsetting the decrease in mobility. The resistance therefore decreases and the voltage across the thermistor drops. The operating point of the thermistor must be selected carefully to avoid damage due to overheating.

Bead type thermistors are used in temperature measurement and in anemometry, whereas the rod and disc types are used in control applications such as surge suppression and time delay. Depending on the dissipation constant and dimensions, the response time of thermistors ranges between 1 and 30 s.

Gunn devices

Commonly-used Gunn devices consist of an epitaxial layer of n-type gallium arsenide grown on a low-resistivity substrate of the same material. Although the two terminals of the device are known as the cathode and anode, there is no p-n junction.

The n-type gallium arsenide material is such that its conduction band contains a higher energy **satellite** band in which the effective mass of the electrons is higher and their mobility lower. Provided that the energy gap between the normal conduction band and the satellite band is considerably smaller than the forbidden gap between the valence and conduction bands, the majority of conduction band electrons can be excited into the satellite band by the application of an electric field of about 350 volts per millimetre. As the critical field ε_c is reached, the average velocity of the electrons is reduced and a negative resistance characteristic AB is produced (Fig. 1.44). When all the available electrons have moved into the satellite band, the velocity increases with the applied field (BC in Fig. 1.44).

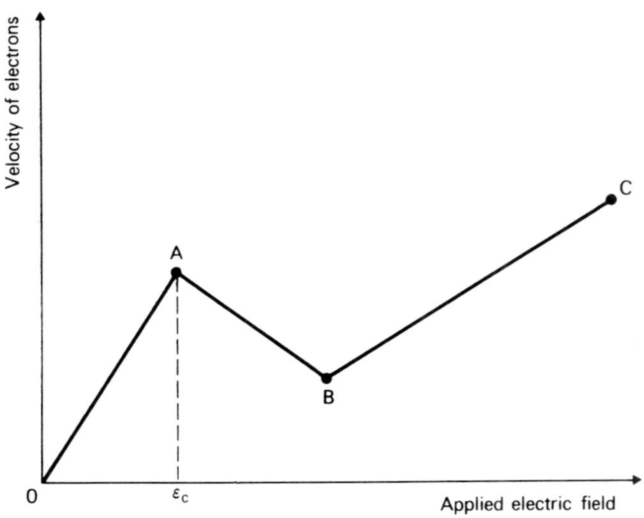

Fig. 1.44 Mechanism of a Gunn device

Unlike conventional negative resistance devices, when a Gunn device is biased to the negative resistance region, high field domains pass between the anode and cathode. This gives rise to a train of pulses whose frequency is determined by the transit time of the domains through the epitaxial layer.

When used to generate microwaves in the 8–12 GHz range, the device is mounted in a tuneable cavity and can deliver several milliwatts of microwave power. Devices with a higher power rating have successfully replaced reflex klystron microwave generators.

References

(1) Phillips, J. C., Energy band interpolation scheme based on a pseudopotential. *Phys. Rev.*, **112**, 685 (1958)
(2) Heine, V., The pseudopotential concept. *Solid State Physics*, **24**, Academic Press (1970)
(3) Kittel, C., *Introduction to Solid State Physics*, 2nd edition. Wiley (1965)
(4) Sze, S. M., *Physics of Semiconductor Devices.* Wiley (1969)
(5) Sevin, L. J. Jr, *Field Effect Transistors.* McGraw-Hill (1965)

Further reading

Ankrum, P. D., *Semiconductor Electronics.* Prentice-Hall (1971)
Brillouin, L., *Wave Propagation in Periodic Structures*, 2nd edition. Dover (1953)
Carroll, J. E., *Physical Models for Semiconductor Devices.* Edward Arnold (1974)
Van der Ziel, A., *Solid State Physical Electronics*, 2nd edition. Prentice-Hall (1968)

2

Microelectronic device technology

2.1 Introduction

Most modern electronic instruments use commercially available monolithic integrated circuits and microelectronic circuit assemblies. Individual semiconductor devices, encapsulated to withstand a specified mechanical stress and temperature, are classed as discrete devices and are used mainly in interface and power circuitry. The following sections outline the basic physical processes involved in the microminiaturisation of components, complex electronic circuits and circuit assemblies. The resultant microelectronic devices, with their small size, high reliability and low cost, may be looked upon as the 'building blocks' of modern analogue and digital circuitry discussed in Chapters 3 and 4.

The present pace of developments in the manufacture of microelectronic circuits suggests a further reduction in the space required for a given electronic function and the cost of implementing it.

2.2 The planar technique

The method used to manufacture microelectronic circuits is an extension of the planar technique, which was initially devised for the manufacture of planar diodes and transistors. Such circuits are made up of planes, or layers, of p-type and n-type silicon or other semiconductor materials. Figure 2.1(a) shows a planar diode structure consisting of two layers, and Fig. 2.1(b) shows the structure of an npn planar transistor made up of three layers, built on a slice of silicon which is subsequently cut up into individual wafers. The electrical connections are 'dye bonded' and the type of encapsulation of a batch of wafers is selected according to the proposed circuit applications (see Section 2.6).

Similarly, the structure of a microelectronic circuit consists of many layers, each of which has a detailed structure. Some of the layers lie within a slice of pure silicon, while others are stacked on the top. The manufacturing process consists of forming this sequence of layers very accurately so that an entire electronic circuit can be assembled on a single silicon slice. Resistors are formed in the silicon slice by diffusing a suitable impurity into a defined region. The value of the resistor depends on the concentration of the impurity, the dimensions of the region at the surface and the depth to which the impurity is diffused. Similarly, capacitors are made by employing the capacity of a reverse biased p-n junction or a MOS structure. The main physical processes involved in the fabrication of microelectronic circuits are:

1) preparation of a pure silicon slice and controlled oxidation;
2) preparation of masks and photolithography;
3) selective doping;
4) thin film deposition, isolation, interconnection and encapsulation.

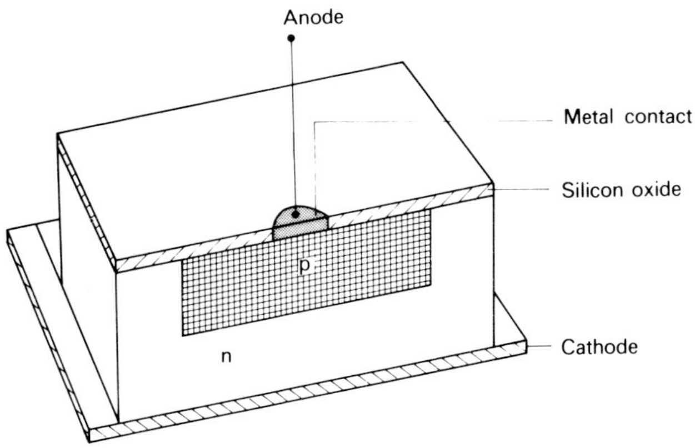

(a)

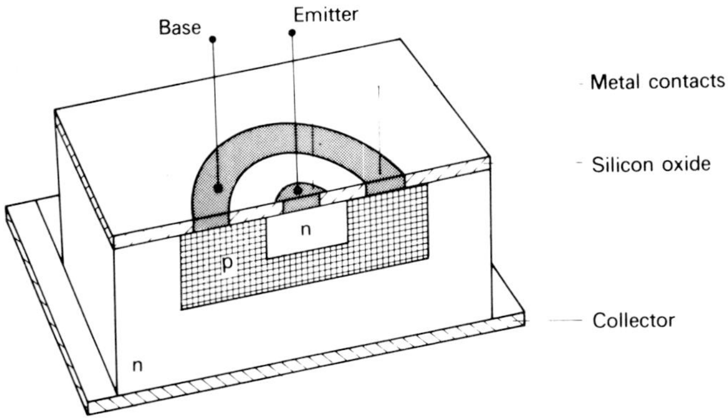

(b)

Fig. 2.1 (a) Planar diode structure; (b) planar transistor structure

2.3 Preparation of a pure silicon slice and controlled oxidation

Raw silicon is chemically purified to one part in 10^7. When it is in molten form and in an inert atmosphere, trivalent or pentavalent impurities are added to produce p- or n-type silicon. A large single crystal is grown from the melt using a perfect single crystal 'seed'. The crystal is ground into a cylindrical shape, and slices about 0.5 mm thick are cut from it using a high-speed diamond saw. A slice to be used for the fabrication of a microcircuit must be ground and highly polished on one side in a 'clean' environment in which the dust level is kept to a minimum. It must be pointed out that any polishing defect, dust or chemical pollution of the sample at this stage can result in the malfunction of the entire microcircuit.

The widespread commercial use of silicon as the material for microelectronic circuits is mainly due to the properties of its oxide. Silicon dioxide is a clear glass with a softening point higher than 1400°C. If the silicon slice is heated in an atmosphere of oxygen or water vapour, a film of silicon dioxide, which adheres well and is hard and durable, forms on the surface of the slice. Moreover, the oxide is an excellent insulator; a property that is utilised for the selective introduction of dopants. Convenient thicknesses of silicon dioxide can be grown at temperatures between 1000°C and 1200°C. The exact thickness can be accurately controlled by selecting the appropriate time and temperature of oxidation. For example, a layer of oxide 0.1 μm thick will grow in one hour at a temperature of 1050°C in an atmosphere of pure oxygen.

Several hundred samples, separated by a few millimetres, are loaded into slots in a quartz 'boat' and slowly moved into the hottest part of a cylindrical high-temperature furnace, through which is passed a purified stream of oxygen-containing gas. The temperature in the process zone is controlled to an accuracy better than one part in 10^3. The temperature and environment of the furnace, as well as the insertion and withdrawal of the samples, are monitored and controlled by a small process-control computer.

2.4 Preparation of masks and photolithography

The first step in the development of a new microelectronic circuit consists of checking its functional characteristics and the physical size and layout of circuit components. Computer simulation is preferred to building and testing a circuit on a 'bread board' because it is easier to manipulate the circuit elements and to optimise the design with a high degree of accuracy. The layout is such that a desired circuit function is achieved using the smallest possible space. The final layout, giving the precise positions of the various circuit elements, specifies the pattern of each layer of the microcircuit.

The computer memory now contains a list of the exact position of every circuit element, and a set of plates called **photomasks** are prepared from this digital information, each mask holding the pattern of a single layer of a microcircuit. The photomask is made by scanning a computer-controlled light spot across a photographic plate in the appropriate pattern. The primary pattern is called the **reticle**, typically ten times the final size of the circuit, and this pattern is checked for errors and corrected. The diminished image of the reticle is projected optically on a suitable plate so as to obtain the final photomask. The latter may consist of a fixed image on an ordinary photographic emulsion plate or a pattern etched in a chromium film on a glass substrate.

The process of engraving the pattern contained by a photomask onto a layer or a substrate is known as **photolithography**. The principle of this process is as follows. In Fig. 2.2(a), an oxidised slice is coated with a photoresist lacquer which polymerises when exposed to ultraviolet light and which then resists attack by acids and solvents. A photomask is placed over the resist-coated slice with opaque regions located at sites where it is required to remove the oxide layer. When illuminated with ultraviolet light, the exposed photoresist polymerises while the unexposed parts under the opaque regions of the photomask are unaffected. These regions can be etched away, leaving windows as shown in Fig. 2.2(b) for selective doping. For example, in Fig. 2.3 p-type material is diffused through a window to join the p-type substrate.

An important requirement of the photolithographic process is that each pattern must be positioned accurately with respect to the layers below it. Several techniques ranging from 'contact printing' to 'projection alignment' are available. The choice in a given situation is usually a compromise between a reasonable lifespan of the mask and adequate resolution to print the smallest circuit elements in the device.

2.5 Selective doping

To construct active elements like MOS and bipolar transistors, it is necessary to introduce

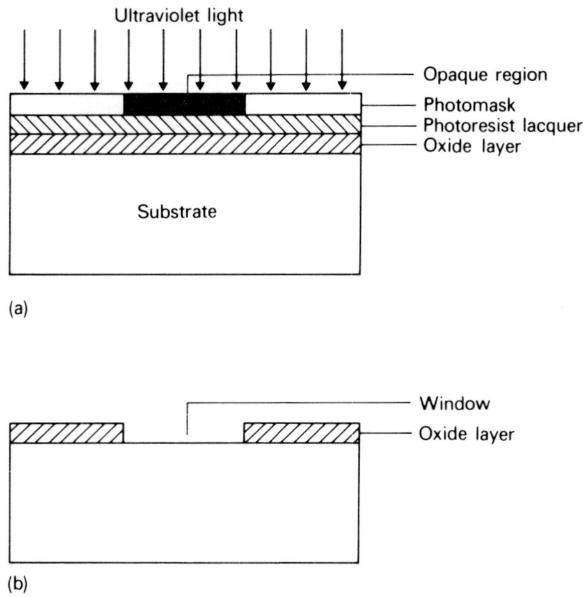

Fig. 2.2 (a) The photolithographic process; (b) 'window' etched in the oxide layer

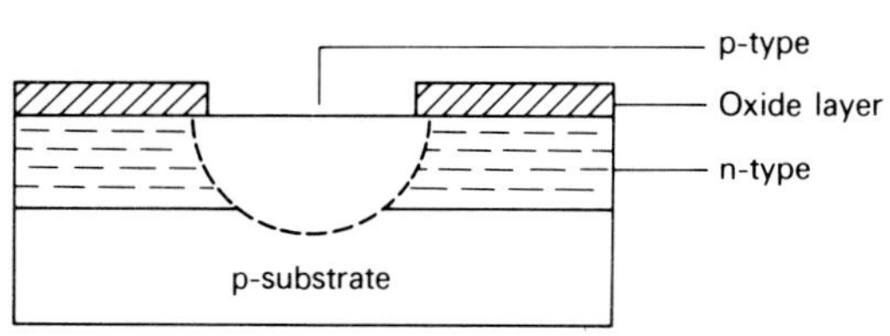

Fig. 2.3 Diffusion of p-type material through the window to join the p-type substrate

impurities selectively so as to create localised n-type and p-type regions by adding appropriate dopant atoms. Selective doping of a silicon crystal may be achieved by one of two techniques:

1) diffusion;
2) ion implantation.

Diffusion

At a temperature of about 100°C, certain impurities such as boron and phosphorus move more slowly through silicon dioxide than through silicon itself. Thin oxide patterns can be used as impurity masks, so that the dopant enters the silicon wherever it is unprotected, diffusing slowly into the bulk of the slice, Fig. 2.3. Solid-state diffusion ceases as soon as the silicon slice is removed from the furnace. The depth to which the impurities diffuse is controlled by the temperature and the time during which the heated silicon is exposed to the dopant atmosphere. For maximum control, diffusions are performed in two steps. In the first step, or **predeposit**, the sample stays in a furnace whose temperature is selected to achieve the best possible control of the amount of impurity introduced, since temperature determines the solubility of the dopant in silicon. In the second step, or **diffusion drive-in**, the sample stays in a furnace at a higher temperature selected to achieve the desired depth of diffusion.

Ion implantation

The dopant ions are accelerated by passing them through a potential difference of about 100 kV. When they strike the silicon slice, they are embedded at different depths depending upon their mass and energy. The silicon slice can therefore be masked by using a patterned oxide layer. For example, in Fig. 2.4(a), by choosing a suitable acceleration voltage the ions can be made to just penetrate the thin region of the oxide layer but not the thick region. By measuring the total charge that accumulates on the crystal, the number of impurities can be determined accurately; hence the choice of the technique in applications where the doping level must be accurately controlled.

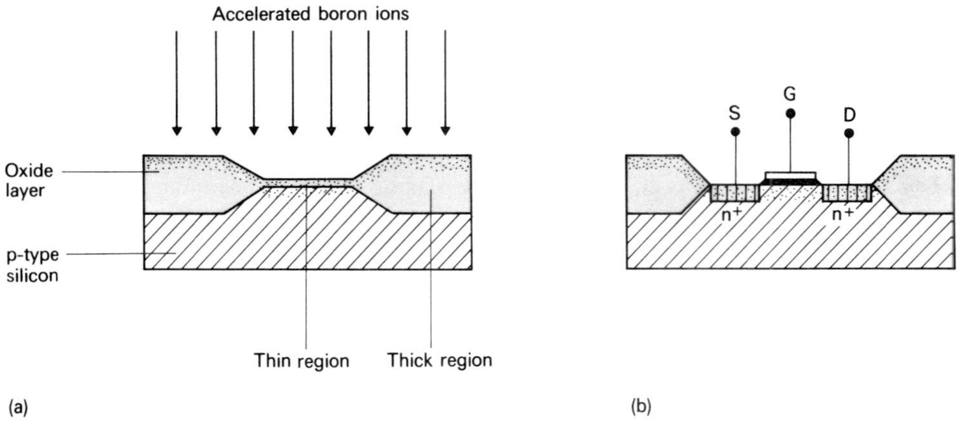

Fig. 2.4 (a) Ion implantation; (b) n-channel MOS transistor

Figure 2.4(b) illustrates the use of both diffusion and ion implantation in the manufacture of MOS transistors. After the implantation of the boron ions, polycrystalline silicon is deposited and patterned to form the gate region of the transistor. A thin layer of the oxide is then removed, and the source and drain regions of the transistor are formed by the diffusion of an n-type impurity.

2.6 Thin-film deposition, isolation, interconnection and encapsulation

Both conducting and insulating thin films, formed in a given pattern, are required in the manufacture of microelectronic devices. Two techniques for the deposition of thin films are currently available:

1) chemical vapour deposition;
2) evaporation.

Silane gas (SiH_4) decomposes into silicon and hydrogen on heating. If the silicon slice is heated in a dilute atmosphere of silane, a uniform film of polycrystalline silicon is formed on its surface. Such films are used in silicon gate-MOS technology. If a source of oxygen such as carbon dioxide is present during the decomposition of silane, then an insulating film of silicon dioxide is formed. Similarly, a thin film of silicon nitride may be formed by decomposing silane in the presence of a nitrogen compound such as ammonia.

In the evaporation technique, the samples to be coated are placed above a crucible containing the coating material, which is usually aluminium. A high vacuum is established and the material is heated by impinging a high-energy electron beam on it. During evaporation, the samples are rotated to ensure a good uniformity of the film, and wobbled with respect to the source to obtain a continuous film. The films are approximately one micrometre thick and are commonly found in the uppermost conducting layers of most microelectronic circuits.

As all the elements of a microelectronic device are in close proximity, each one of them must be electrically isolated in order to minimise unwanted coupling. The only connections between the elements are via the metallised pattern on the surface, which is introduced when the microcircuit is assembled. The two most common methods of isolation are diode isolation and oxide isolation.

Figure 2.3 illustrates the **diode isolation** technique; p-type material is diffused through a window to join the p-type substrate. This leaves the n-type regions each separated from the substrate by a p-n junction. The p-n junctions are all biased in the reverse direction by connecting the p-type substrate to a potential more negative than any part of the circuit. Each junction presents a very high resistance which isolates the element formed in its n-type region.

In the **oxide isolation** technique, a layer of silicon oxide is formed around each element so that it is isolated and supported on a polycrystalline silicon substrate.

Extensive computer-controlled electrical testing is carried out to ensure that all the circuits function properly. The defective ones are marked before separating the silicon slice into individual 'chips' by a 'scribing and breaking' process similar to the one used in glass cutting. The good circuits are bonded into packages and are connected by fine wires to the electrodes leading out of the package. The package is then sealed and subjected to an exhaustive series of electrical tests, to make sure that it functions perfectly and reliably.

There are many types of 'packaging' or encapsulation, all much larger and stronger than the chips themselves. The commonly-used types of encapsulation for discrete devices are shown in Fig. 2.5, and the three main types used for microcircuits are shown in Fig. 2.6.

2.7 Microcircuit functions

The first generation of microelectronic circuits consisting of a few simple circuits, with about ten transistors per chip, are referred to as **small-scale integrated circuits** (SSI circuits). To form a required functional circuit, the interconnections of the assembly had to be done by external wiring.

Digital devices comprising more than 20 transistors but less than 200 transistors per chip are referred to as **medium-scale integrated circuits** (MSI circuits). A MSI circuit assembly may consist of several chips on a ceramic substrate interconnected by a thick film pattern, or it might be a monolithic device.

Large-scale integrated circuits (LSI circuits) are constructed from MOS transistors which have been made smaller and packed closer. Such circuits have more than 200 transistors per chip and are invariably monolithic devices built on large silicon chips. A typical microprocessor LSI chip can contain several thousand transistors and perform complex

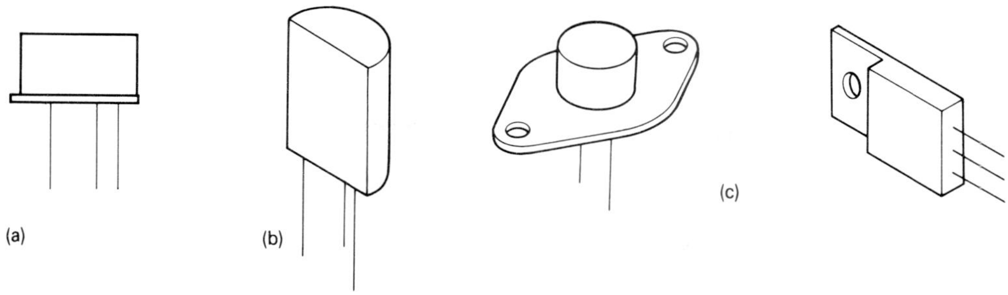

Fig. 2.5 Discrete device encapsulations

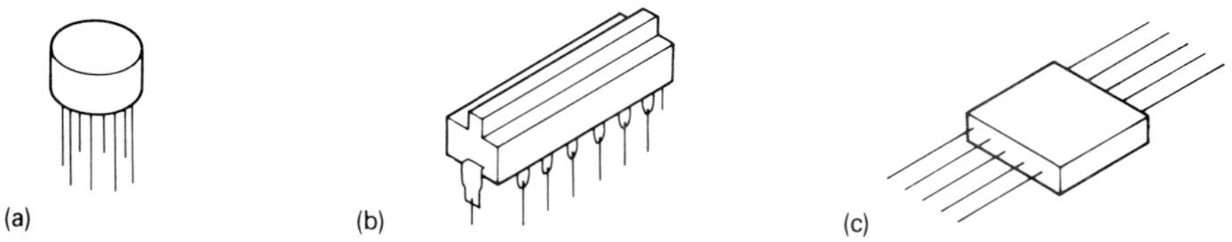

Fig. 2.6 Microcircuit encapsulations

circuit functions. As the limit of optical resolution is reached, new techniques like X-ray lithography are being used to achieve a higher packing density.

One microelectronic functional element that merits description is the **charge-coupled device** (CCD). This may be regarded as an extension of the enhancement mode MOS transistor and is capable of high packing densities. For example, Fig. 2.7 shows the mechanism of a charge-coupled shift register consisting of a source, a drain and a long row of gates. In a p-channel device, a 'charge packet' consisting of a concentration of holes is injected via the gate input, V_g, which is a steady negative voltage deriving the holes from the source. If two trains of pulses are applied to alternate gates, with the trailing edge of one pulse coinciding with the rising edge of the other, then a sequence of charge packets can be transferred from gate to gate and finally collected at the drain.

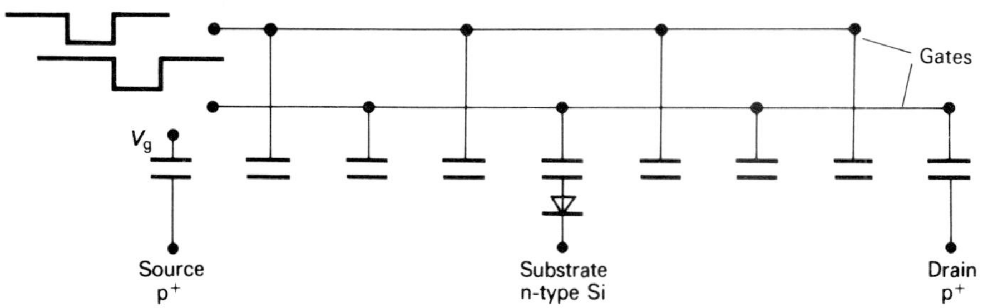

Fig. 2.7 Mechanism of a charge-coupled shift register

A basic memory element may be defined by the presence or absence of holes under a gate. A memory device in which digits are stored as long strings of bits may be fabricated from a charge-coupled device with a large number of gates. Since the packing density is high, such devices are suitable for LSI manufacture. Lastly, instead of injecting the charge packets at the source, they may be generated photoelectrically, the charge being proportional to the incident light. A CCD array can therefore be used for digitising and storing optical images.

Further reading

Carr, W. N., and Mize, J. P., *MOS/LSI Design and Application.* McGraw-Hill (1972)
Hibberd, Robert G., *Integrated Circuits.* McGraw-Hill (1969)
Keonjian, Edward, *Microelectronics.* McGraw-Hill (1963)

3

Linear circuit design and building blocks of analogue circuitry

3.1 Introduction

Circuits which are required exclusively to handle signals of small amplitude are usually analysed with the help of equivalent circuits. These involve the use of general network theory and the small-signal parameters of active devices; these parameters may be supplied by the manufacturer or obtained by measurement. Devices that are capable of handling large signals, usually power amplifiers, have different design criteria. In general, circuits that do not have a purely logical or digital function may be classed as analogue circuits. A wide variety of elaborate analogue circuits are available in integrated circuit form and may be looked upon as functional building blocks.

In the following sections a brief outline of small-signal analysis is given, along with large-signal circuit design criteria. Brief descriptions of some distinct types of analogue integrated circuits are included, together with their uses, so that users can choose suitable building blocks and can interface the units according to their specific needs.

3.2 'Black box' representation

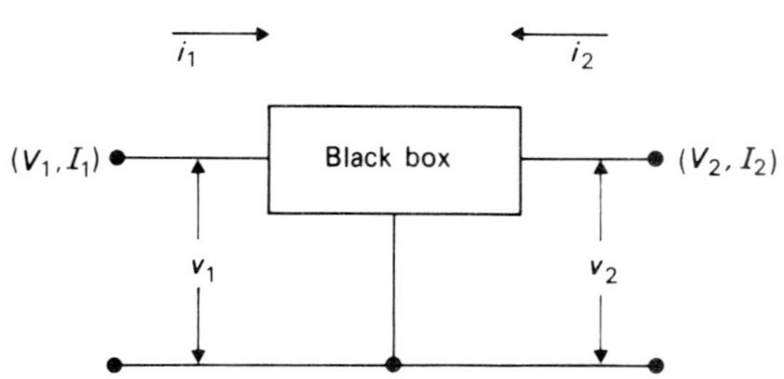

Fig. 3.1 'Black box' representation

At low frequencies the effects associated with the inductance and capacitance of an active device are negligible. Thus, in the audio-frequency range the relation between input and output voltages and currents may be considered to be frequency independent. The active device can therefore be represented by a 'black box', Fig. 3.1, and general equations can be written relating the input voltage and current (V_1, I_1) and output voltage and current (V_2, I_2). Thus, at a low frequency, if $V_1 = f_1(I_1, I_2)$ and $V_2 = f_2(I_1, I_2)$, then the small-signal variations of V_1 and V_2, i.e. v_1 and v_2, can be expressed in terms of the variations in I_1 and I_2, i.e. i_1 and i_2, as

$$v_1 = \frac{\delta f_1}{\delta I_1} i_1 + \frac{\delta f_1}{\delta I_2} i_2 \tag{3.1}$$

$$v_2 = \frac{\delta f_2}{\delta I_1} i_1 + \frac{\delta f_2}{\delta I_2} i_2 \tag{3.2}$$

The coefficients have the dimensions of impedance, so that the two simultaneous equations can be written in terms of measurable quantities as

$$v_1 = z_{11} i_1 + z_{12} i_2 \tag{3.3}$$

$$v_2 = z_{21} i_1 + z_{22} i_2 \tag{3.4}$$

These equations will characterise the linear active device within the box.

Of the four variables v_1, i_1, v_2, i_2, any two may be chosen as the independent ones, so that there are six possible pairs of equations which may be used to represent the black box. One of the most commonly-used sets comprises equations 3.3 and 3.4 above; other popular sets are equations 3.5, 3.6 and equations 3.7, 3.8 below.

$$i_1 = y_{11} v_1 + y_{12} v_2 \tag{3.5}$$

$$i_2 = y_{21} v_1 + y_{22} v_2 \tag{3.6}$$

$$v_1 = h_{11} i_1 + h_{12} v_2 \tag{3.7}$$

$$i_2 = h_{21} i_1 + h_{22} v_2 \tag{3.8}$$

The parameters z_{mn}, y_{mn} and h_{mn} will be discussed later.

In circuit analysis, when dealing with interconnected black boxes, it is convenient to use matrix algebra for manipulating large arrays of simultaneous equations. For example, equations 3.3 and 3.4 may be expressed in matrix form as

$$\begin{pmatrix} v_1 \\ v_2 \end{pmatrix} = \begin{pmatrix} z_{11} & z_{12} \\ z_{21} & z_{22} \end{pmatrix} \begin{pmatrix} i_1 \\ i_2 \end{pmatrix} \tag{3.9}$$

The z-matrix containing the **impedance** or ***z*-parameters** characterises the black box. Similarly, the **admittance** or ***y*-parameters** of equations 3.5 and 3.6 can be written as

$$\begin{pmatrix} i_1 \\ i_2 \end{pmatrix} = \begin{pmatrix} y_{11} & y_{12} \\ y_{21} & y_{22} \end{pmatrix} \begin{pmatrix} v_1 \\ v_2 \end{pmatrix} \tag{3.9a}$$

and the **hybrid** or ***h*-parameters** of equations 3.7 and 3.8 can be written as

$$\begin{pmatrix} v_1 \\ i_2 \end{pmatrix} = \begin{pmatrix} h_{11} & h_{12} \\ h_{21} & h_{22} \end{pmatrix} \begin{pmatrix} i_1 \\ v_2 \end{pmatrix} \tag{3.9b}$$

The remaining three pairs of equations can be written as follows.

The **inverse hybrid** or ***g*-parameters**,

$$\begin{pmatrix} i_1 \\ v_2 \end{pmatrix} = \begin{pmatrix} g_{11} & g_{12} \\ g_{21} & g_{22} \end{pmatrix} \begin{pmatrix} v_1 \\ i_2 \end{pmatrix} \tag{3.9c}$$

The **transmission** or ***a*-parameters**,

$$\begin{pmatrix} v_1 \\ i_1 \end{pmatrix} = \begin{pmatrix} a_{11} & a_{12} \\ a_{21} & a_{22} \end{pmatrix} \begin{pmatrix} v_2 \\ i_2 \end{pmatrix} \tag{3.9d}$$

The **inverse transmission** or ***b*-parameters**,

$$\begin{pmatrix} v_2 \\ i_2 \end{pmatrix} = \begin{pmatrix} b_{11} & b_{12} \\ b_{21} & b_{22} \end{pmatrix} \begin{pmatrix} v_1 \\ i_1 \end{pmatrix} \tag{3.9e}$$

3.3 Equivalent circuits

For each of the pairs of equations described above it is possible to draw an equivalent electrical circuit which can be considered to represent the contents of the box. Two theorems that are useful in the modelling process are the Thevenin and Norton theorems.

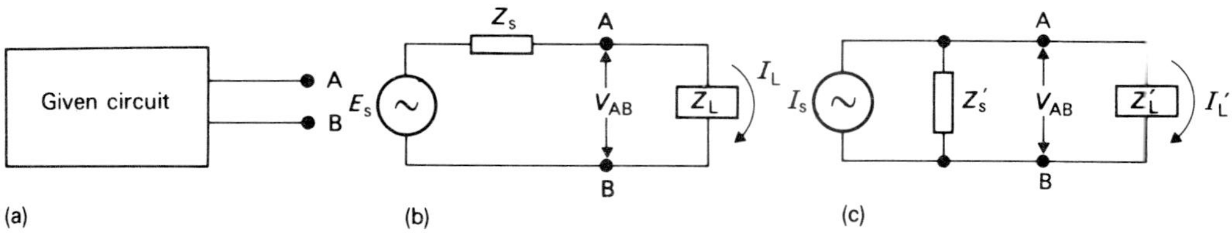

Fig. 3.2 (a) A given circuit; (b) Thevenin representation and (c) Norton representation of circuit in (a)

The **Thevenin theorem** enables a given circuit, Fig. 3.2(a), to be replaced by an ideal voltage generator E_s in series with an impedance Z_s, Fig. 3.2(b). The **Norton theorem** enables the circuit to be represented by an ideal current generator I_s shunted by an impedance Z_s', Fig. 3.2(c). In the Thevenin representation, if a load Z_L demands a current I_L then the terminal voltage $V_{AB} = (E_s - Z_s I_L)$ so that when $I_L = 0$, $E_s = V_{AB}$, the open-circuit voltage. In the Norton representation, $I_L' = I_s - (V_{AB}/Z_s')$ so that $I_s = I_L'$ when $V_{AB} = 0$, i.e. the short-circuit output current.

If $Z_s \ll Z_L$, then it is a low-impedance or constant voltage source and Thevenin representation is used. If $Z_s \gg Z_L$, then it is a high-impedance or constant current source and Norton representation is more suitable.

Using the above theorems, the small-signal parameters of equations (3.3, 3.4), (3.5, 3.6), (3.7, 3.8) can be interpreted as follows.

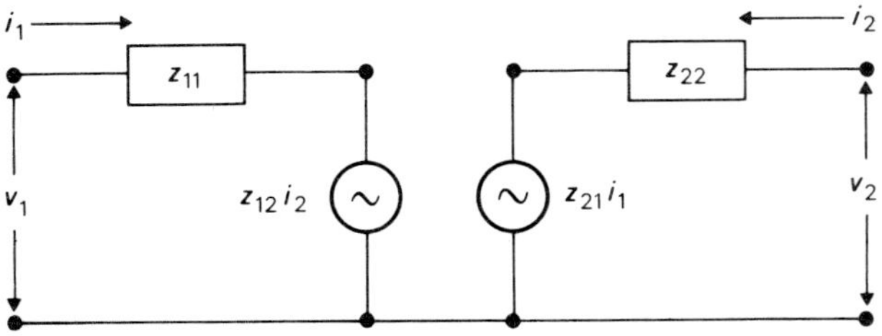

Fig. 3.3 *z*-parameter equivalent circuit

The circuit shown in Fig. 3.3 satisfies equations 3.3 and 3.4. If $z_{11} = (v_1/i_1)_{i_2=0}$ then the input impedance, with the output open-circuited, can be determined by measurement. The remaining three parameters are as follows:

z_{12}, the open-circuit reverse transfer impedance;
z_{21}, the open-circuit forward transfer impedance;
z_{22}, the open-circuit output impedance.

In the measurement of any of these parameters, either i_1 or i_2 must be zero, i.e. the z-parameters are determined from open-circuit measurements.

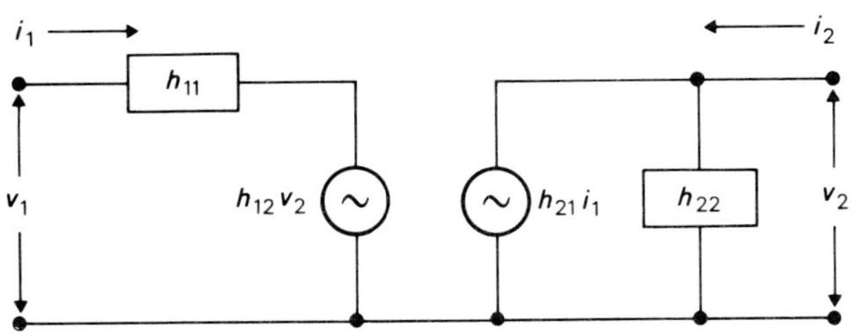

Fig. 3.4 *h*-parameter equivalent circuit

The circuit of Fig. 3.4 satisfies equations 3.7, 3.8. If the output terminals are AC short-circuited, then $v_2 = 0$ and $h_{11} = v_1/i_1$, which is the input impedance with output short-circuited, also denoted by h_i. Also, $h_{21} = i_2/i_1$ is the current amplification with output short-circuited, denoted by h_f.

With an open-circuited input, $i_1 = 0$ and $h_{12} = v_1/v_2$, which is the reverse voltage feedback ratio with open-circuited input, denoted by h_r. Also, $h_{22} = i_2/v_2$, which is the output admittance with open-circuited input, denoted by h_o.

The h-parameters are a combination of impedance and admittance parameters, ideally suited for describing the low input impedance and the high output impedance of bipolar transistors. In transistor circuit analysis a second subscript, 'e' for common-emitter, 'b' for common-base, and 'c' for common-collector is added to the h-parameters to denote the circuit configuration. For example, h_{fe} implies that it is the value of h_f when the transistor is used in the common-emitter mode.

It is possible to derive performance equations using the general parameters h_i, h_r, h_f, h_o, and finally to substitute the value of the parameter corresponding to the configuration used. Interrelations between the h-parameters corresponding to the three configurations have been worked out mathematically and have been given in tabular form.[1] Thus

$$h_{fe} = -\left(\frac{h_{fb}}{1 + h_{fb}}\right) \tag{3.10}$$

where h_{fe} or β is the current gain in the common-emitter configuration and h_{fb} or α is the current gain in the common-base configuration.

Similarly, the circuit of Fig. 3.5 satisfies equations 3.5 and 3.6. The y-parameters are obtained from short-circuit measurements. For short-circuited output, $v_2 = 0$ so that $y_{11} = i_1/v_1$, which is the input admittance, and $y_2 = i_2/v_1$, which is the forward transfer admittance. For short-circuited input, $v_1 = 0$ so that $y_{22} = i_2/v_2$, which is the output admittance, and $y_{12} = i_1/v_2$, which is the reverse transfer admittance.

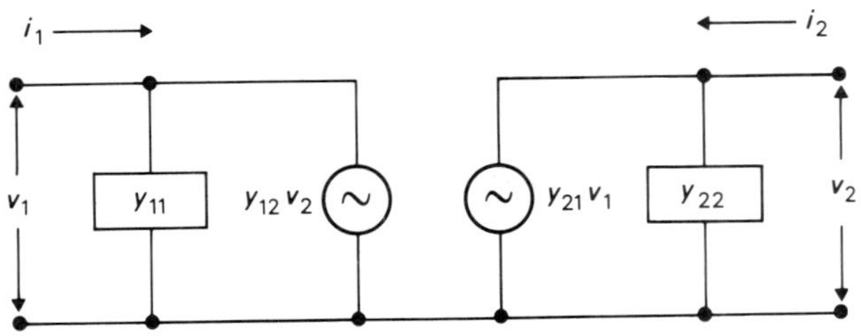

Fig. 3.5 *y*-parameter equivalent circuit

3.4 Interrelation between the various parameters

Since each of the six pairs of equations is a different way of characterising the behaviour of the same black box, transformation of one set into another is often required. These transformations are easily effected by matrix methods and can be found in tabular form.[2] For example, if the black box shown in Fig. 3.6 is represented in terms of *a*-parameters, then the defining equations can be written in the following matrix form:

$$\begin{pmatrix} v_1 \\ i_1 \end{pmatrix} = \begin{pmatrix} a_{11} & a_{12} \\ a_{21} & a_{22} \end{pmatrix} \begin{pmatrix} v_2 \\ i_2 \end{pmatrix} \tag{3.11}$$

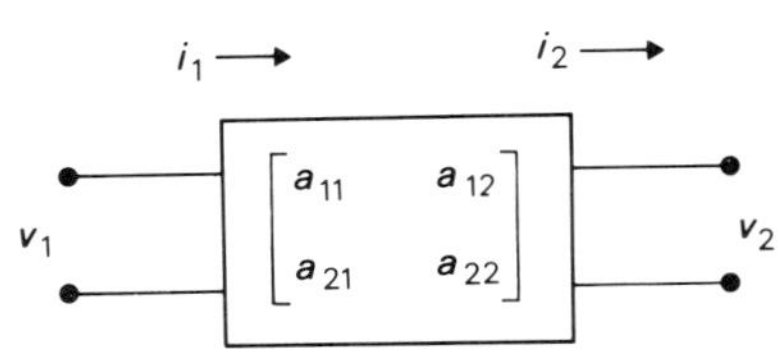

Fig. 3.6 *a*-parameter equivalent circuit

In terms of the *h*-parameters, the same box can be represented as:

$$\begin{aligned} v_1 &= h_{11}i_1 + h_{12}v_2 \\ -i_2 &= h_{21}i_1 + h_{22}v_2 \end{aligned} \tag{3.12}$$

Therefore

$$v_1 = -\left(\frac{h_{11}h_{22} - h_{12}h_{21}}{h_{21}}\right)v_2 - \frac{h_{11}}{h_{21}}i_2$$

or $$v_1 = -\frac{\Delta h}{h_{21}}v_2 - \frac{h_{11}}{h_{21}}i_2$$

where

$$\Delta h = \begin{vmatrix} h_{11} & h_{12} \\ h_{21} & h_{22} \end{vmatrix}$$

Similarly,

$$i_1 = -\frac{h_{22}}{h_{21}}v_2 - \frac{1}{h_{21}}i_2$$

Therefore

$$\begin{pmatrix} v_1 \\ i_1 \end{pmatrix} = -\frac{1}{h_{21}}\begin{pmatrix} \Delta h & h_{11} \\ h_{22} & 1 \end{pmatrix}\begin{pmatrix} v_2 \\ i_2 \end{pmatrix} \tag{3.13}$$

Comparing equations 3.11 and 3.13, the transmission matrix can be expressed as

$$\begin{pmatrix} a_{11} & a_{12} \\ a_{21} & a_{22} \end{pmatrix} = -\frac{1}{h_{21}}\begin{pmatrix} \Delta h & h_{11} \\ h_{22} & 1 \end{pmatrix} \tag{3.14}$$

where the matrix elements can be identified as

$$a_{11} = -\frac{\Delta h}{h_{21}} \qquad a_{12} = -\frac{h_{11}}{h_{21}}$$

$$a_{21} = -\frac{h_{22}}{h_{21}} \qquad a_{22} = -\frac{1}{h_{21}}$$

In practice these are found from tables of matrix interrelations.

3.5 Analysis of commonly-used linear circuits

The concept of equivalent circuits is useful for formulating a *qualitative* picture of circuit behaviour. When *quantitative* information is required, the use of matrix methods results in considerable saving in time and labour. The following simple examples will illustrate the technique at low frequencies as well as at high frequencies.

Single-stage transistor amplifier

Assuming that the correct DC bias conditions exist, the output and input of the transistor amplifier, Fig. 3.7(a), can generally be given a black box representation from which the output impedance, the input impedance, the current gain and the voltage gain can be determined as follows. In Fig. 3.7(b), if z_g is the internal impedance of a source v_1 then the input and output are related via the transistor transmission matrix of equation 3.11. Hence

$$\begin{pmatrix} v_2 \\ i_2 \end{pmatrix} = \begin{pmatrix} a_{11} & -a_{12} \\ -a_{21} & a_{22} \end{pmatrix}^{-1} \times \begin{pmatrix} 1 & z_g \\ 0 & 1 \end{pmatrix} \times \begin{pmatrix} v_1 \\ i_1 \end{pmatrix} \tag{3.15}$$

$$\begin{pmatrix} v_2 \\ i_2 \end{pmatrix} = \frac{1}{\Delta}\begin{pmatrix} a_{22} & +a_{12} \\ +a_{21} & a_{11} \end{pmatrix} \times \begin{pmatrix} 1 & z_g \\ 0 & 1 \end{pmatrix} \times \begin{pmatrix} v_1 \\ i_1 \end{pmatrix} \tag{3.15a}$$

where

$$\Delta = \begin{vmatrix} a_{11} & a_{12} \\ a_{21} & a_{22} \end{vmatrix}$$

which is the determinant of the transmission matrix.

From the triple matrix product of equation 3.15*a*,

$$v_2 = \frac{a_{22}}{\Delta}v_1 + \frac{a_{22}z_g + a_{12}}{\Delta}i_1 \tag{3.16}$$

$$i_2 = \frac{a_{21}}{\Delta}v_1 + \frac{a_{21}z_g + a_{11}}{\Delta}i_1 \tag{3.17}$$

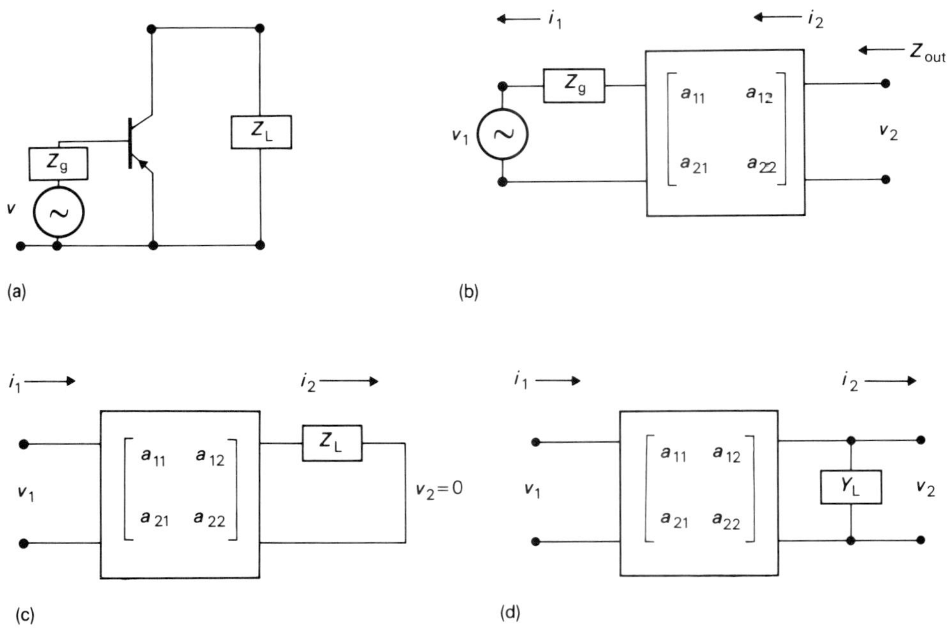

Fig. 3.7 (a) Single-stage transistor amplifier; (b) 'black box' representation with source impedance Z_g; (c) load represented as an impedance Z_L; (d) load represented as an admittance Y_L

The output impedance $z_{out} = (v_2/i_2)_{v_1=0}$, so

$$z_{out} = \frac{a_{22}z_g + a_{12}}{a_{21}z_g + a_{11}} \tag{3.18}$$

In terms of the h-parameters, $a_{11} = -\Delta h/h_{21}$; $a_{12} = -h_{11}/h_{21}$; $a_{21} = -h_{22}/h_{21}$, $a_{22} = -1/h_{21}$, and

$$z_{out} = \frac{z_g + h_{11}}{h_{22}z_g + \Delta h} \tag{3.19}$$

In Fig. 3.7(c), if Z_L is the load impedance, then

$$\begin{pmatrix} v_1 \\ i_1 \end{pmatrix} = \begin{pmatrix} a_{11} & a_{12} \\ a_{21} & a_{22} \end{pmatrix} \times \begin{pmatrix} 1 & Z_L \\ 0 & 1 \end{pmatrix} \times \begin{pmatrix} v_2 \\ i_2 \end{pmatrix}$$

$$= \begin{pmatrix} a_{11} & a_{11}Z_L + a_{12} \\ a_{21} & a_{21}Z_L + a_{22} \end{pmatrix} \begin{pmatrix} v_2 \\ i_2 \end{pmatrix} \tag{3.20}$$

From the matrix product of equation 3.20,

$$v_1 = a_{11}v_2 + (a_{11}Z_L + a_{12})i_2 \tag{3.21}$$

$$i_1 = a_{21}v_2 + (a_{21}Z_L + a_{22})i_2 \tag{3.22}$$

The input impedance $Z_{in} = (v_1/i_1)_{v_2=0}$, so

$$Z_{in} = \frac{a_{11}Z_L + a_{12}}{a_{21}Z_L + a_{22}} \tag{3.23}$$

As in equation 3.19, in terms of the h-parameters,

$$Z_{\text{in}} = \left(\frac{\Delta h Z_{\text{L}} + h_{11}}{h_{22} Z_{\text{L}} + 1}\right) \tag{3.24}$$

From equations 3.21 and 3.22, the current gain A_{i} of the amplifier is given by

$$A_{\text{i}} = \left(\frac{i_2}{i_1}\right)_{v_2 = 0} = \frac{1}{a_{21} Z_{\text{L}} + a_{22}} \tag{3.25}$$

In terms of the h-parameters,

$$A_{\text{i}} = -\frac{h_{21}}{h_{22} Z_{\text{L}} + 1} \tag{3.26}$$

Lastly, in Fig. 3.7(d), if the load is represented as an admittance Y_{L}, then

$$\begin{pmatrix} v_1 \\ i_1 \end{pmatrix} = \begin{pmatrix} a_{11} & a_{12} \\ a_{21} & a_{22} \end{pmatrix} \times \begin{pmatrix} 1 & 0 \\ Y_{\text{L}} & 1 \end{pmatrix} \times \begin{pmatrix} v_2 \\ i_2 \end{pmatrix}$$

$$= \begin{pmatrix} a_{11} & a_{12}\, Y_{\text{L}} & a_{12} \\ a_{21} & a_{22}\, Y_{\text{L}} & a_{22} \end{pmatrix} \begin{pmatrix} v_2 \\ i_2 \end{pmatrix} \tag{3.27}$$

From the matrix product of equation 3.27,

$$v_1 = (a_{11} + a_{12}\, Y_{\text{L}}) v_2 + a_{12} i_2 \tag{3.28}$$

$$i_1 = (a_{21} + a_{22}\, Y_{\text{L}}) v_2 + a_{22} i_2 \tag{3.29}$$

Hence from equation 3.28 the voltage gain A_{v} of the amplifier is given by

$$\left(\frac{v_2}{v_1}\right)_{i_2 = 0} = \frac{1}{a_{11} + a_{12}\, Y_{\text{L}}} \tag{3.30}$$

In terms of h-parameters,

$$A_{\text{v}} = \frac{h_{21}}{\Delta h + h_{11}\, Y_{\text{L}}} = \frac{h_{21} Z_{\text{L}}}{\Delta h Z_{\text{L}} + h_{11}} \tag{3.31}$$

The equations for the output impedance, 3.19, the input impedance, 3.24, the current gain, 3.26 and the voltage gain, 3.31, are applicable to any of the three modes in which a transistor is used, provided that the correct values of the h-parameters are chosen. For example, for a transistor used in the common-emitter mode, the current gain of equation 3.26 will be given by

$$A_{\text{i}} = \frac{h_{\text{fe}}}{h_{\text{oe}} Z_{\text{L}} + 1} \tag{3.32}$$

Impedance matching

The equivalent circuit technique can easily be applied to the case of matching a high-frequency amplifier having an output impedance Z_{o} via a suitable passive network to a load Z_{L}, as shown in Fig. 3.8(a). The T network is described by a Z matrix after including the imaginary parts of Z_0 and Z_1 as in Figs 3.8(b) and 3.8(c).

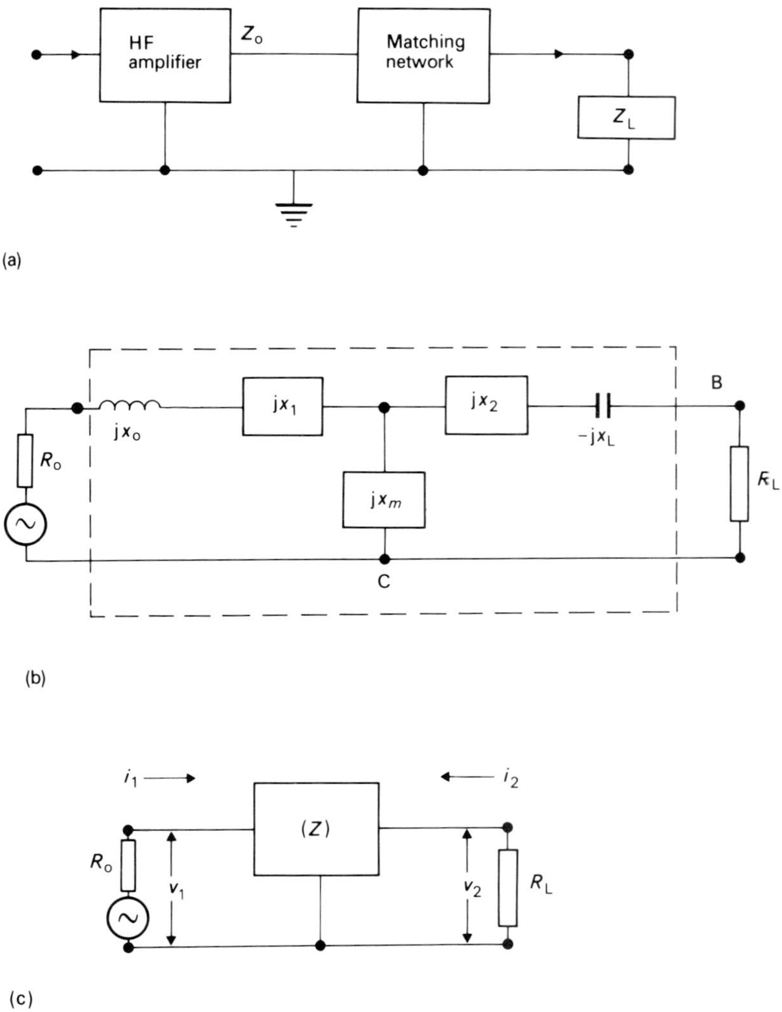

Fig. 3.8 (a) HF amplifier with output impedance Z_o; (b) T-matching network; (c) Z matrix representation

$$(Z) = \begin{pmatrix} Z_{11} & Z_{12} \\ Z_{21} & Z_{22} \end{pmatrix}$$

$$= \begin{pmatrix} j(x_0 + x_1 + x_m) & jx_m \\ jx_m & j(-x_L + x_2 + x_m) \end{pmatrix} \tag{3.33}$$

From Fig. 3.8(c), since $v_1 = -i_1 R_o$,

$$\begin{bmatrix} 0 \\ v_2 \end{bmatrix} = \begin{pmatrix} Z_{11} + R_o & Z_{12} \\ Z_{21} & Z_{22} \end{pmatrix} \begin{bmatrix} i_1 \\ i_2 \end{bmatrix} \tag{3.34}$$

so that

$$Z'_o = \frac{v_2}{i_2} = \frac{\Delta Z + Z_{22} R_o}{Z_{11} + R_o} \tag{3.35}$$

which is the output impedance of the box, where

$$\Delta Z = \begin{vmatrix} Z_{11} & Z_{12} \\ Z_{21} & Z_{22} \end{vmatrix}$$

Also

$$v_2 = -i_2 R_{\rm L}$$

$$\begin{bmatrix} v_1 \\ 0 \end{bmatrix} = \begin{pmatrix} Z_{11} & Z_{12} \\ Z_{21} & Z_{22}+R_{\rm L} \end{pmatrix} \begin{bmatrix} i_1 \\ i_2 \end{bmatrix} \tag{3.36}$$

$$Z_{\rm in} = \frac{v_1}{i_1} = \frac{\Delta Z + Z_{11} R_{\rm L}}{Z_{22}+R_{\rm L}} \tag{3.37}$$

which is the input impedance of the box. For matching, $Z_{\rm o}' = R_{\rm L}$ and $Z_{\rm in} = R_{\rm o}$ and equations 3.35, 3.37 must be satisfied simultaneoulsy, therefore

$$R_{\rm L} = \frac{\Delta Z + Z_{22} R_{\rm o}}{Z_{11}+R_{\rm o}} \qquad R_{\rm o} = \frac{\Delta Z + Z_{11} R_{\rm L}}{Z_{22}+R_{\rm L}}$$

$$R_{\rm L} = \sqrt{\Delta Z (Z_{11}/Z_{22})} \tag{3.38}$$

$$R_{\rm o} = \sqrt{\Delta Z (Z_{22}/Z_{11})} \tag{3.39}$$

$$R_{\rm o} R_{\rm L} = \Delta Z = (Z_{11} Z_{22} - Z_{21} Z_{12})$$

Therefore

$$R_{\rm o} R_{\rm L} = -(x_0 + x_1 + x_m)(-x_{\rm L} + x_2 + x_m) + x_m^2$$

For conjugate matching, the algebraic sum of the reactances between the points A and B and the points B and C in Fig. 3.8(c) must be zero. Hence

$$x_0 + x_1 + x_m = 0 \tag{3.40}$$

$$-x_{\rm L} + x_2 + x_m = 0 \tag{3.41}$$

giving $R_{\rm o} R_{\rm L} = x_m^2$. Therefore,

$$x_1 = (-x_0 - x_m) = -x_0 - \sqrt{R_{\rm o} R_{\rm L}} \tag{3.42}$$

$$x_2 = (x_{\rm L} - x_m) = x_{\rm L} - \sqrt{R_{\rm o} R_{\rm L}} \tag{3.43}$$

Darlington pair

The Darlington pair configuration of Fig. 3.9 is often used in cases where the current gain and input impedance obtainable from a single transistor is not adequate. The first transistor $\rm T_A$ acts as a common-collector current amplifier for the second transistor $\rm T_B$. The composite parameters of such a pair may be determined by matrix methods. If the y-matrix of $\rm T_A$ alone with respect to terminal 2 is given by

$$\begin{pmatrix} y_{\rm iA} & y_{\rm rA} \\ y_{\rm fA} & y_{\rm oA} \end{pmatrix}$$

then the indefinite or floating matrix with respect to terminal 4 is obtained by creating an

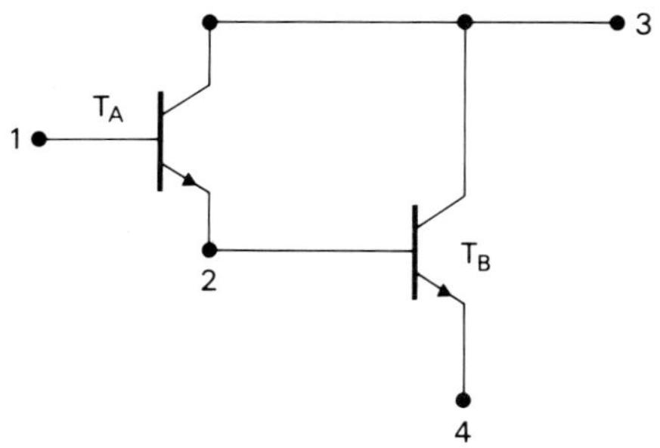

Fig. 3.9 Darlington pair

extra row and column, in which the elements are chosen so as to make the sum of the elements in each row and each column equal to zero, i.e.

$$\begin{pmatrix} y_{iA} & -(y_{iA}+y_{rA}) & y_{rA} \\ -(y_{fA}+y_{iA}) & \sum y_A & -(y_{oA}+y_{rA}) \\ y_{fA} & -(y_{fA}+y_{oA}) & y_{oA} \end{pmatrix}$$

where

$$\sum y_A = y_{iA}+y_{rA}+y_{fA}+y_{oA}$$

The y-matrix of the network with respect to terminal 4 with transistor T_A removed is

$$\begin{pmatrix} 0 & 0 & 0 \\ 0 & y_{iB} & y_{rB} \\ 0 & y_{fB} & y_{oB} \end{pmatrix}$$

The y-matrix of the Darlington pair is the sum of the above two matrices,

$$\begin{pmatrix} y_{iA} & -(y_{iA}+y_{rA}) & y_{rA} \\ -(y_{fA}+y_{iA}) & (y_{iB}+\sum y_A) & -(y_{oA}+y_{rA}-y_{rB}) \\ y_{fA} & (y_{fB}-y_{fA}-y_{oA}) & (y_{oA}+y_{oB}) \end{pmatrix}$$

As the terminal is practically an open circuit, row 2 and column 2 may be eliminated and the matrix rewritten as

$$\frac{1}{y_{iB}+\sum y_A}\begin{pmatrix} \begin{matrix} y_{iA}(y_{iB}+\sum y_A)- \\ (y_{fA}+y_{iA})(y_{iA}+y_{rA}) \end{matrix} & \begin{matrix} y_{rA}(y_{iB}+\sum y_A)- \\ (y_{iA}+y_{rA})(y_{oA}+y_{rA}-y_{rB}) \end{matrix} \\ \begin{matrix} y_{fA}(y_{iB}+\sum y_A)+ \\ (y_{fA}+y_{iA})(y_{fB}-y_{fA}-y_{oA}) \end{matrix} & \begin{matrix} (y_{iB}+\sum y_A)(y_{oA}+y_{oB})+ \\ (y_{fB}-y_{fA}-y_{oA})(y_{oA}+y_{rA}-y_{rB}) \end{matrix} \end{pmatrix}$$

If the magnitudes of y_{rA} and y_{rB} are zero, and those of y_{fA} and y_{fB} are large, then the matrix may be approximated to

$$\frac{1}{y_{fA}}\begin{pmatrix} y_{iA}(y_{iB}+y_{oA}) & -y_{iA}y_{oA} \\ (y_{fA}y_{fB}+y_{fA}y_{iB}+y_{fB}y_{iA}) & (y_{oA}y_{fB}+y_{oB}y_{fA}) \end{pmatrix}$$

The short-circuit current amplification will be

$$\left(\frac{I_3}{I_1}\right)_{v_3=0} \approx \frac{y_{fA}y_{fB}}{y_{iA}(y_{iB}+y_{oA})}$$

n-port network

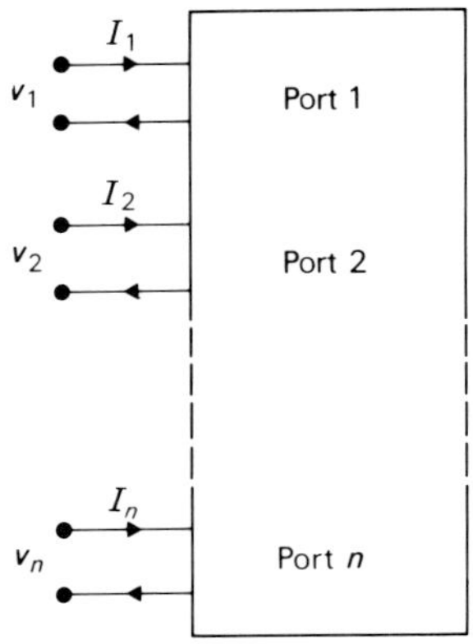

Fig. 3.10 n-port network

The techniques described here may be extended to n-port networks; port voltages must be linearly dependent on port currents and vice versa. The open-circuit voltage produced at port i by a current source applied at port j when all other ports are open-circuited may be described by the following set of equations.

$$\begin{aligned} v_1 &= Z_{11}I_1 + Z_{12}I_2 + \ldots + Z_{1n}I_n \\ v_2 &= Z_{21}I_1 + Z_{22}I_2 + \ldots + Z_{2n}I_n \\ v_n &= Z_{n1}I_1 + Z_{n2}I_2 + \ldots + Z_{nn}I_n \end{aligned}$$

where $Z_{ij} = (v_i/v_j)_{I_1, I_2, \ldots, I_{j+1}, I_n = 0}$. These equations may be expressed in matrix form as

$$\begin{pmatrix} v_1 \\ v_2 \\ v_n \end{pmatrix} = \begin{pmatrix} Z_{11} & Z_{12} & \ldots & Z_{1n} \\ Z_{21} & Z_{22} & \ldots & Z_{2n} \\ Z_{n1} & Z_{n2} & \ldots & Z_{nn} \end{pmatrix} \begin{pmatrix} I_1 \\ I_2 \\ I_n \end{pmatrix}$$

3.6 High-frequency circuits in general

Even at the upper end of the audio-frequency spectrum, the reactive components, such as junction capacities of transistors and other circuit elements, begin to have an appreciable effect on circuit performance. In the black box representation of Fig. 3.1, v_1 and v_2 are functions of I_1 and I_2 and of frequency ω. Thus, for example,

$$v_1 = f'_1(I_1, I_2, \omega) \quad \text{and} \quad v_2 = f'_2(I_1, I_2, \omega)$$

The z-parameters of equations 3.3 and 3.4 will be complex numbers, usually expressed in terms of an α cut-off frequency, defined as that frequency at which $|\alpha|$ is 3 dB below its low-frequency value.

The above mathematical concept helps to show the general problems of small-signal circuit analysis at high frequencies. It is often possible to modify a particular low-frequency equivalent circuit to take into account the effect of increasing frequency. For example, the base-collector capacity C_{bc} of a bipolar transistor, although smaller than the base-emitter capacity C_{be}, is in parallel with the high resistance of the reverse biased base-collector junction, as shown in Fig. 3.11(a). The capacity C_{be} causes signal feedback from output to

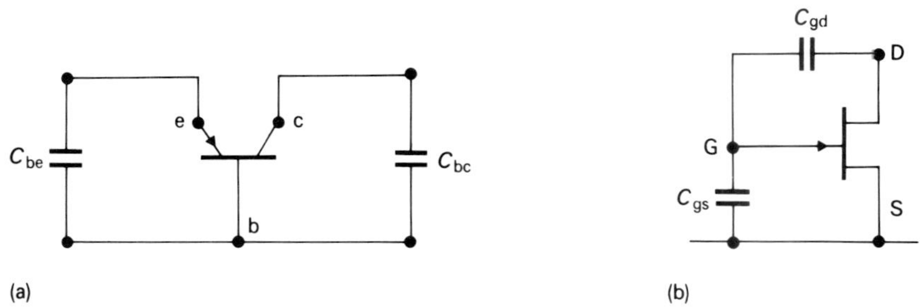

Fig. 3.11 (a) Interelectrode capacitances of a bipolar junction transistor; (b) interelectrode capacitances of a JFET

input, and in amplifiers with an inductive load, the phase shift round the feedback loop at high frequencies is often sufficient to give rise to instabilities (see Chapter 6).

Similarly, in the FET, Fig. 3.11(b), both the gate-source capacitance C_{gs} and the gate-drain capacitance C_{gd} are important, because the input resistance R_g is large. If C_{gd} is ignored, then the high-frequency effect of C_{gs} is defined by the resistance R of the signal source. However, if C_{gs} is ignored, then C_{gd} and R form a voltage feedback path. For resistive loading C_{gd} will enhance the input capacity, but with inductive loading the feedback can become positive, with the consequent risk of instability.

Finally, it may be pointed out that the effects of lead inductance and stray capacities are important at operating frequencies above 50 MHz. Such effects are minimal in the integrated circuits themselves but arise mainly in the interface circuitry. Neutralisation, i.e. the use of deliberate feedback loops to cancel the effect of undesired coupling, is often used in high-frequency linear circuitry.

3.7 Large-signal circuit design criteria

Large-signal amplifiers, commonly known as **power amplifiers**, are used to drive energy conversion devices which may be acoustic, mechanical, radio-frequency, etc. There are several modes of operation, originally defined for vacuum tube circuits. For example, in class A operation the power output stage conducts over the entire cycle. In class B operation each stage conducts over one half of the entire input cycle, so that two identical stages, biased near cut-off, are required to amplify a sinusoidal signal. In class C operation a single stage conducts over less than one half of a signal, as in pulse amplifiers. Each of these modes has distinct advantages and disadvantages. For instance, a class A transistor-amplifier requires an operating point that is characterised by a high collector current, whereas a class B or push-pull amplifier requires a low standby current.

In using IC power amplifiers or power transistors, heat sinks are often required to facilitate heat transfer from the device. For example, most of the power dissipation occurs at the collector junctions of power transistors and the maximum tolerable value of the junction temperature is usually known. The temperature T_j at which a junction is operating depends on the surrounding temperature T_s, the thermal resistance R_T in °C per watt of the heat path from the junction to the surroundings, and the electrical power W being converted into heat at the junction. Thus

$$T_j = (T_s + R_T W)\,^\circ\text{C} \tag{3.44}$$

and $W = V_c I_c$ where V_c and I_c are the junction voltage and current respectively. Hence when

$$\frac{\partial W}{\partial T_j} = \frac{1}{R_T} \tag{3.45}$$

the rate of increase of collector power dissipation with junction temperature exceeds the ability of the device to remove the heat, and **thermal runaway** will occur. For thermal stability, $(\partial W/\partial T_j) < (1/R_T)$.

The primary criterion of large signal devices is that their power conversion efficiency, defined as

$$\frac{\text{AC power delivered to the load}}{\text{DC power supplied}}$$

must be as high as possible. For example, the power output of a common emitter transistor stage is limited by the following factors:

1) maximum allowable collector-emitter current;
2) maximum allowable power dissipation;
3) maximum allowable collector-emitter voltage;
4) preventing distortion by avoiding the saturation and cut-off regions of the transistor characteristics.

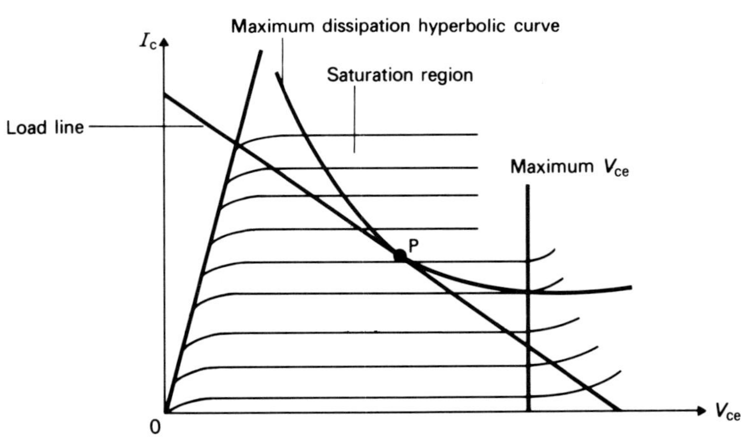

Fig. 3.12 Operating limits of a common-emitter power stage

Figure 3.12 shows the V_{ce}–I_c characteristics, with the maximum power dissipation hyperbolic curve defined by $V_{ce}I_c$ = constant for a given junction temperature. The quiescent point P and the load are selected so that the load line is tangential to the maximum collector dissipation hyperbola at P. To obtain an operating point such as P, a large standby collector current must be supplied so that the efficiency of such a device is inherently low. One of the techniques for increasing the efficiency is to use two transistors in push-pull configuration, as illustrated in Fig. 3.13. Two complementary transistors conduct only in response to an applied signal, say $v \sin \omega t$, one during the positive half of a cycle and the other during the negative half. Two supply rails $+V$ and $-V$ are used, and the load R_L is driven with respect to

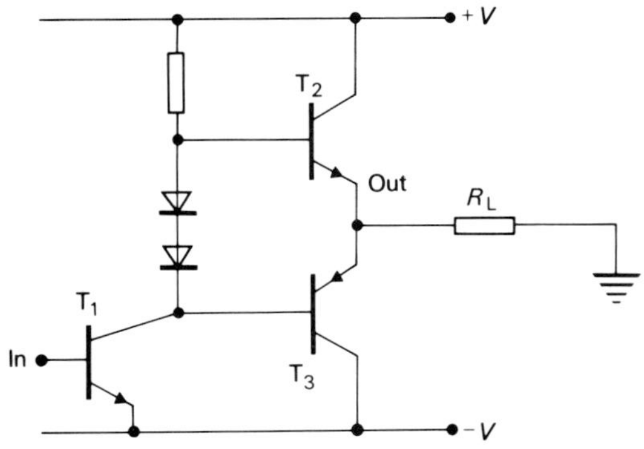

Fig. 3.13 Push-pull power stage

earth. In the ideal case, a sinusoidal input voltage will give rise to half sine wave current pulses in each transistor and a complete sine wave voltage will develop across the load.

However, load power $= K^2V^2/2R_L$ where $0 < K < 1$. During half a conduction cycle, the mean current drawn from the supply will be given by

$$\frac{1}{\pi}\int_0^{\pi}\frac{KV}{R_L}\sin\theta\,d\theta = \frac{2}{\pi}\left(\frac{KV}{R_L}\right)$$

Since the circuit is symmetrical, the same mean current will flow during the second half cycle.

$$\text{Supply power} = \frac{2}{\pi}\left(\frac{KV^2}{R_L}\right)$$

and efficiency is given by

$$\frac{K^2V^2}{2R_L}\times\frac{\pi R_L}{2KV^2} = \frac{K\pi}{4}$$

When $K = 1$ the maximum efficiency is equal to 78.5%.

Lastly, the inputs and outputs of large-signal circuits usually require some kind of protection to prevent damage to the transistors. For example, excessive input voltage will damage the base-emitter junction of the input transistor T_1, Fig. 3.13, and a short circuit in R_L will result in heating because of excessive collector currents in T_2 and T_3. There is a wide variety of protection circuits, but all of them perform two basic functions:

1) the required transistor currents are sensed;
2) when the specified values are exceeded, solid-state switches such as switching transistors or thyristors are activated so as to divert the excess currents.

It is important to locate the cause of the overvoltage or short-circuit whenever the protection circuitry is triggered.

3.8 Building blocks of analogue circuitry

Analogue integrated circuits, monolithic and hybrid, can be broadly classified as follows:

Analogue ICs

- Operational amplifiers
- Voltage regulators
- Voltage comparators
- Other ICs

Operational amplifiers

Over the last ten years, operational amplifiers which were originally designed as sub-assemblies for analogue computers have become a basic building block of all analogue circuitry. Modern IC operational amplifiers approach the following idealised properties:

i) 'infinite' open loop gain (in practice $\approx 10^5$);
ii) 'infinite' input resistance (several megohms);
iii) 'zero' output resistance (a few ohms);
iv) no phase shift of the signal with frequency;
v) zero output voltage for zero input voltage (no offset).

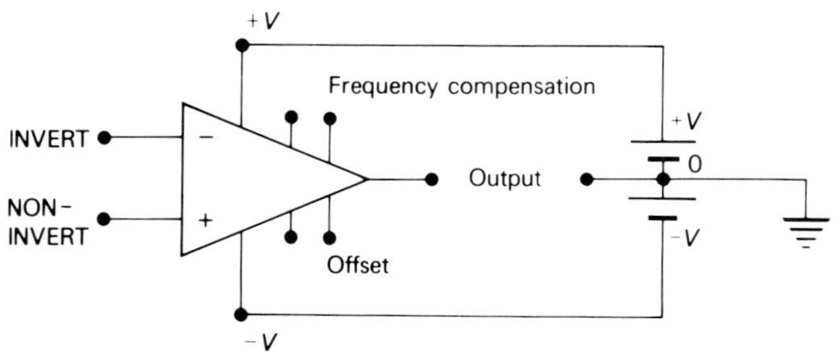

Fig. 3.14 A typical operational amplifier

Operational amplifiers are essentially direct-coupled high gain differential amplifiers designed to operate from symmetrical positive and negative power supplies, Fig. 3.14. The output voltage can be positive or negative with respect to earth (common terminal of $+V$ and $-V$; with no input the output terminal is ideally at earth potential. Two input terminals are available; one (NON-INVERT) is in phase with the output and the other (INVERT) is 180° out of phase with it. The performance of an IC operational amplifier is specified by the manufacturer, using the following terminology.

Input impedance is the impedance measured at one input with the other input at earth potential. Its value is several megohms for bipolar transistors and above 100 megohms for FETs. The input capacitance and resistance are sometimes given.

Input bias current is temperature-dependent and is of the order of a few μA for bipolar transistors and a few pA for FETs.

Input offset current is the difference between the input currents at zero output voltage; it is temperature-dependent. Its value is of the order of 0.5 μA for bipolar transistors and about 1 pA for FETs.

Open loop gain is the low-frequency differential voltage gain measured at a given temperature and load. Its value is of the order of 10^5.

Common mode rejection ratio is the ratio of the gain with a differential signal to that with equal signal at each input. Its value is of the order of 90 dB.

Slewing rate is the maximum rate of change of the output voltage in response to a step input voltage difference. It defines the rate at which the output can track change in the signal; its value is of the order of $1/V\,\mu s^{-1}$ to $50\,\mu s^{-1}$.

Supply voltage rejection ratio With the inputs short-circuited, this is the change in output voltage produced by a change in the supply voltage. Its value is of the order of $10\,\mu V\,V^{-1}$.

Output protection denotes the presence of an active switching device for diverting the output current when the latter exceeds a specified value.

Latch up When the absolute value of the voltage at the inverting input is exceeded, the circuit goes into saturation. Input protection circuits are used to prevent 'latch up' and to avoid transistor damage.

Output voltage swing is the maximum excursion of the output voltage; its value is usually about ± 15 V for general-purpose amplifiers.

Settling time is the time required for the output voltage to settle to within a specified percentage of its final value, following a step change in the signal. It is important in applications involving signal switching, for example in analogue/digital (A/D) conversion and multiplexing.

Bandwidth usually ranges from DC to a frequency at which the small-signal open loop gain falls to unity (≈ 1 MHz). Some amplifiers have internal compensation while others have provision for connecting external compensating networks.

The following types of IC operational amplifiers are typical examples, selected from a wide and ever-increasing range for performing specific electronic functions. The IC numbers are usually given along with the manufacturer's individual designations for the device family, temperature ranges, packing, etc. For example, LM101/LM201/LM301 is the notation of National Semiconductors Ltd for linear monolithic amplifiers, the 1–2–3 numbering system denoting the temperature ranges, i.e. 1 for $-55°C$ to $+125°C$, 2 for $-25°C$ to $+85°C$, and 3 for $0°C$ to $+70°C$.

General-purpose type

General-purpose operational amplifiers such as the SN 72702A (Texas Instruments) have a differential input and a single-ended emitter-follower output. Terminals are provided for adding external components to compensate the amplifier for stable operation under various feedback and load conditions. High-performance amplifiers such as the 741 have internal frequency compensation that ensures stability without external components, over a frequency range of several hundred kHz. In most cases, terminals are provided for adjusting the output offset voltage to zero. Thus 741N is a low-noise version of 741.

High voltage type

High-voltage operational amplifiers such as the LH0003 (National Semiconductors Ltd) are designed to operate from supply voltages up to ± 40 V, although they work satisfactorily at voltages as low as ± 5 V. Their high gain and wide range of operating voltages make them ideal for applications requiring a large output swing and low power dissipation.

FET input operational amplifiers

The MOS input and output types such as 3130S feature a low input bias current of about 5 pA, a large output voltage swing with a short input time which is constant despite the high input resistance of about $1.5 \times 10^{12}\,\Omega$. The JFET type such as the μAF 355 TC (SGS Ltd) feature an input bias current of about 30 pA with a low input offset voltage, large open loop gain and input resistance of about $10^{12}\,\Omega$. The input current noise is lower than that of the corresponding bipolar types.

Dual and quad operational amplifiers

More than one operational amplifier can be encapsulated in a single package. For example, the 747 is a dual version of the 741 with a channel separation of 120 dB, and the 348 includes four 741 operational amplifiers in one package.

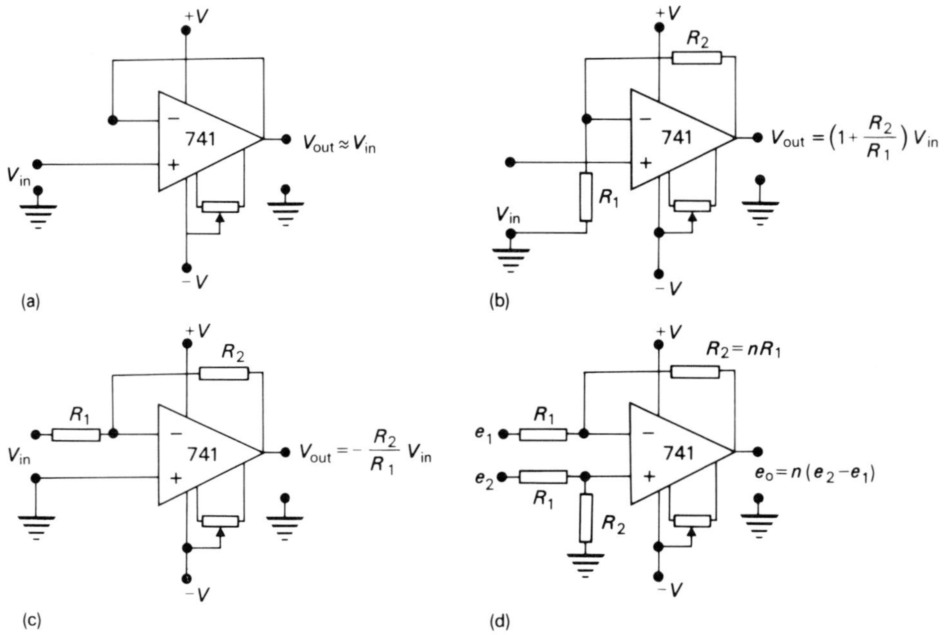

Fig. 3.15 Typical operational amplifier applications; (a) voltage follower; (b) non-inverting amplifier; (c) inverting amplifier; (d) differential input, single-ended output.

Figure 3.15 shows some commonly-used DC and low-frequency circuit applications of IC operational amplifiers, and Fig. 3.16 shows their use for simple analogue operations. Further circuit applications are described by Graeme[3] and Clayton.[4]

The use of operational amplifiers for DC amplification of signals from a source such as a thermocouple is limited by the temperature dependence of the offset voltage and current. The DC zero drift in the output is appreciable, especially in cases where several directly-coupled stages are used. Whenever a DC closed loop gain of about 10^3 is required, chopper stabilisation is perhaps the best practical solution.

Voltage regulators

IC voltage regulators basically consist of a temperature-compensated voltage reference V_r, Fig. 3.17, a differential amplifier and a series voltage-control element such as a transistor. Voltage regulation is achieved by comparing a fraction of the output voltage with the internal preset reference V_r, amplifying the error signal and utilising it via the closed feedback loop to correct for any variations in V_{out}. In practice, monolithic IC voltage regulators contain complex circuitry and are available as two terminal types, for 0 to $+v$ or for 0 to $-v$, and three terminal types, Fig. 3.18, in which the positive and negative output voltages may have different values. A wide variety of IC voltage regulators, differing mainly in power handling and voltage output capabilities, are available. The regulated output should ideally be

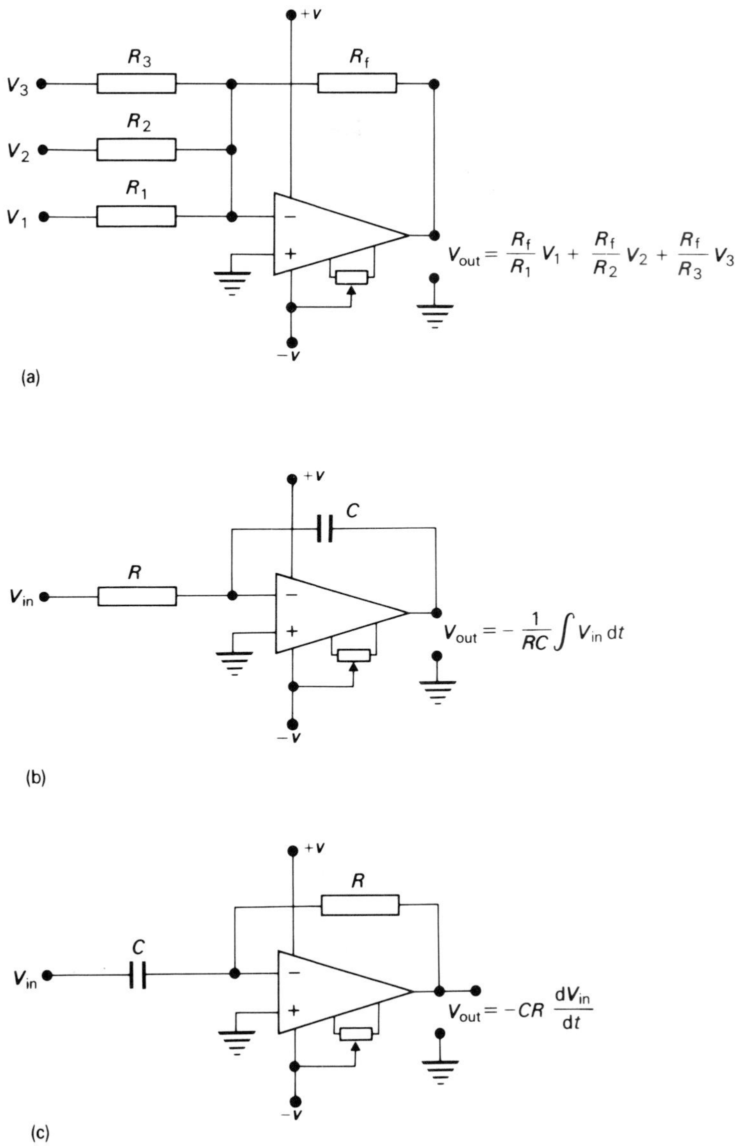

Fig. 3.16 Simple analogous operations: (a) summation; (b) integration; (c) differentiation

independent of the load current, temperature and variations in the AC mains voltage. Most regulators feature an internal current-limiting and thermal shut-down protection. Typical parameters in the operating temperature range of a device are as follows: ripple rejection = 0.01 %; full load regulation with current limiting = 0.1 %; line regulation = 0.1 %.

The unregulated input voltage must be at least a few volts greater than the regulated output voltage, which may be adjustable or which may be fixed. In electronic instrumentation, monolithic regulators are used to distribute unregulated voltage supplies to various sections of a circuit, to provide regulation for individual stages and to improve their isolation and decoupling.

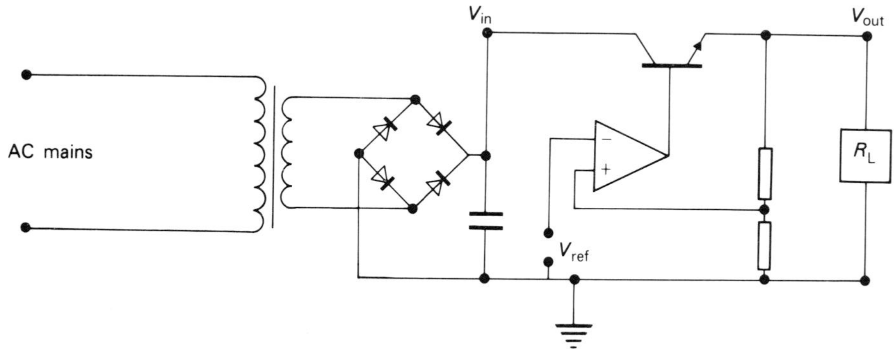

Fig. 3.17 Principle of two-terminal voltage regulator

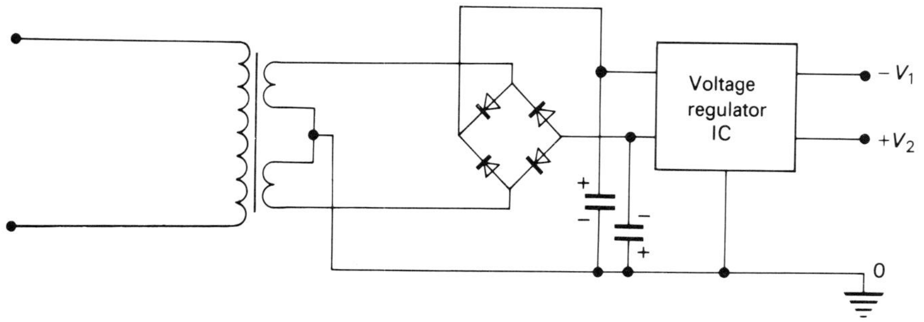

Fig. 3.18 Three-terminal voltage regulator

Voltage comparators

These are differential amplifiers arranged to take two input signals and may have a differential or single-ended output. One of the inputs is connected to a reference voltage and the other to the analogue signal voltage which is to be compared to the reference. Comparators such as the 710 have a single-ended output level that changes from -0.5 V to $+3$ V when the signal voltage exceeds the reference. They are much faster than operational amplifiers and the output levels are compatible with most TTL (transistor-transistor logic) and other integrated circuits. Typical applications include level detectors, pulse height discriminators and voltage comparators in high-speed A/D converters.

Dual comparators such as the 711 have separate differential inputs and a single-ended output. They are mostly used as double-ended limit detectors, the upper and lower limits being set via the two input references, Fig. 3.19. Quad versions such as the 139 series are designed to interface directly with TTL and MOS logic circuits. It is often necessary to convert a very slowly varying analogue signal into one that has an abrupt waveform occurring at a precise value of the input voltage. This function can obviously be implemented by using voltage comparators. However, a 'Schmitt trigger' integrated circuit such as the 7413, which may be looked upon as a regenerative comparator, performs this function by rapidly switching to the ON state when the input signal reaches a preset voltage reference level. Moreover, the element remains in this state as long as the signal exceeds the reference level

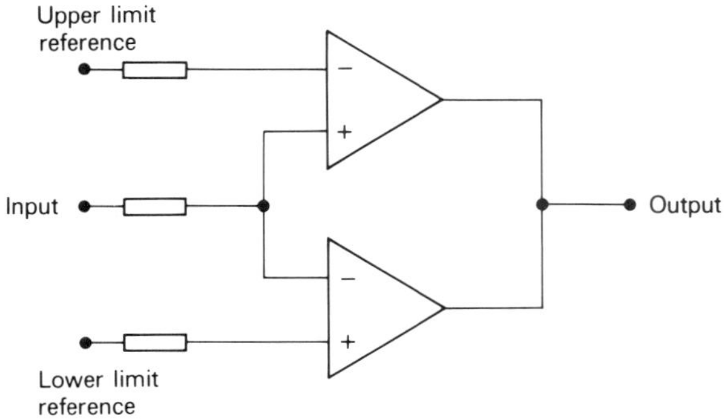

Fig. 3.19 Dual voltage comparator

and returns to the original state only when the signal falls below it. The switching action from one state to the other takes place in less than 10 ns and the circuit is commonly used for pulse shaping and sine to square wave conversion.[5]

Other ICs

Integrated circuits belonging to this category perform combinations of analogue and digital functions. The following are some typical examples.

Sense amplifiers

A sense amplifier basically consists of a differential input amplifier, a differential voltage level detector and circuitry for adjusting the output pulse level. Sense amplifiers are mostly used for processing signals from core memories. The millivolt level memory sense signal is amplified and compared with an internal reference. A logic '1' output is produced only if the signal exceeds the reference, various logic connections being provided to suit specific applications. The input threshold voltage is adjustable and the output is TTL compatible.

Timers

Monolithic timing circuits such as the 555 consist of a transistor switch T driven by a flip-flop FF which in turn is controlled by two voltage comparators C_1 and C_2, with internal preset reference voltages V_{r_2} and V_{r_1}, Fig. 3.20, derived from the external supply voltage V_{cc}.

The switch is connected across an external timing condenser C charging via the resistor R. A trigger pulse at the input of comparator C_1 sets the flip-flop and releases the short-circuit across C, which then charges until the voltage across it reaches the trigger level of comparator C_2, when the flip-flop is reset and the condenser discharged rapidly via T. Once triggered, the output stays high for an interval of time τ defined by R and C.

Since the charging rate and the threshold levels of the comparators are both directly proportional to the supply voltage V_{cc}, the timing interval is independent of V_{cc}. A pulse at the reset terminal during the timing cycle will discharge the timing condenser and cause the cycle to start over again soon afterwards. The circuit can be connected to operate on either monostable or free-running mode and its most common applications are in timing over long intervals, in delay generation, and in pulse generation. Dual versions are also available.

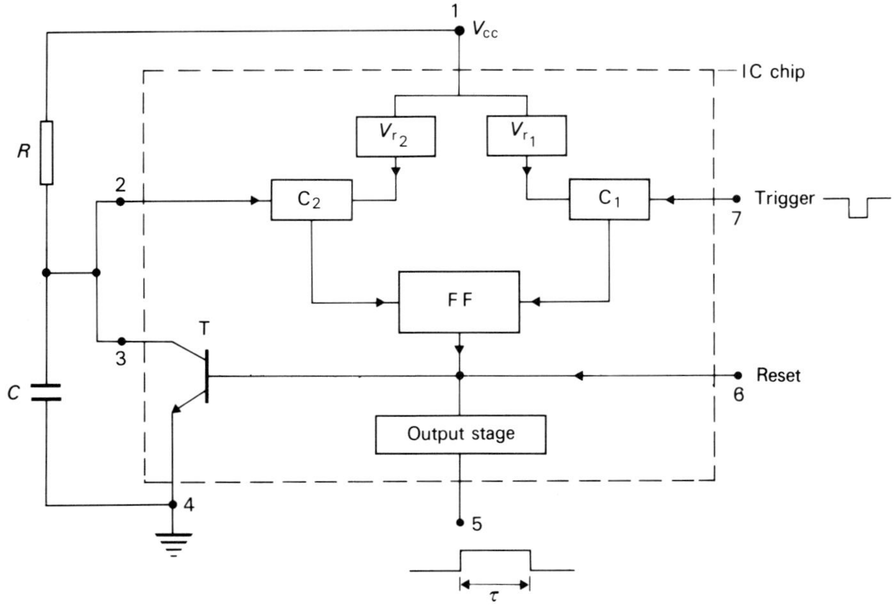

Fig. 3.20 A timer IC with external terminals 1, 2, 3, . . . etc.

Phase-locked loops (PLL)

Phase-locked loop technology dates back more than 40 years but its complexity made it impractical and uneconomical in the majority of applications until the advent of monolithic circuits. Basically it is a servo loop consisting of a phase detector, a low-pass filter and a voltage-controlled oscillator (VCO), Fig. 3.21. The input signal is phase compared to the output of the VCO, and the error signal at the output of the phase detector is filtered, amplified and fed back to the input of the VCO, the centre frequency of which is adjusted to coincide with that of the incoming signal. Any changes in the input signal frequency produce a change in the phase detector output voltage, which in turn alters the VCO frequency so as to preserve the loop in a locked condition.

There are two main types of PLL packages. The general-purpose ones such as the 560B, 562B and 565 (Signetics Ltd) contain an oscillator whose frequency can be set with external

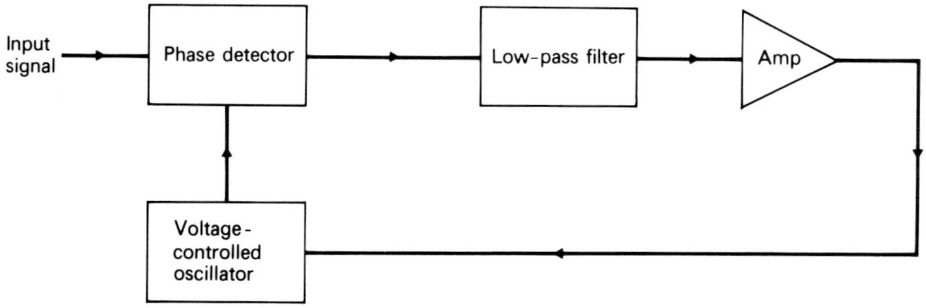

Fig. 3.21 A phase-locked loop (PLL)

components, a phase detector and an amplifier preceded by a low-pass filter with external components. When locked to an incoming signal these packages provide two outputs: a voltage proportional to the frequency of the output and a square wave VCO output which, during lock, has the same frequency as the input signal. All general-purpose devices have a linear frequency-to-voltage transfer characteristic.

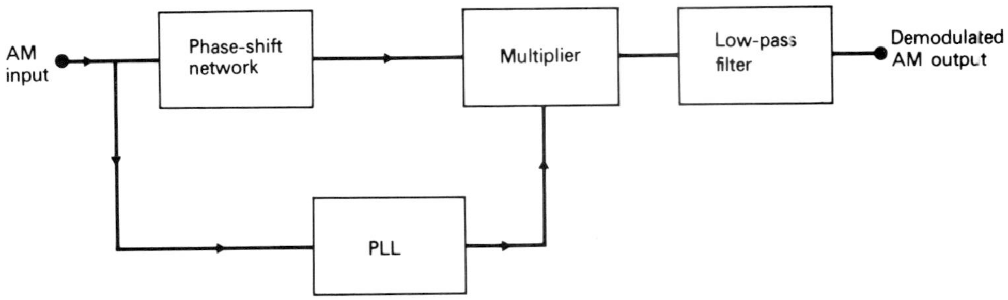

Fig. 3.22 A combined PLL and multiplier IC

Devices such as the 561B (Signetics Ltd) contain a complete PLL as described above and an additional multiplier for AM demodulation. The PLL locks on the carrier of the AM signal, so that the VCO output has the same frequency as the carrier, is noise-free and is shifted in phase. The demodulated AM output is obtained by multiplying the VCO output with the noisy input and passing the product through a suitable low-pass filter, Fig. 3.22. As this method is essentially a coherent detection technique involving the averaging of the product of two signals (see Chapter 10), it has a higher degree of noise immunity than the conventional peak detector type AM demodulators. The PLL has an inherently high degree of selectivity centred about the free-running VCO frequency, and is widely used in tracking weak signals from satellites.

A large number of applications circuits drawn from various sources may be found elsewhere.[4, 6, 7]

Circuits for radio and television

Apart from functional blocks using PLL, such as AM–FM demodulators and stereo decoders, there is a wide variety of integrated circuit subsystems, designed for television and radio, which require a minimum number of external components for their operation. The components in question are physically too large to be incorporated in integrated form and consist mainly of tuned circuits and RC networks. The following are some of the typical integrated circuits belonging to this category.

The AM radio receiver subsystem type CA 3123E (RCA Ltd) is a monolithic integrated circuit that provides an RF amplifier, IF amplifier, mixer, oscillator, automatic gain control (AGC) detector and voltage regulator on a single chip. It is intended for use in superheterodyne AM receivers. The TBA 651 (Mullard Ltd) is a similar integrated circuit.

The 1351 (National Semiconductors Ltd) is a monolithic integrated circuit containing a limiter, FM detector and an audio-amplifier capable of driving an external output stage.

The 3028 is an RF–IF amplifier operating from DC to 120 MHz, intended for communications equipment.

The 1845 is a signal processing system for television receivers, and performs the functions of automatic gain control and 'synch' separations.

The 3064 is designed for fine tuning in television receivers and includes an IF amplifier, a differential peak detector and an AGC circuit.

Several integrated circuit subsystems for colour television receivers are also currently available. For example, the 3066 is an elaborate colour signal processor whereas the 3067 is primarily designed for colour signal demodulation.

MOS transistors, mainly the p-channel enhancement types, are being introduced into bipolar integrated circuits in order to obtain a high input impedance $\approx 10^{14}\,\Omega$. However, MOS linear circuits by themselves are suitable for fabrication in integrated form, because MOS structures for active elements and load resistors are small and simple. MOS large-scale integration is used in the construction of specialised devices such as electrometers, analogue switches and control systems which complement the current range of linear circuits with bipolar transistors.

Microwave integrated circuits are another group of linear circuits which mainly use the hybrid approach. Strip lines are formed by thin film metallisation on a ceramic substrate and the semiconductor devices are added as separate chips. Modules with preamplifiers, mixers, frequency multipliers, phase shifters and power amplifiers are widely used in radar and communications circuitry.

Analogue multipliers

A building block fundamental to analogue signal processing is the four-quadrant multiplier, so called because both the signals to be multiplied may be either positive or negative. There is a large number of schemes for multiplying two time-varying analogue signals, but most of them are meant for signals of a single polarity, either positive or negative. Two commonly-used types of four-quadrant multiplier are the Hall effect multiplier and the translinear multiplier.

In the **Hall effect multiplier**, one input v_x is made proportional to the excitation current and the other v_y proportional to the current producing the magnetic field. The output Hall voltage is proportional to $v_x v_y$.

The **translinear multiplier** utilises a combination of differential amplifiers with diodes in their collector circuits and constant current sources. The differential output currents can be shown to be proportional to the product of the input currents i_x and i_y, provided that the diodes and transistors are sufficiently forward biased for their current to be represented by $i = Ae^{qv/kT}$ where v is the base-emitter voltage in the case of the transistors. The MC1495 or MC1595 (Motorola Ltd) are examples of such inexpensive four-quadrant multiplier ICs. External resistances are used to adjust the bias values of circuits and to convert the input and output currents into voltages.

References

(1) Fitchen, F. C., *Transistor Circuit Design and Analysis*, 2nd edition. Van Nostrand (1970)
(2) Zelinger, G., *Basic Matrix Analysis and Synthesis*. Pergamon Press (1966)
(3) Graeme, J. G., *Applications of Operational Amplifiers*. McGraw-Hill (1973)
(4) Clayton, C. B., *Linear Integrated Circuit Applications*. Macmillan (1975)
(5) Characteristics and applications of the SN 7413 N dual Schmitt trigger, *Application Report B81*. Texas Instruments (November 1969)
(6) Gardner. Floyd M., *Phaselock Techniques*. John Wiley and Sons (1966)
(7) Signetics Applications Handbook

Further reading

Carter, G. W., and Richardson, A., *Techniques of Circuit Analysis.* OUP (1972)
Graeme, J. G., and Tobey, G. E., *Operational Amplifiers.* McGraw-Hill (1971)
National Semiconductors, *Linear Integrated Circuit Applications*
National Semiconductors, *Linear Integrated Circuit Data Handbook*

4

Logic design and building blocks of digital circuitry

4.1 Introduction

Digital circuitry is currently used in one form or another in a vast majority of modern electronic instruments. The popularity of digital techniques is mainly due to the fact that they are capable of giving greater measurement accuracy, stability and dynamic range than their analogue counterparts. Low-cost digital circuits are readily available in IC form, and LSI devices can replace entire circuit assemblies so that digital functions can be easily implemented.

The following sections deal with two main aspects of digital circuitry:

a) the application of Boolean algebra in the design or description of logic systems;
b) the main features and functions of a wide range of digital ICs which may be looked upon as 'building blocks' of complex digital systems.

4.2 Boolean algebra and the physical implementation of its concepts

The three main Boolean operations, in terms of which all others can be expressed, are as follows:

1) Boolean **addition**, Fig. 4.1(a), written as $(A + B)$ and interpreted as A **or** B;
2) Boolean **multiplication**, Fig. 4.1(b), written as $(A.B)$ and interpreted as A **and** B;
3) Boolean **inversion,** Fig. 4.1(c), written as A and interpreted as the **complement** of A.

The logical value of any given Boolean function of variables A, B, C etc. can be represented on a **truth table** in terms of the values of all the variables. For example, the INVERT, OR and AND functions of two variables, A and B, are shown in Table 4.1.

Other commonly-used Boolean functions implemented by **gates** or gating circuits are:

4) EXCLUSIVE OR, Fig. 4.1(d), written as $A \oplus B$, which gives an output 1 if either input is at 1 but not if both inputs are at 1, as shown in Table 4.2;
5) NAND, Fig. 4.1(e), written as $\overline{A . B}$, is an inversion of AND and stands for NOT AND, the function being shown in Table 4.2;
6) NOR, Fig. 4.1(f), written as $\overline{A + B}$, is an inversion of OR and stands for NOT OR, the function being shown in Table 4.2.

The following identities and theorems are useful in analysing Boolean functions, simplifying them and converting them to forms suitable for physical implementation with electronic gating circuits.

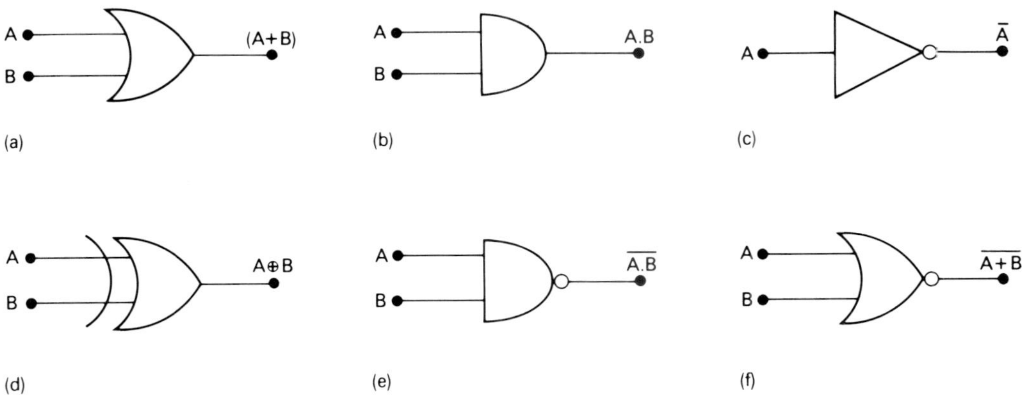

Fig. 4.1 (a) Two-input OR gate; (b) two-input AND gate; (c) inverter or NOT gate; (d) two-input EXCLUSIVE OR gate; (e) two-input NAND gate; (f) two-input NOR gate

Table 4.1 Truth table of OR and AND functions

A	$\overline{A}$	B	$\overline{B}$	A + B	A . B
0	1	0	1	0	0
0	1	1	0	1	0
1	0	0	1	1	0
1	0	1	0	1	1

Table 4.2 Truth table of EXCLUSIVE OR, NAND and NOR functions

A	B	$A \oplus B$	$\overline{A.B}$	$\overline{A+B}$
0	0	0	1	1
0	1	1	1	0
1	0	1	1	0
1	1	0	0	0

1 AND and OR functions $A.A = A;\ A + A = A$
$A.1 = A;\ A + 1 = 1$
$A.0 = 0;\ A + 0 = A$

2 Redundancy In a Boolean expression containing a sum of products, a product that contains all the factors of another product is redundant. For example, $A.B + A.B.C + A.B.D = A.B.$

3 Commutation $A + B = B + A$ and $A.B = B.A$

4 Association $A + (B + C) = (A + B) + C$
$A.(B.C) = (A.B).C$

5 Distribution $A + (B.C) = (A + B).(A + C)$
$A.(B + C) = A.B + A.C$

6 De Morgan's theorems In general they may be written as follows.

$$A + B + C \ldots + N = \overline{\overline{A}\,.\,\overline{B}\,.\overline{C}\,\ldots\ldots\overline{N}}$$
$$A.B.C.\ldots.N = \overline{\overline{A} + \overline{B} + \overline{C} + \ldots + \overline{N}}$$

where any finite number of variables A to N may be included. The transformation involves three steps:

a) AND is changed to OR and OR to AND;
b) each variable is replaced by its complement, A by $\overline{A}$ and $\overline{A}$ by A, etc;
c) the logical state of the entire expression is replaced by its complementary value.

De Morgan's theorems may be verified by examining the truth tables of Boolean expressions such as Table 4.3.

Table 4.3 Truth table of $A+B=\overline{\overline{A}.\overline{B}}$

A	B	$\overline{A}$	$\overline{B}$	$\overline{A}.\overline{B}$	$\overline{\overline{A}.\overline{B}}$	A+B
0	0	1	1	1	0	0
0	1	1	0	0	1	1
1	0	0	1	0	1	1
1	1	0	0	0	1	1

De Morgan's theorems are useful in simplifying Boolean expressions or in transforming a part of an expression in order to get that part in a form suitable for implementation with readily available ICs. For example, the expression $(A+\overline{B.C})$ may be implemented without using an AND gate since $A+\overline{B.C}=A+\overline{B}+\overline{C}$ which requires only OR gates and inverters. Alternatively, the expression may be written as $A+\overline{B.C}=\overline{\overline{A}.B.C}$ which requires only AND gates and inverters.

If an IC gate has fewer inputs than the number required, then several units usually available in the same package have to be cascaded, Fig. 4.2. Also, if a gate has too many inputs, then two or more inputs must be connected together to give an input which is effectively single.

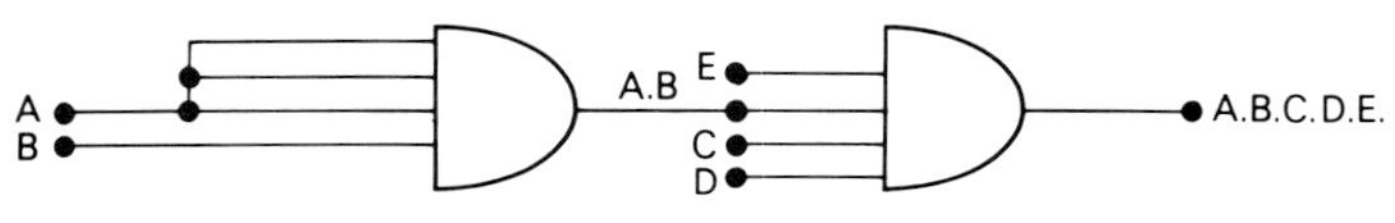

Fig. 4.2 Implementation of the function (A.B.C.D) with two four-input AND gates

Lastly, the implementation of a Boolean function before and after simplification is illustrated in the following example.

Consider the Boolean function $\overline{\overline{A.\overline{B}}+C}+\overline{\overline{A}+\overline{\overline{B}.\overline{C}}}$ implemented in Fig. 4.3. On simplification,

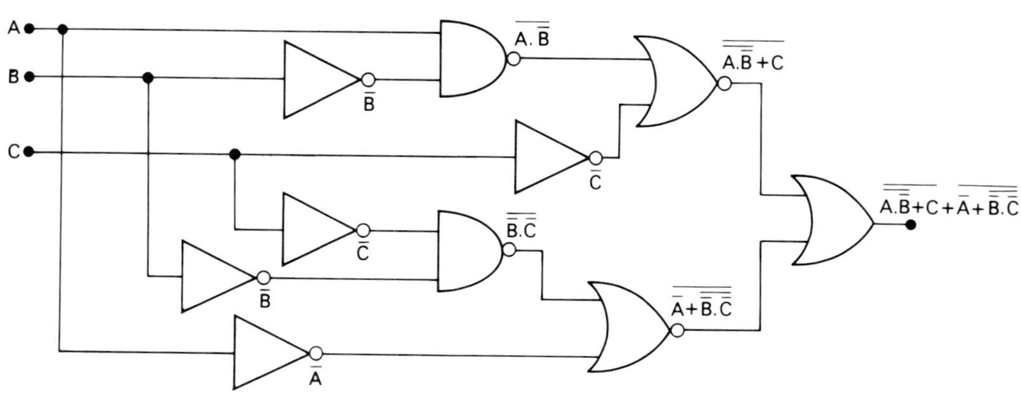

Fig. 4.3 Implementation of the function $\overline{\overline{A.\overline{B}}+C}+\overline{\overline{A}+\overline{\overline{B}.\overline{C}}}$

$$\overline{A.\overline{B}+\overline{C}} = A.\overline{B}.C \quad \text{and} \quad \overline{\overline{A}+\overline{B}.\overline{C}} = A.\overline{B}.\overline{C}$$

Hence

$$\begin{aligned}\overline{\overline{A.\overline{B}}+\overline{C}}+\overline{\overline{A}+\overline{\overline{B}.\overline{C}}} &= A.\overline{B}.C + A.\overline{B}.\overline{C} \\ &= A.\overline{B}(C+\overline{C}) \\ &= A.\overline{B}\end{aligned}$$

Figure 4.4 is the implementation of the simplified version which contains a single gate and an inverter.

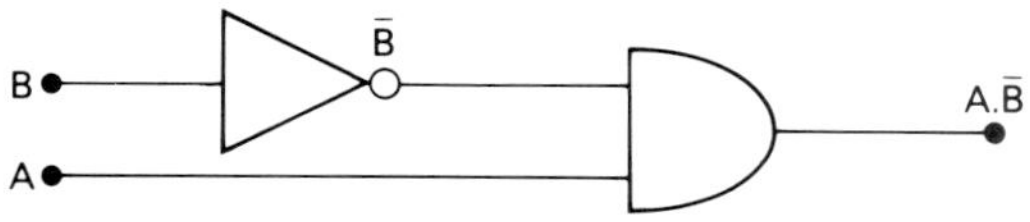

Fig. 4.4 Implementation of the function of Fig. 4.3 after simplification

4.3 Minimisation of Boolean expressions

Simplification of Boolean expressions is of great importance in digital systems using large numbers of logic elements, because it is the only way of determining the most economical implementation. With the low cost and ready availability of ICs, the minimisation of the number of chips required to implement a given logic function is more important than the Boolean minimisation of the number of logic elements. However, the formal minimisation procedure is useful in handling Boolean expressions and may be outlined as follows.

Any Boolean function may be expressed in at least two ways:

1) as a logical sum of logical product terms called **minterms** or **P terms**;
2) as a logical product of logical sum terms called **maxterms** or **S terms**.

The function

$$f(A, B) = A.\overline{B} + B.\overline{A}$$

may be written as

$$f^*(A, B) = (\overline{A}+B).(\overline{B}+A)$$

Each maxterm or minterm in a function of n variables must itself be a logical sum or product, respectively, of those n variables. For any number of variables n, there are 2^n possible minterms or maxterms. For example, functions of two variables A and B can contain only four minterms $\overline{A}\overline{B}$, $\overline{A}B$, $A\overline{B}$ and AB or four maxterms $(A+B)$, $(A+\overline{B})$, $(\overline{A}+B)$ and $(\overline{A}+\overline{B})$. The following are some of the important minterm and maxterm relationships.

1) The complement of any minterm is a maxterm and *vice-versa*, i.e.

$$S_i = \overline{P}_i \quad \text{and} \quad P_i = \overline{S}_i$$

For example,

$$\overline{A}+B+\overline{C}+\overline{D} = \overline{A.\overline{B}.C.D}$$

2) The logical sum of all the 2^n minterms of any function of n variables is unity, and the logical product of all its maxterms is zero, i.e.

$$\sum_{i=0}^{2^n-1} \mathrm{P}_i = 1 \quad \text{and} \quad \prod_{i=0}^{2^n-1} \mathrm{S}_i = 0$$

For example,

$$\overline{\mathrm{A}}\,.\,\overline{\mathrm{B}} + \overline{\mathrm{A}}\,.\,\mathrm{B} + \mathrm{A}\overline{\mathrm{B}} + \mathrm{AB} = 1$$

and

$$(\mathrm{A}+\mathrm{B})\,.\,(\mathrm{A}+\overline{\mathrm{B}})\,.\,(\overline{\mathrm{A}}+\mathrm{B})\,.\,\overline{\mathrm{A}}+\overline{\mathrm{B}}) = 0$$

3) The logical product of any two different minterms in a given function is zero and the logical sum of any two different maxterms is unity, i.e.

$$\mathrm{P}_j\mathrm{P}_k = 0 \quad \text{if} \quad j \neq k$$

and

$$\mathrm{S}_j + \mathrm{S}_k = 1 \quad \text{if} \quad j \neq k$$

For example,

$$(\overline{\mathrm{A}}\,.\,\mathrm{B}\,.\,\overline{\mathrm{C}})(\mathrm{A}\,.\,\mathrm{B}\,.\,\overline{\mathrm{C}}) = 0$$

and

$$(\mathrm{A}+\overline{\mathrm{B}}+\mathrm{C}) + (\overline{\mathrm{A}}+\mathrm{B}+\overline{\mathrm{C}}) = 1$$

Depending on the application, minimisation of a Boolean function consists in expressing that function as a sum of products with the least number of inputs or the least number of terms. One of the commonly-used methods of achieving this is by inspecting its minterms represented on a **Karnaugh map**.

Consider a function Q of four variables such as

$$\mathrm{Q} = \overline{\mathrm{A}}\,.\,\overline{([(\mathrm{B}+\mathrm{C})\,.\,\mathrm{D}]\,.\,\overline{\mathrm{A}})}$$

Table 4.4 Truth table of $\mathrm{Q} = \overline{\mathrm{A}}\,.\,\overline{([(\mathrm{B}+\mathrm{C})\,.\,\mathrm{D}]\,.\,\overline{\mathrm{A}})}$

A	B	C	D	(B + C)	(B + C) . D	[(B + C) . D] . $\overline{\mathrm{A}}$	$\overline{(\mathrm{B}+\mathrm{C})\,.\,\mathrm{D}\,.\,\overline{\mathrm{A}}}$	Q
0	0	0	0	0	0	0	1	1
0	0	0	1	0	0	0	1	1
0	0	1	0	1	0	0	1	1
0	0	1	1	1	1	1	0	0
0	1	0	0	1	0	0	1	1
0	1	0	1	1	1	1	0	0
0	1	1	0	1	0	0	1	1
0	1	1	1	1	1	1	0	0
1	0	0	0	0	0	0	1	0
1	0	0	1	0	0	0	1	0
1	0	1	0	1	0	0	1	0
1	0	1	1	1	1	0	1	0
1	1	0	0	1	0	0	1	0
1	1	0	1	1	1	0	1	0
1	1	1	0	1	0	0	1	0
1	1	1	1	1	1	0	1	0

This function has 16 possible combinations shown in the truth table 4.4, Q being valid for all input *combinations* of the inputs that give Q = 1, i.e.

$$Q = \overline{A}.\overline{B}.\overline{C}.\overline{D} + \overline{A}.\overline{B}.\overline{C}.D + \overline{A}.\overline{B}.C.\overline{D} + \overline{A}.B.\overline{C}.\overline{D} + \overline{A}.B.C.\overline{D}$$

This can be represented on a Karnaugh map, Fig. 4.5, using Table 4.5 which represents all possible combinations of the variables A, B, C and D. In order to obtain a minimised expression using the Karnaugh map of Fig. 4.5, the adjacent 1's are grouped together and the sum of the logical expressions is written as

$$Q = \overline{A}.\overline{D} + \overline{A}.\overline{B}.\overline{C}$$

CD ↓ \ AB →	00	01	11	10
00	1	1	0	0
01	1	0	0	0
11	0	0	0	0
10	1	1	0	0

Fig. 4.5 Karnaugh map representing the function $Q = \overline{A}.(\overline{[B+C).D].\overline{A}})$

Table 4.5 Possible combinations of four variables A, B, C and D

CD ↓ \ AB →	00	01	11	10
00	$\overline{A}.\overline{B}.\overline{C}.\overline{D}$	$\overline{A}.B.\overline{C}.\overline{D}$	$A.B.\overline{C}.\overline{D}$	$A.\overline{B}.\overline{C}.\overline{D}$
01	$\overline{A}.\overline{B}.\overline{C}.D$	$\overline{A}.B.\overline{C}.D$	$A.B.\overline{C}.D$	$A.\overline{B}.\overline{C}.D$
11	$\overline{A}.\overline{B}.C.D$	$\overline{A}.B.C.D$	$A.B.C.D$	$A.\overline{B}.C.D$
10	$\overline{A}.\overline{B}.C.\overline{D}$	$\overline{A}.B.C.\overline{D}$	$A.B.C.\overline{D}$	$A.\overline{B}.C.\overline{D}$

In general, the process of minimisation of logical expressions using a Karnaugh map may be summarised as follows.

1) The function is expressed as a sum of products.
2) Each term is represented on a Karnaugh map whose size is defined by the number of variables.
3) All the 1's are grouped together so as to form the largest possible block of cells in binary groups of 2, 4, 8, etc. Overlapping groups are desirable to reduce the expression to its simplest form.
4) The sum of all the logical expressions gives the minimised function.

The Karnaugh map of a function helps to simplify it by grouping the adjacent cells so that each group can be described by the simplest possible combination of variables. The map may be considered to be continuous from the right-hand edge to the left-hand edge. For example, a function of three variables, such as

$$F = A.\overline{B}.C + \overline{A}.\overline{B}.\overline{C} + A.B.C + A.\overline{B}.\overline{C}$$

Table 4.6 Truth table of $F = A.\overline{B}.C + \overline{A}.\overline{B}.\overline{C} + A.B.C + A.\overline{B}.\overline{C}$

A	B	C	$A.\overline{B}.C$	$\overline{A}.\overline{B}.\overline{C}$	$A.B.C$	$A.\overline{B}.\overline{C}$	F
0	0	0	0	1	0	0	1
0	0	1	0	0	0	0	0
0	1	0	0	0	0	0	0
0	1	1	0	0	0	0	0
1	0	0	0	0	0	1	1
1	0	1	1	0	0	0	1
1	1	0	0	0	0	0	0
1	1	1	0	0	1	0	1

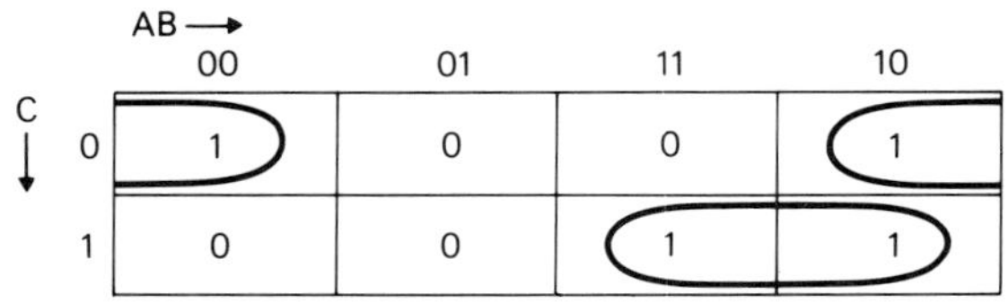

Fig. 4.6 Karnaugh map representing the function F = A.B.C + A.B.C + A.B.C + A.B.C.

Table 4.7 Possible combinations of three variables A, B and C

AB		00	01	11	10
C	0	$\overline{A}.\overline{B}.\overline{C}$	$\overline{A}.B.\overline{C}$	$A.B.\overline{C}$	$A.\overline{B}.\overline{C}$
	1	$\overline{A}.\overline{B}.C$	$\overline{A}.B.C$	$A.B.C$	$A.\overline{B}.C$

has eight possible combinations shown in truth table 4.6; F is valid for all input combinations that give F = 1 and can be represented on the Karnaugh map by Fig. 4.6 using Table 4.7. The marked grouping will give the simplified function

$$F = A.C + \overline{B}.\overline{C}$$

4.4 Binary numbers and codes

In Boolean algebra the variables are allowed to have two values described as 'true or false', 'high or low', '1 or 0', etc. The two states form the basis of the binary number system in which the **base** or **radix** is 2, instead of 10 in the decimal system. Thus a binary number 111001 may be written as follows:

Most significant bit (MSB)   Least significant bit (LSB)

$$\begin{array}{cccccc} 1 & 1 & 1 & 0 & 0 & 1 \\ \downarrow & \downarrow & \downarrow & \downarrow & \downarrow & \downarrow \\ 1\times 2^5 & +1\times 2^4 & +1\times 2^3 & +0\times 2^2 & +0\times 2^1 & +1\times 2^0 \\ \downarrow & \downarrow & \downarrow & \downarrow & \downarrow & \downarrow \\ 32 & +\ 16 & +\ 8 & +\ 0 & +\ 0 & +\ 1 \end{array} = 57$$

The simplicity of using only two states in the binary system is achieved at the cost of using additional binary digits. In general, to represent n decimal digits in a system of radix r requires

$$\frac{n}{\log_{10} r} = m \quad \text{digits}$$

The 0 and 1 **binary digits** or **bits** can be represented by voltage or current pulses, so that a binary number can be expressed as a train of pulses suitable for manipulation by digital electronic circuits. Prior to processing the pulse trains are often coded, the commonly-used codes being:

i) binary coded decimal (BCD);
ii) octal;
iii) hexadecimal.

Binary coded decimal

In the **BCD** code, each decimal digit is represented by four binary digits or bits: $A = 2^0 = 1$, $B = 2^1 = 2$, $C = 2^2 = 4$, and $D = 2^3 = 8$. Thus a decimal number 728 may be written as

D	C	B	A	D	C	B	A	D	C	B	A
0	1	1	1	0	0	1	0	1	0	0	0
		↓				↓				↓	
		7				2				8	

Octal code

In the **octal** code, the base or radix is 8 and the permitted symbols are 0, 1, 2, 3, . . . , 7. Thus a decimal number 24 may be written as

$$\begin{array}{ccccccc} 3 & & 0 & & & \rightarrow & \text{octal} \\ \downarrow & & \downarrow & & & & \\ 3 \times 8^1 & + & 0 \times 8^0 & & & & \\ \downarrow & & \downarrow & & & & \\ 24 & + & 0 & = & 24 & \rightarrow & \text{decimal} \end{array}$$

Binary coding of octal numbers requires only three bits, $A = 2^0 = 1$, $B = 2^1 = 2$, $C = 2^2 = 4$. The binary number is arranged in groups of three bits and each group is an octal coded number. Thus a decimal number 1206 may be written as a binary 010010110110 and arranged in groups of three digits as follows:

$$\begin{array}{cccc} \overbrace{0 \quad 1 \quad 0} & \overbrace{0 \quad 1 \quad 0} & \overbrace{1 \quad 1 \quad 0} & \overbrace{1 \quad 1 \quad 0} \\ 0\times2^2+1\times2^1+0\times2^0 & 0\times2^2+1\times2^1+0\times2^0 & 1\times2^2+1\times2^1+0\times2^0 & 1\times2^2+1\times2^1+0\times2^0 \\ \downarrow & \downarrow & \downarrow & \downarrow \\ 2 & 2 & 6 & 6 \end{array}$$

The octal number is thus 2266.

Hexadecimal

In the **hexadecimal** or hex code, the base or radix is $16 = 2^4$ so that each hexadecimal digit corresponds to a group of four bits according to the code,

binary	0000	0001	0010	0011	0100	0101	0110	0111
hex	0	1	2	3	4	5	6	7
binary	1000	1001	1010	1011	1100	1101	1110	1111
hex	8	9	A	B	C	D	E	F

Thus an 8 bit binary number can be written as a pair of hex digits. For example,

Binary → Hex
00110100 → 34
00111100 → 1E
11111100 → FC

To distinguish hex numbers from decimal numbers or words, they are preceded by the sign '?' or followed by an H in cases where the first digit of the number is in the range 0–9. Examples of the hex notation are ?34, 34H, ?FC, 1EH.

4.5 Digital IC terminology

Logic functions can be implemented by a wide variety of commercially-available digital integrated circuits which use bipolar transistors, in saturated or unsaturated states, or MOSFETs as active switching elements. The main factors that dictate their choice for a given application are specified by manufacturers using the following terminology.

Logic levels are the maximum and minimum voltage values which define a binary 0 or 1 and must always be verified during test procedures to ensure that the integrated circuit operates satisfactorily within the specified limits.

Propagation delays are time delays inherent in all switching circuits, and can be traced back to storage effects (see Chapter 1), the nature of the devices and the mode in which they are used. It is, however, important to note that when switching circuits are cascaded, the propagation delays accumulate and limit the operating speed of the entire circuit.

Power dissipation is the average value in milliwatts per logic unit and depends upon the power consumption in the ON and OFF states. The value ranges from a few microwatts in certain types of MOS circuits to tens of milliwatts per unit in high-speed nonsaturated logic circuits. In most bipolar transistor circuitry, the power dissipation is related to the speed of switching and a compromise in the choice of the IC is often required. To meet the power dissipation requirements of some circuits, cooling arrangements are often necessary to ensure proper operation.

Noise immunity is the capability of a digital circuit to reject noise voltages that would cause a circuit to switch at an improper time to an undesirable state. The noise immunity of most logic circuits is between 10 and 50% of the supply voltage.

'Fan-in' and 'Fan-out' 'Fan in' of a device indicates the number of inputs allowed to it. If a device such as a gate with a fan-in restriction is used to implement a Boolean expression, then it is equivalent to specifying the maximum size of the products and sums of the expression. The 'fan-out' of a device indicates the number of loads $L_1, \ldots, L_n$ of a specified magnitude that a device output will handle without any significant deterioration of its performance and without excessive heat generation.

4.6 Families of digital ICs

On account of their ready availability and low cost, the following 'families' of digital integrated circuits, each family with distinct characteristics, have gained popularity and allow the user to choose an appropriate device for a given application.

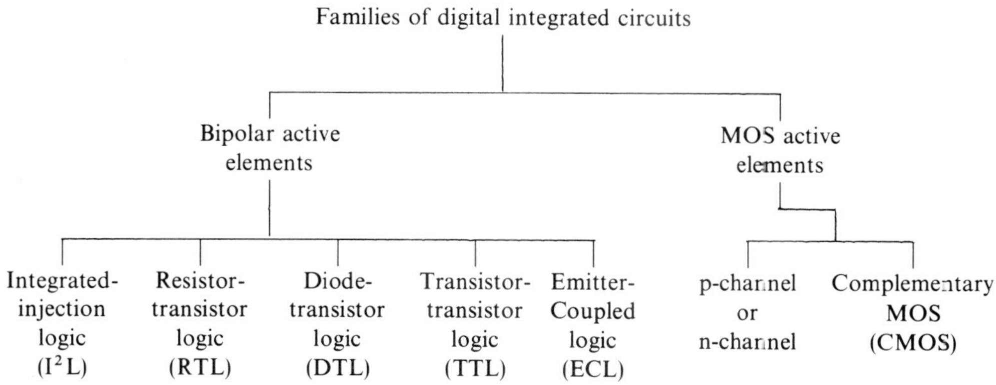

Transistor-transistor logic

These circuits use a multiple-emitter transistor such as Q_1 in Fig. 4.7 connected to the base of a transistor Q_2 in a phase-splitting circuit which drives the output stage Q_3, Q_4. When all inputs are at level 1, Q_2 and Q_4 are ON, Q_3 is OFF and the output is 0. When one or more inputs are at level 0, Q_2 and Q_4 are OFF, Q_3 is ON, and the output is 1. The gate is a NAND element with a low output impedance and a fast speed. There are several versions of TTL circuits made for special applications. Some of these are now discussed.

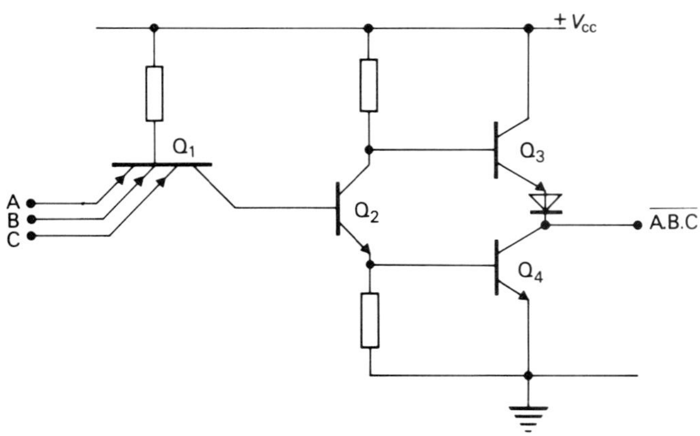

Fig. 4.7 TTL three-input NAND gate

Low power TTL
These circuits have approximately $\frac{1}{10}$ the power consumption of normal TTL circuits, but a propagation delay of 30 to 40 ns. They are ideal for applications where a high speed of operation can be sacrificed for low power consumption.

High power TTL
The propagation delay in these circuits is reduced to about 6 ns at the expense of the power consumption, which is about twice that of a normal TTL circuit.

Schottky TTL

In these circuits a Schottky diode is connected between the base and collector of each transistor (see Chapter 1). The transistors do not saturate, there are no charge storage problems and propagation delays as low as 3 ns can be achieved, with a power dissipation which is lower than the high-power TTL. A special low-power Schottky TTL, with a power dissipation of 2 mW per unit and a propagation delay of 10 ns, is also available.

Three-state TTL

In addition to the binary 0 (low) and 1 (high) outputs, this type of TTL has a third state which depends on the state of the control electrode. In this state, if the control electrode is at logic 1, the output terminal represents a very high impedance and is therefore equivalent to disconnecting the output circuit from the output pin. These circuits are used in multiplexed or 'bussed' data transmission and were originally introduced by the National Semiconductor Corporation as 'Tri-State Logic' circuits. One of several bidirectional data bus lines, each carrying one bit of a multiple bit data word (or **byte**) using tristate logic gates, is shown in Fig. 4.8. The bus line may be simply a cable several feet long. Any number of tristate outputs can be connected in parallel to form a common bus line, since all the gates not transmitting data can have their control electrodes at logic 1 so that their outputs are effectively disconnected. Only the gate C_3 required to transmit a bit will be enabled, and the information will be received by R'_1. The same bus can be used to transmit one bit from either C'_1, C'_2 or C'_3 at some other time, and this information will be received by R_1. Thus only a single data transmission can take place at any given time in either direction, from any one of the several transmitters to several receivers.

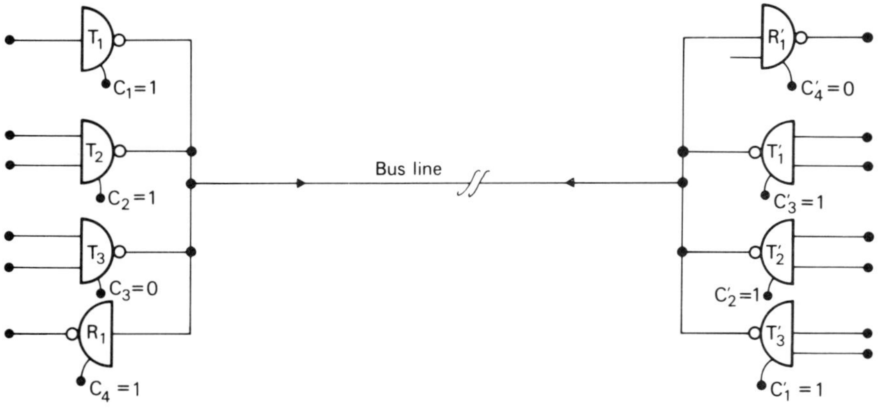

Fig. 4.8 Three-state TTL bidirectional data line

Emitter-coupled logic

These circuits attain switching speeds of a fraction of a nanosecond by always working the transistors in the nonsaturated region. The basic circuit is an emitter-coupled amplifier or long-tailed pair such as Q_1Q_2 in Fig. 4.9. A reference voltage V_r is applied to Q_2 and the inputs to Q_1. When all the inputs are at level 0, Q_1 carries less current than Q_2 and the potential of the Q_1 collector is higher than that of the Q_2 collector. If any one of the inputs is switched to level 1 then Q_1 carries more current than Q_2 and the Q_1 collector is lower than the

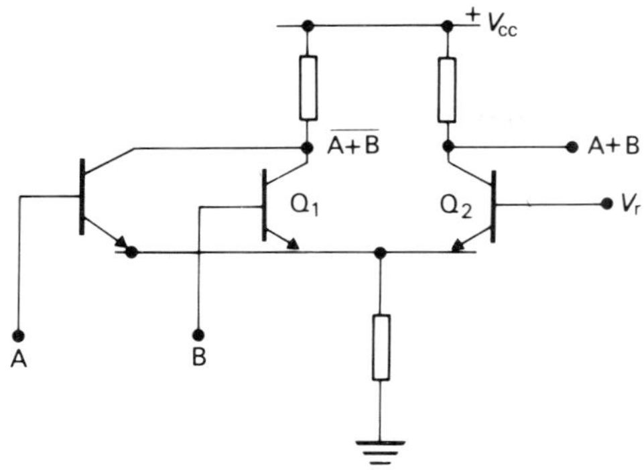

Fig. 4.9 ECL two-input OR-NOR gate

Q_2 collector. The value of V_r is arranged to be approximately midway between the two logic levels.

Resistor-transistor and diode-transistor logic

RTL was the earliest form of logic available in IC form. A transistor was supplied for each input and only a few resistors with small values were required, as shown in Fig. 4.10. In DTL, Fig. 4.11, the inputs are isolated with diodes and the collector of the output transistor T is at level 0 only when all the inputs are at level 1.

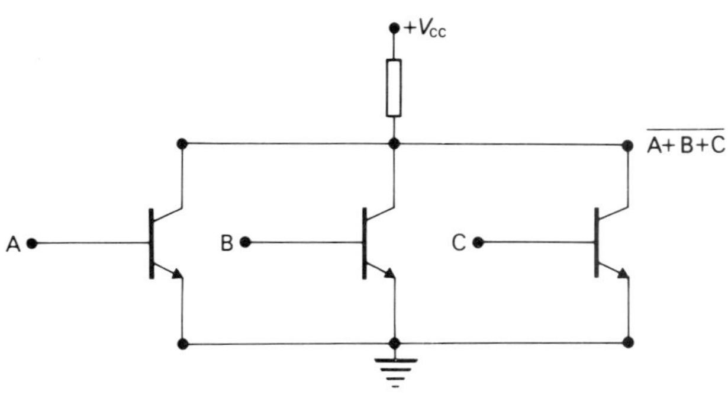

Fig. 4.10 RTL three-input NOR gate

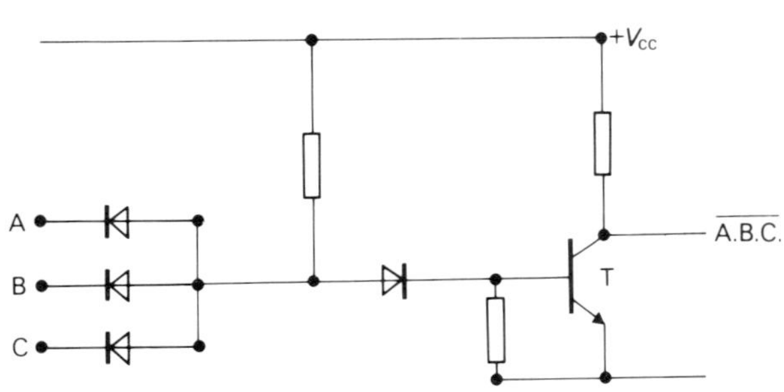

Fig. 4.11 DTL three-input NAND gate

MOS and complementary MOS (CMOS)

MOS logic circuits (p-type or n-type) that perform logic functions with voltage levels are known as **static logic circuits**. Dynamic logic circuits take advantage of the capacitive nature of the input of MOS devices, and operate by transferring stored charges, corresponding to logic levels, from one circuit to the next using high-speed clock signals. They consume less power than the static circuits and achieve higher operating speeds.

CMOS transistor (CMOST) logic circuits use both p-channel and n-channel enhancement mode MOSFETs, and the switching action of such a complementary pair is illustrated in Fig. 4.12. When C, the common gate input, is at logic 0 (low or near earth potential) T_2 is OFF but T_1 is ON. Hence the output $\overline{C}$ will be at logic 1 (high or near supply voltage V_{dd}). When C is at logic 1 (high or near V_{dd}), T_2 is ON but T_1 is OFF so that the output $\overline{C}$ will be at logic 0 (low or near earth potential). Low power dissipation is achieved because in neither of the logic states is there a continuous path through any of the devices from the supply voltage to earth. However, both the devices are ON momentarily together during the switching of the output state, thereby causing a small current flow. The power consumption of a CMOS device, therefore, increases with its operating frequency. The CMOS logic circuit family features a low power dissipation, excellent noise immunity, wide power supply voltage variations, high fan-out and a moderately high speed of operation. Above all, a high packing density can be achieved and this family is therefore the obvious choice for LSI circuitry such as is used in microprocessors. CMOS technology excels in the electronic watch industry.

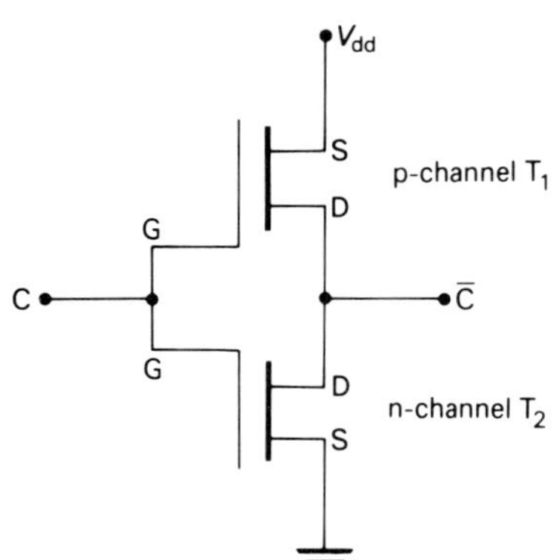

Fig. 4.12 Switching action of a CMOST pair

Integrated-injection logic (I^2L)

The basic I^2L element is a pnp–npn pair of transistors on a silicon substrate which serves as the emitter of the npn transistor, Fig. 4.13. The current in this transistor flows upwards through the base to multiple collectors. The substrate is also the base of the pnp transistor, which has a lateral current flow. By this arrangement, the isolation islands normally required in bipolar circuitry are eliminated and packing densities similar to MOS circuits may be achieved. I^2L gates are faster than MOS gates but not as fast as some TTL circuits.

4.7 Combinational logic circuits and some commonly-used MSI functional circuit assemblies

Combinational logic circuits are digital circuits which are made up of gates and inverters. In

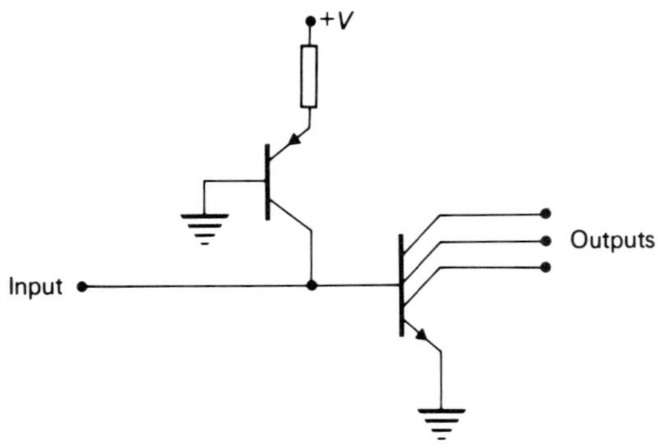

Fig. 4.13 Basic I^2L element made up of a pnp–npn pair of transistors

principle, combinational circuits can be constructed with three basic building blocks; i.e. AND, OR and INVERT gates. However, circuits constructed entirely with NAND or NOR gates are generally more economical and convenient to use. Two main features of combinational logic circuits are as follows:

1) gate fan-in restriction, or the maximum number of inputs to a gate;
2) state hazards or unwanted transient changes in the output.

Fan-in restriction

When the fan-in of the gates to be used for implementing a Boolean function is specified, the given function must be rearranged to satisfy the restriction. For example, the function $f = AB + A\overline{C} + A\overline{D}$ can be implemented directly with six NAND gates, but one of them has three inputs, Fig. 4.14. If the fan-in restriction is two, then the function must be rearranged as $f = AB + A(\overline{C} + \overline{D})$ which requires only four NAND gates, Fig. 4.15.

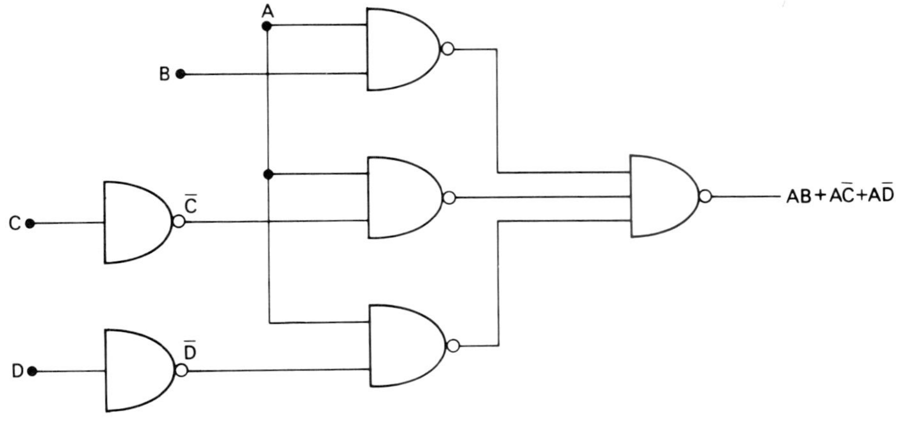

Fig. 4.14 Direct implementation of the function $A.B + A.\overline{C} + A.\overline{D}$ with six NAND gates

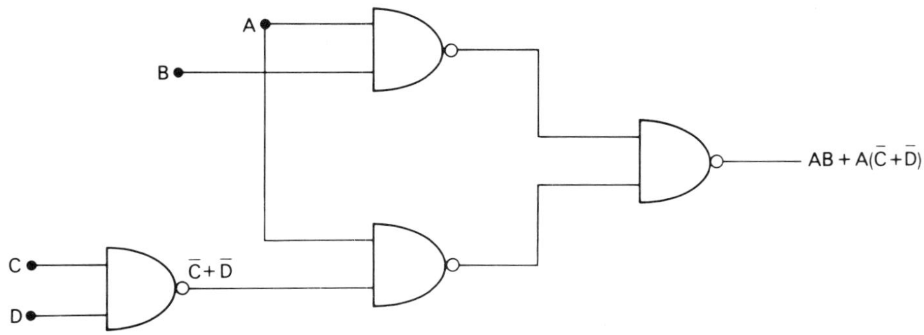

Fig. 4.15 Implementation of the function $A.B + A.\bar{C} + A.\bar{D}$ with a fan-in restriction of two on the NAND gates

There is a systematic method based on the use of merging tables[1,2] for obtaining the best possible arrangement of a function to be implemented using gates with a specified fan-in.

State hazards

In combinational logic circuits, a change in a single variable from 0 to 1 or from 1 to 0 may cause an unwanted transient change in the output. A common example is the spurious pulse caused by the finite propagation delay of practical gates. In Fig. 4.16 the output $(A + \bar{A})$ falls to zero for an instant equal to the propagation delay of the INVERT gate.

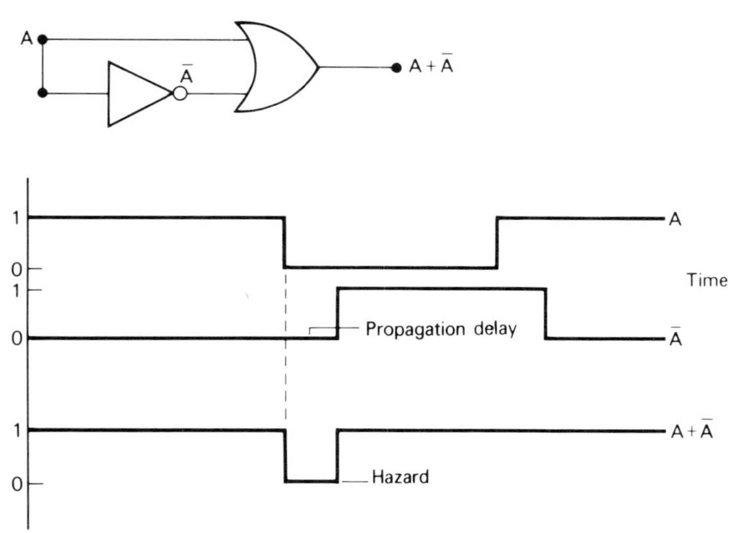

Fig. 4.16 Hazard caused by propagation delay

Another example of a state hazard is shown in Fig. 4.17. A hazard is said to exist when the input signal combinations change in such a way that they cause a change in the adjacent cells in the Karnaugh map that are not grouped together. By grouping the cells as in Fig. 4.18 the hazard can be eliminated, at the expense of including a redundant AND gate in the network.

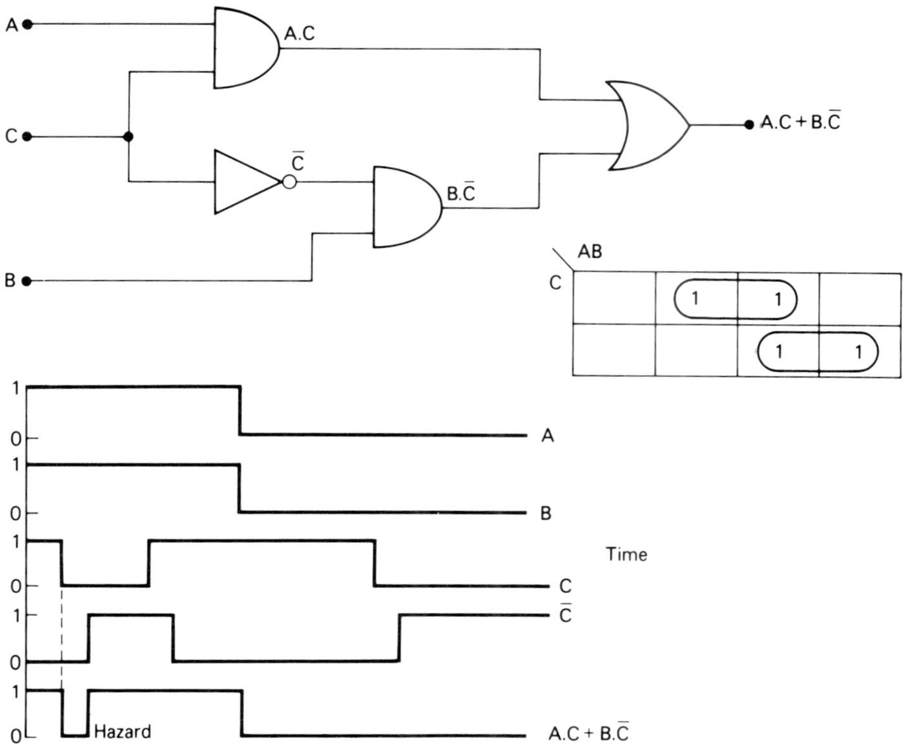

Fig. 4.17 Implementation of the function $f = A.C + B.\overline{C}$

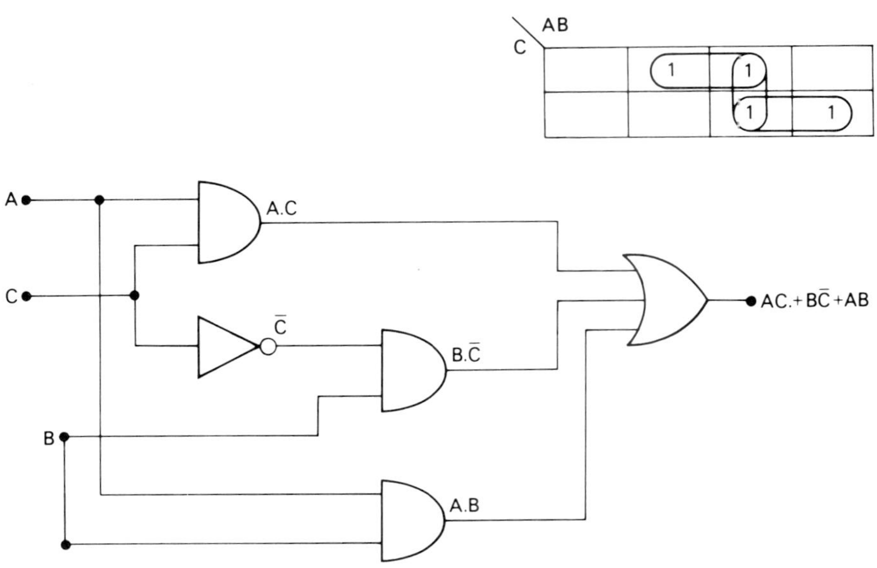

Fig. 4.18 Hazard-free version of Fig. 4.17

Despite the large number of possible combinational circuits, most digital operations can be implemented by using the following MSI functional circuit assemblies:

1) decoders;
2) encoders;
3) multiplexers;
4) demultiplexers;
5) EXCLUSIVE OR;
6) code converters.

Decoders

A single binary word N digits in length can represent 2^N different elements of information. A combination of basic logic elements such as OR, INVERT, and AND gates are used to decode or to detect the presence of a specific binary number or word. For example, a two-bit binary word contains $2^2 = 4$ distinct states and can be decoded using four two-input NAND gates, Fig. 4.19. The output shows a level 1 only when both its inputs are at level 0, thereby defining a state uniquely as shown in Table 4.8.

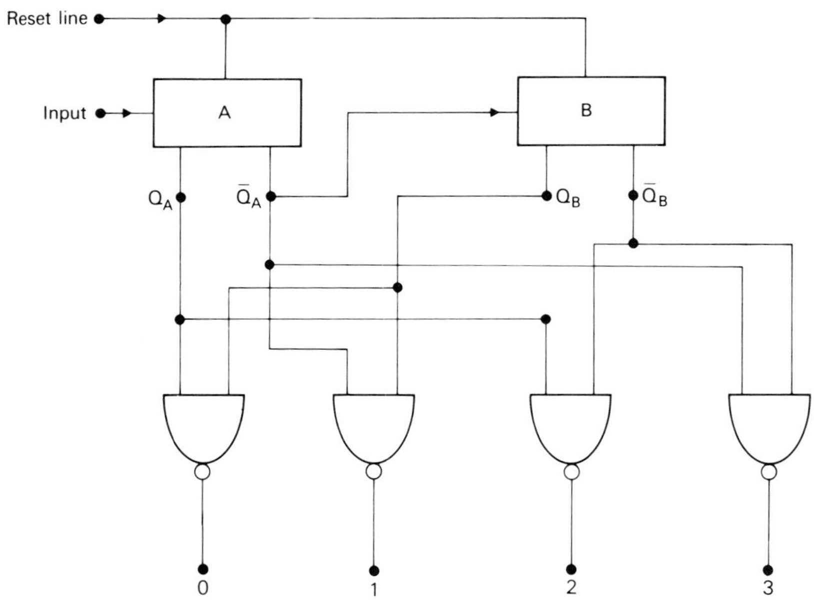

Fig. 4.19 A two-bit binary word decoder using four two-input NAND gates

A variety of codes can be generated for binary representation of decimal numbers. Each code is designed for a specific purpose and therefore may not necessarily use the full capacity of the binary system. For example, the popular BCD code uses only 10 out of the possible 16 states of a four-bit word. Commonly-used decoders are available as MSI packages, and it is no longer necessary to construct a decoder from individual logic elements. The following applications illustrate the use of some of these decoders.

Table 4.8 The four states 0, 1, 2, 3 distinctly defined by the outputs of the four NAND gates

State	Q_A	$\bar{Q}_A$	Q_B	$\bar{Q}_B$	*Gate inputs*
0	0	1	0	1	$Q_A \cdot Q_B$
1	1	0	0	1	$\bar{Q}_A \cdot Q_B$
2	0	1	1	0	$Q_A \cdot \bar{Q}_B$
3	1	0	1	0	$\bar{Q}_A \cdot \bar{Q}_B$

BCD to decimal decoders

The 74141 BCD to decimal 'decoder-driver' is mainly used for decoding the output of BCD decade counters such as the 7490 and driving cold cathode tubes, Fig. 4.20, such as the Burroughs NIXIE. The 'driver' stage of the decoder consists of ten open collector transistors, each collector being connected to one cathode of the NIXIE tube. The gas discharge strikes between the anode and any one cathode, the drive transistor of which happens to be in a conducting state. Decoders such as the 7442 have high-performance npn transistor output stages suitable for driving relays or indicator lamps.

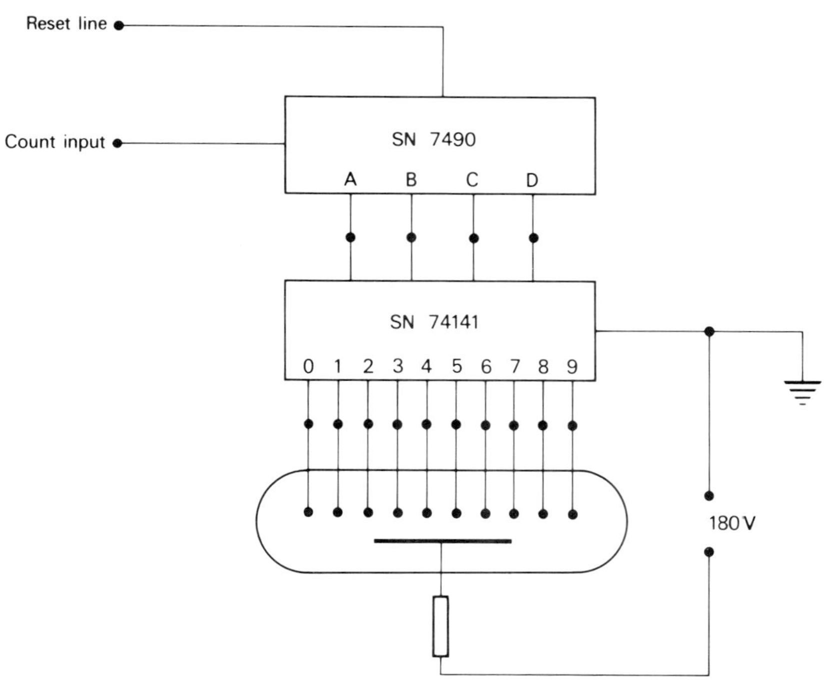

Fig. 4.20 BCD to decimal 'decoder-driver' of a cold cathode display tube

BCD to seven-segment decoders

A seven-segment readout displays the decimal numbers 0 to 9 and occasionally special letter combinations, by illuminating two or more segments in a specially arranged seven-segment

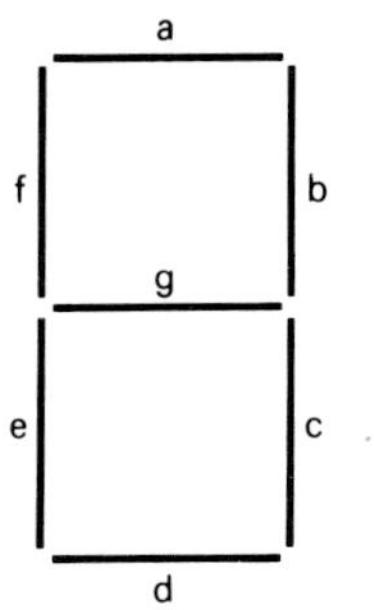

(a)

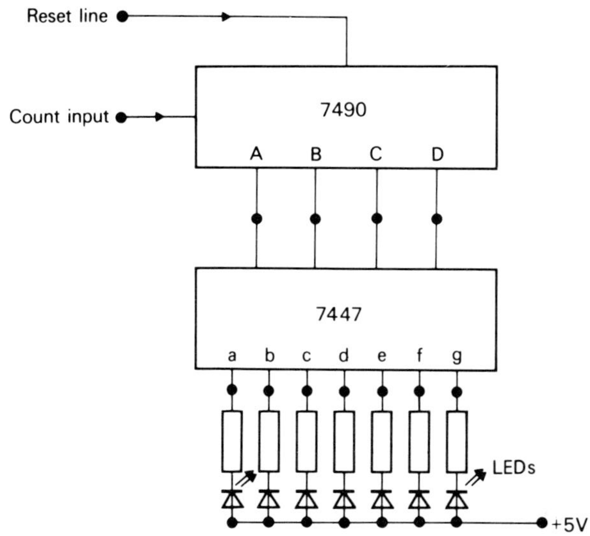

(b)

Fig. 4.21 (a) A seven-segment LED display; (b) a BCD to seven-segment 'decoder-driver'

pattern such as Fig. 4.21(a). The segments may be constructed of a wide variety of light-emitting elements such as light-emitting diodes (LEDs), incandescent filaments or liquid crystal segments. The function of decoders such as the 7447 is to activate the appropriate segments to form the numbers from 0 to 9 via open collector output transistors, each of which is capable of handling about 20 mA, Fig. 4.21(b).

Octal and hex decoders

An octal decoder accepts a parallel three-bit input word and decodes all eight output states representing the decimal numbers 0 to 7. Figure 4.22 shows the decimal weights of the inputs

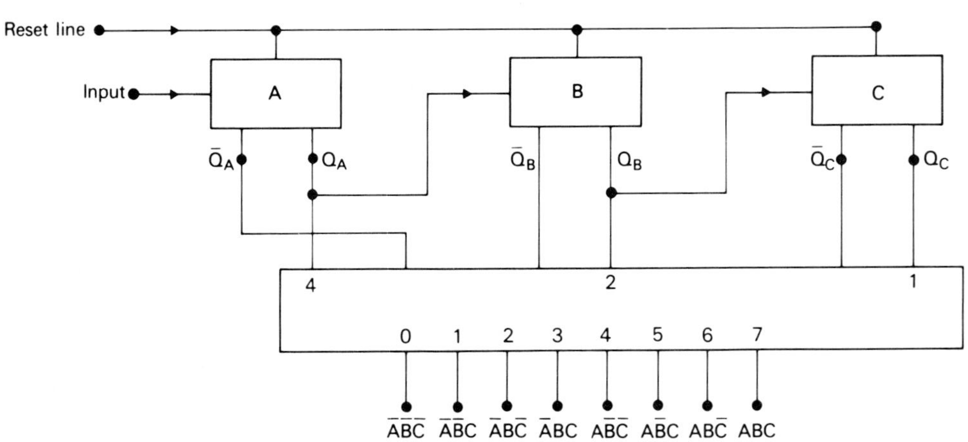

Fig. 4.22 An octal decoder

$Q_A = 4$, $Q_B = 2$, $Q_C = 1$ and the decimal equivalents of the decoded outputs ($A.B.C = 4+2+1 = 7$). Similarly, a hex decoder takes a four-bit input word and decodes all the sixteen states represented by that word.

Finally, in decoding binary sequences with redundant states, simplification using a Karnaugh map is possible. For example, the logic circuitry required for decoding 8-4-2-1 weighted BCD numbers into decimal is deduced from the Karnaugh map in Fig. 4.23. Only the numbers zero and unity require four binary variables, whereas other cells can be grouped with one or more X.

Decimal number	BCD number A (8)	B (4)	C (2)	D (1)	Simplified logic representation
0	0	0	0	0	$\bar{A}.\bar{B}.\bar{C}.\bar{D}$
1	0	0	0	1	$\bar{A}.\bar{B}.\bar{C}.D$
2	0	0	1	0	$\bar{B}.C.\bar{D}$
3	0	0	1	1	$\bar{B}.C.D$
4	0	1	0	0	$B.\bar{C}.\bar{D}$
5	0	1	0	1	$B.\bar{C}.D$
6	0	1	1	0	$B.C.\bar{D}$
7	0	1	1	1	$B.C.D$
8	1	0	0	0	$A.\bar{D}$
9	1	0	0	1	$A.D$
10-15	x	x	x	x	

(a)

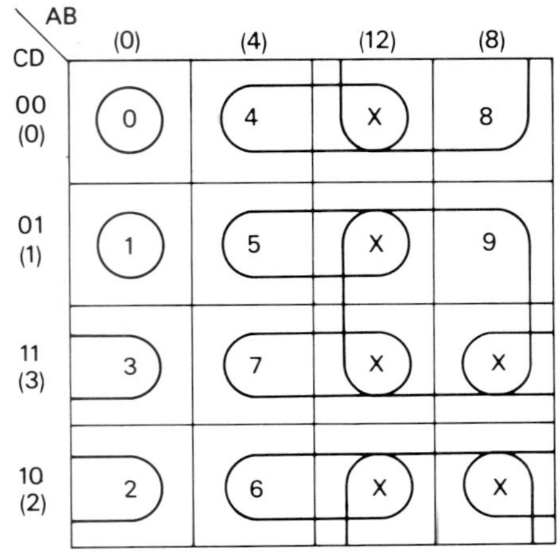

(b)

Fig. 4.23 (a) BCD to decimal decoder; (b) Karnaugh map

Encoders

These are combinational logic circuits that accept one or more inputs and generate specific codes. A simple example of an encoder in which a decimal keyboard signal is converted into a four-bit BCD code is shown in Fig. 4.24. Thus when switch 7 of the keyboard is pressed, the inputs of gates C, B and A are brought to logic 0 whereas the output of gate D will be at logic 1 so that the number generated will be 0111, which is decimal 7. TTL MSI encoders such as the 74147, encoding nine data lines to four output lines (8-4-2-1 BCD), and the 74148, encoding eight data lines to three output lines (4-2-1 binary), are 'priority encoders' implying that when two or more inputs are low simultaneously, the input with the highest priority is represented at the output.

Multiplexers

A multiplexer is a circuit used to select one at a time out of a number of inputs, just like a mechanical selector switch, the basic difference being the high operating speed of the multiplexer and its facility for automatic selection of the inputs. In analogue multiplexers, bipolar transistors or MOSFET switches are used, whereas digital multiplexers are constructed from logic gates because their primary function is the selection of binary data.

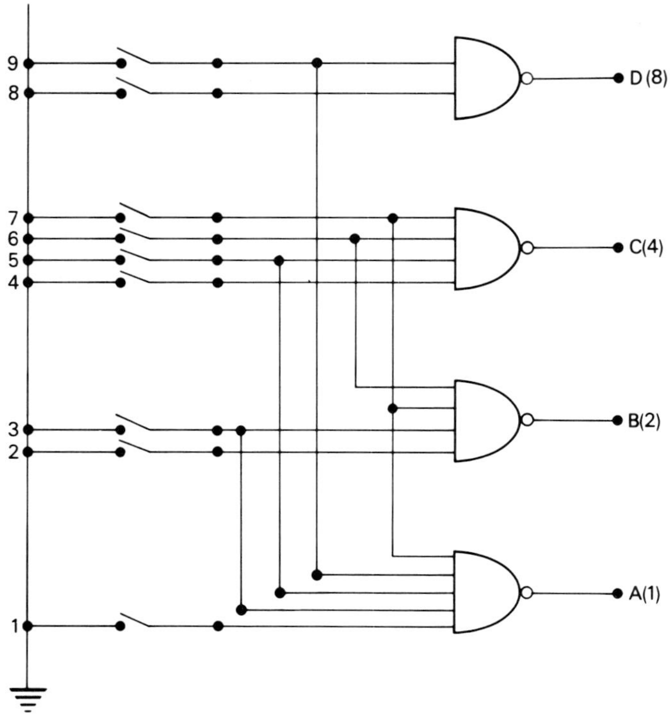

Fig. 4.24 A decimal keyboard to four-bit BCD encoder using NAND gates

For example, two four-bit words such as $A_1B_1C_1D_1$ and $A_2B_2C_2D_2$ can be multiplexed into one four-bit channel, Fig. 4.25, by using the 'select' and 'enable' controls. When the enable line is at logic 1, the output of the inverter I_1 is at logic 0 and all the eight AND gates are inhibited so that neither word can appear at the output. When the enable line is at logic 0, then the select input specifies which one of the two words appears at the output. Thus, when the select input is at logic 1, gates 2, 4, 6 and 8 will be enabled and the word $A_1B_1C_1D_1$ will appear at the output. Similarly, when the select input is at logic 0, gates 1, 3, 5 and 7 will be enabled and the word $A_2B_2C_2D_2$ will appear at the output.

MSI multiplexers such as the 74150, with 16 data inputs and four selector inputs that can be actuated by a four-bit binary counter such as the 7493, are commonly used for parallel-to-serial data conversion, the principle of which is illustrated in Fig. 4.26. A four-bit binary word 1010, set in a register, is serialised by four clock pulses applied to the two flip-flops FF_1 and FF_2 that control the select inputs in a sequence of 00, 01, 10, 11.

Demultiplexers

A demultiplexer has a single input and multiple outputs, the input being connected to one output at a time, in a sequence defined by the 'select' inputs.

MSI demultiplexers such as the 74154, with 16 data outputs and four select inputs that can be actuated by a four-bit binary counter such as the 7493, are commonly used for serial-to-parallel data conversion, the principle of which is illustrated in Fig. 4.27. A register consisting of four flip-flops is initially reset, and the set inputs are connected to four three-input NAND gates.

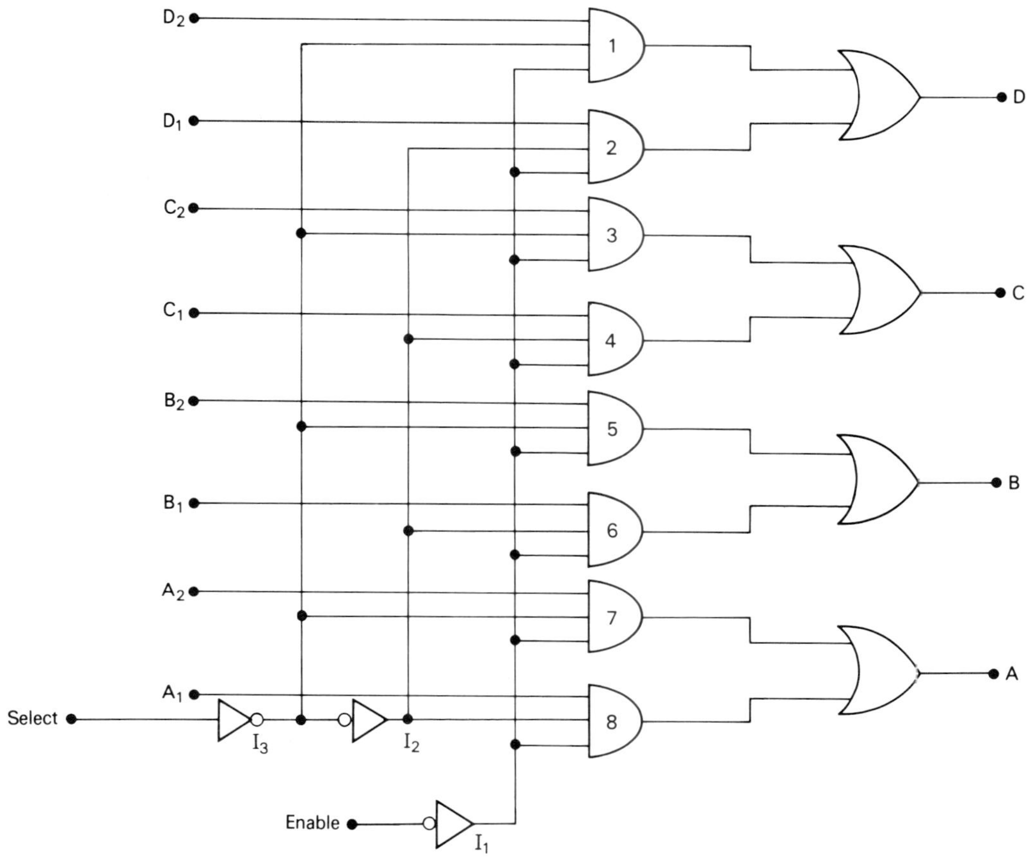

Fig. 4.25 Two four-bit words ($A_1B_1C_1D_1$) and ($A_2B_2C_2D_2$) multiplexed into one four-bit ABCD channel

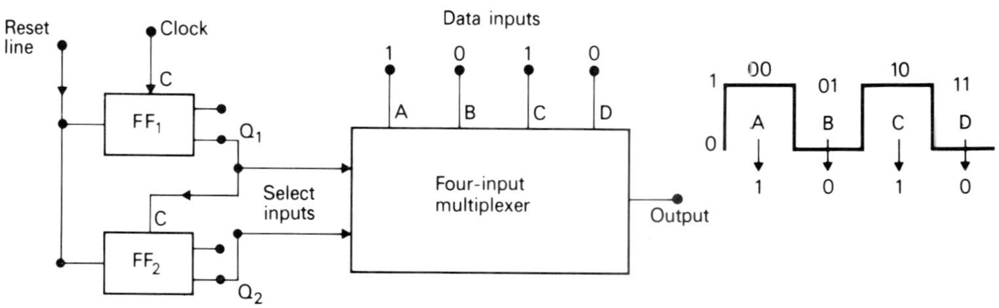

Fig. 4.26 Parallel-to-serial data conversion

The serial input pulses are connected to one set of inputs of the gates, which are enabled one at a time by four clock pulses which control the select inputs in a sequence via a two-bit counter. Thus, any gate that is enabled has its two select inputs at logic 1, so that the data input is shifted into the register.

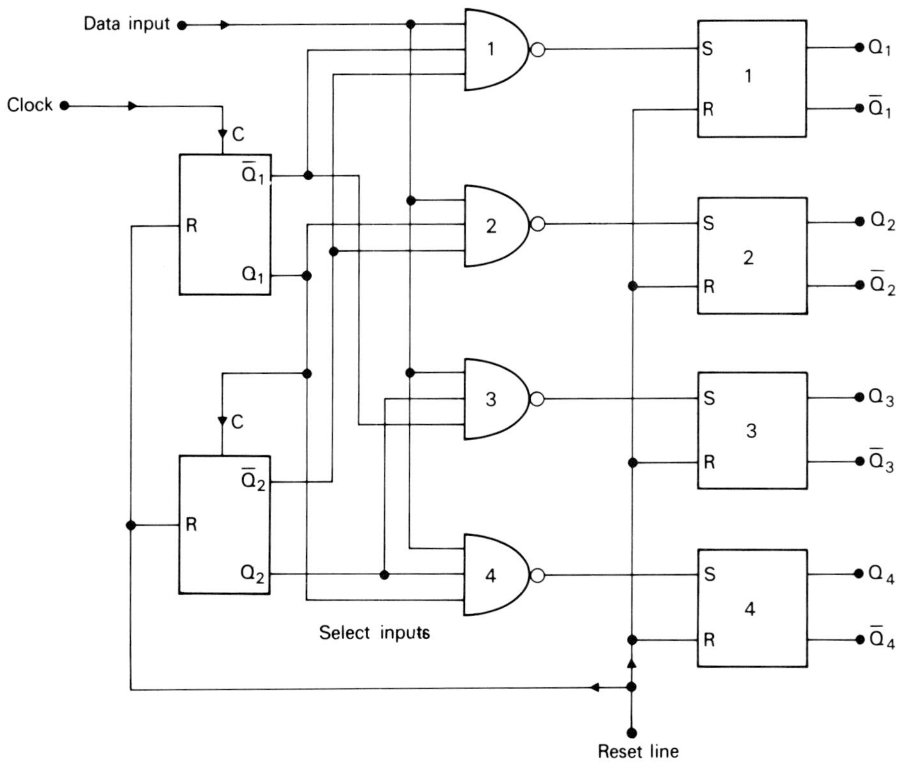

Fig. 4.27 Serial-to-parallel data conversion

MSI decoders such as the 7442 BCD to decimal decoders can be used as demultiplexers or data distributors, either in conjunction with counters having a BCD output, Fig. 4.20, or on their own, with additional circuitry to use the ABC inputs as data select inputs and the D input as the data input line.

EXCLUSIVE OR

Unlike the standard logic OR element, the EXCLUSIVE OR circuit produces a logic 1 output when either, but not both, of its inputs is at logic 1. The EXCLUSIVE OR function can be implemented by using combinations of AND, OR and INVERT logic elements in several ways. The basic Boolean representation is $C = \overline{A}.B + A.\overline{B}$, where A and B are the inputs with complements $\overline{A}$ and $\overline{B}$ and C is the output. For example, the EXCLUSIVE OR function can be implemented simply by using two AND gates fed by complementary inputs A, $\overline{A}$ and B, $\overline{B}$ in conjunction with an OR gate, Fig. 4.28(a). Alternatively, the same function can be implemented by a quad NAND gate such as the 7400, Fig. 4.28(b), in which no complementary inputs are required.

In practice, it is no longer necessary to construct EXCLUSIVE OR gates from individual logic elements, because MSI circuits such as the 7486 are readily available. The complementary version of the EXCLUSIVE OR gate is the EXCLUSIVE NOR. The following common applications illustrate the use of the EXCLUSIVE OR logic elements.

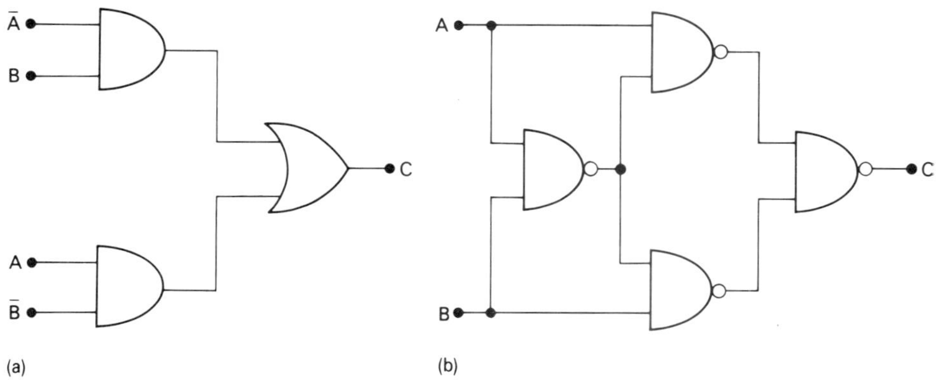

Fig. 4.28 EXCLUSIVE OR function implemented by (a) two AND gates and one OR gate; (b) a quad NAND gate

Binary adders

The logic function of the EXCLUSIVE OR element is well suited for binary addition, the carry operation being done with the help of an AND gate. For example, in Fig. 4.29, the AND gate will give a binary 1 output when both A and B are at binary 1. The circuit is referred to as a **half adder**, and is suitable for a single-bit addition. For adding multiple-bit numbers, there must be provision for adding in the carry from the previous least significant bit as shown in Fig. 4.30, the circuit being commonly known as a **full adder**.

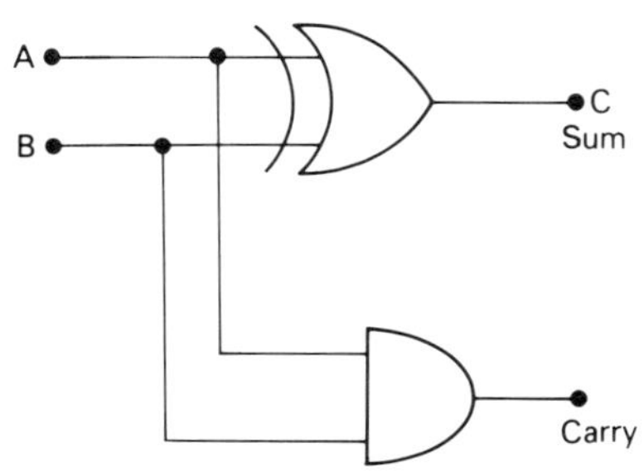

A	0	0	1	1
B	0	1	0	1
C	0	1	1	10

Carry

Fig. 4.29 Half adder using an EXCLUSIVE OR gate and an AND gate

A four-bit parallel adder can be constructed from full adders as shown in Fig. 4.31, but in practice such an adder is available as an MSI circuit such as the 7483.

The four-bit number $A_1A_2A_3A_4$ is added to the four-bit number $B_1B_2B_3B_4$, giving a four-bit sum $S_1S_2S_3S_4$ and a fifth most significant carry bit.

Parity generators/checkers

These are combinational logic circuits used to check errors in stored or transmitted binary data. The binary word to be transmitted or stored is examined, and the number of binary 1's present in that word is determined. A parity bit is generated, based on this information, so that the total number of binary 1's in the word, including the parity bit, is either odd or even.

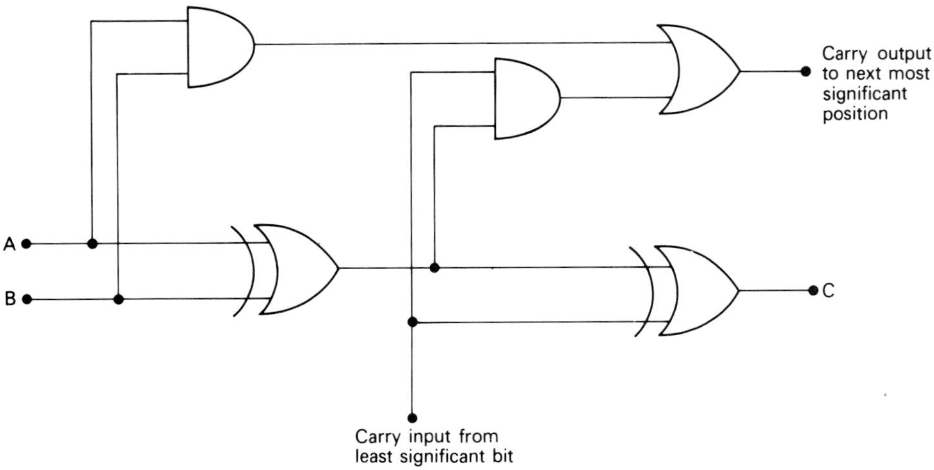

Fig. 4.30 A full adder made up of two half adders (Fig. 4.29) and an OR gate

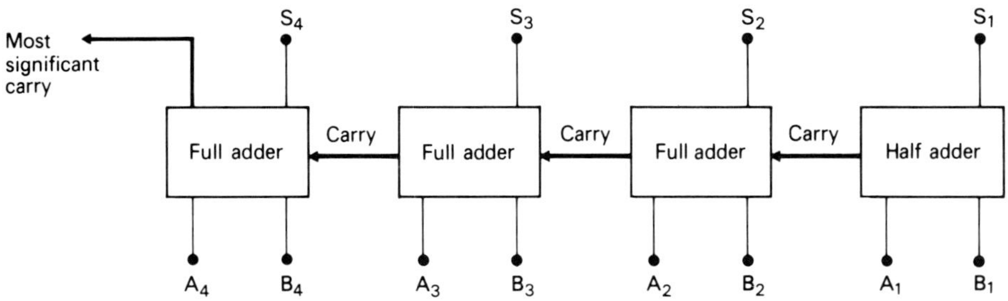

Fig. 4.31 A four-bit parallel adder

Table 4.9 'Odd' and 'even' parity generation

A	B	C	D	*Odd parity*	*Even parity*
0	1	1	1	0	1
1	0	0	1	1	0
1	1	0	1	0	1
1	1	1	1	1	0
0	1	0	0	0	1

Table 4.9 shows the output of an even and odd parity generator when various four-bit binary words are fed to it. This extra parity bit is stored or transmitted along with the information of interest. When a stored word is retrieved or a transmitted word is received, a parity checker generates a parity bit and compares it with the parity bit that was stored or transmitted with the original information. If the two parity bits are identical then no error exists. A difference in the parity bits indicates an error.

The basic implementation of a parity generator with EXCLUSIVE OR gates is shown in Fig. 4.32(a), and that of a checker in Fig. 4.32(b).

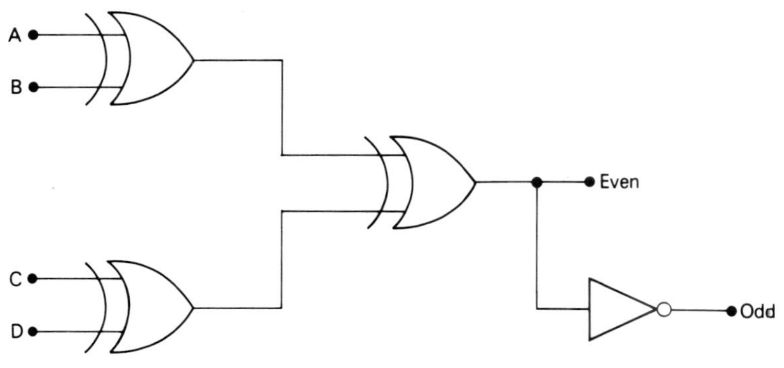

(a)

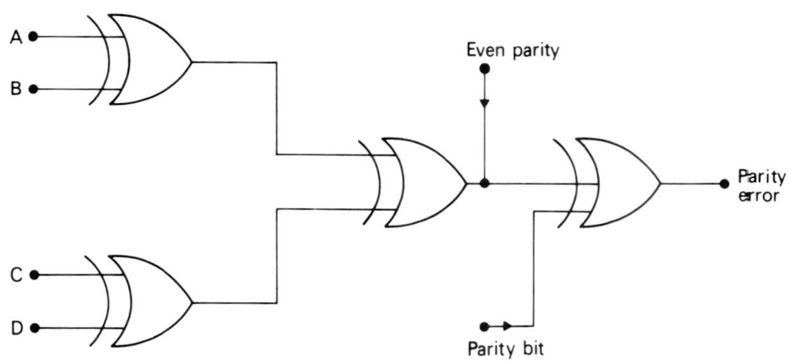

(b)

Fig. 4.32 (a) A parity generator; (b) a parity checker

In practice, MSI circuits such as the 74180 are capable of implementing eight-bit odd/even parity generator/checker functions.

Binary comparators
These are combinational logic circuits used for comparing the magnitude of parallel binary input words. The output of the circuit is a binary 1 only if the magnitude of the two words being compared is exactly the same. Fig. 4.33 shows this function implemented by four EXCLUSIVE NOR gates and a four-input AND gate. The output of the AND gate will be a logic 1 only when the two four-bit binary numbers $A_1A_2A_3A_4$ and $B_1B_2B_3B_4$ are exactly equal.

MSI four-bit magnitude comparators such as the 7485, in addition to providing an output that indicates the equality of the two four-bit words, also generate two additional output signals. One output gives a binary 1 if the input word A is numerically *greater* than the word B

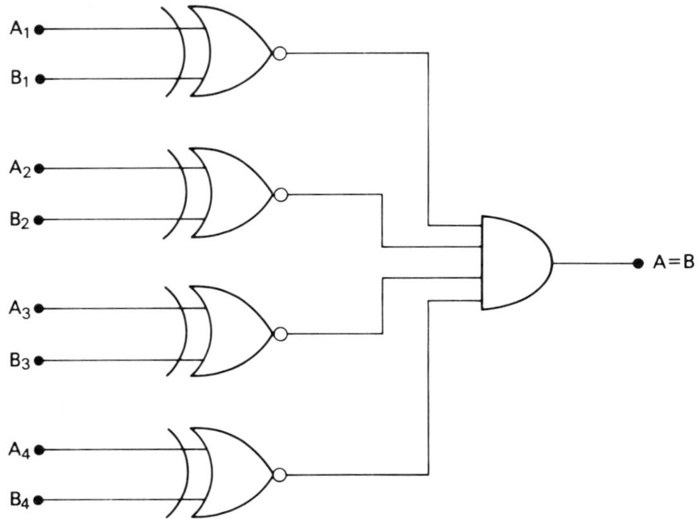

Fig. 4.33 A binary comparator using four EXCLUSIVE NOR gates and a four-input AND gate

(A > B output); the other output gives a binary 1 if the input word A is numerically *less* than the word B (A < B output).

Code converters

There are many applications in digital systems where two or more different binary codes are used. The most common example is the case where data is available in binary form and it is required to convert it to BCD form, or vice versa. Code converters are mostly combinational logic circuits which convert one type of binary code into another so that various digital circuits in a system can be made compatible. Most code converters like BCD to binary, binary to BCD, Gray to binary and binary to Gray (see p. 116), are available as MSI circuits.

4.8 Sequential logic systems

In any analysis of a practical digital system, it is necessary to include a sequential function of Boolean variables, because there is a time delay between stimulus and response of most circuits. For example, an elementary sequential function is a time delay Δ, whose action is to inhibit the response of a given operation A until a time Δt has elapsed. Circuits used for timing, sequencing and storage functions may be divided into three distinct categories:

1) flip-flops;
2) counters;
3) shift registers.

Flip-flops

A flip-flop is a **bistable**, i.e. a circuit with two stable states, whose basic function is 'memory' or storage of a single bit of binary data. It operates in one of two modes, asynchronous or unclocked, and synchronous or clocked. The RS latch is an example of an asynchronous operation. It has two control inputs and two outputs. The R (reset) and S (set) are the asynchronous controls since the output changes immediately when either control input changes. Figure 4.34 shows the RS latch implemented with two NAND gates. Table 4.10 is the truth table, showing that the state R = S = 0, Q = $\overline{Q}$ = 1 is not allowed. The 'clear' and 'preset' are overriding controls, and their function is shown in Table 4.11.

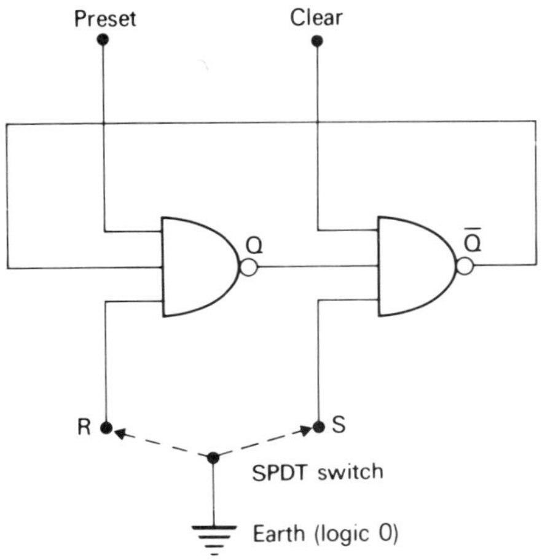

Fig. 4.34 RS latch using two NAND gates and a single-pole double throw (SPDT) switch

Table 4.10 Truth table of a RS latch

R	S	Q	$\overline{Q}$
0	0	Not allowed	
0	1	1	0
1	0	0	1
1	1	No change	

Table 4.11 Action of the 'preset' and 'clear' inputs

Preset	*Clear*	Q	$\overline{Q}$
0	0	1	1
0	1	1	0
1	0	0	1
1	1	No change	

It is often desirable to have some kind of a clock input to the flip-flop so that the device can be operated synchronously with all others in the system. In general, the different types of synchronous data inputs determine the name of the flip-flop, while the different mechanisms for clocking may be used with any type of flip-flop. A synchronous data input is one that does not cause an immediate change in the output. It requires the occurrence of another input, the clock pulse, to generate the change of state. There are four basic types of flip-flops, which differ in the types of input stimuli although their outputs are the same for all four types, representing a normal output Q and its complement $\overline{Q}$ simultaneously.

1) RS flip-flop;
2) toggle or T-type flip-flop;
3) JK flip-flop;
4) D-type flip-flop.

RS flip-flop

The only difference between the RS flip-flop and the RS latch is that the former has a clock input for synchronous operation. It can be implemented with NAND gates as shown in Fig. 4.35. With R = S = 0, Q = $\overline{Q}$ = 1 the clock has no effect on the output. For R = 1, S = 0, the Q output is zero after clocking. In this state, as long as the S input is held at zero level, it does not matter whether the R input is at 0 or 1. The function of the clear and preset inputs is the same as before (see Table 4.11).

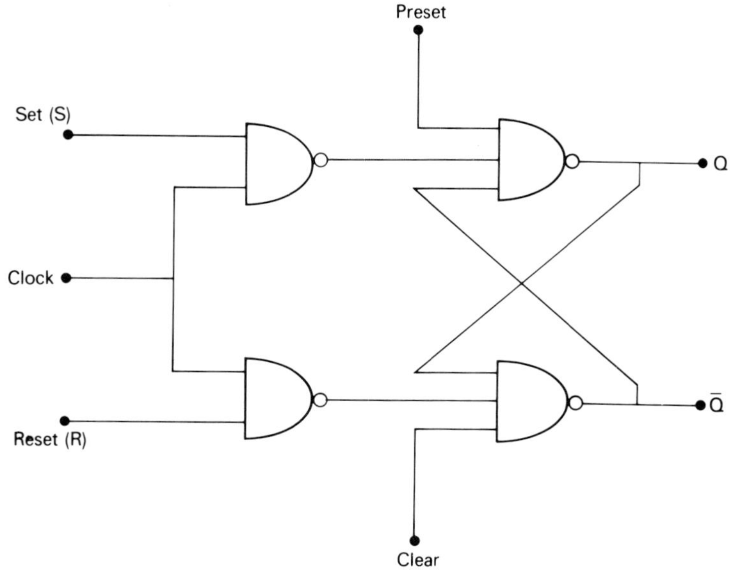

Fig. 4.35 RS flip-flop implemented with NAND gates

T-type flip-flop

This type has a single data input T and a clock input. If the T input is in state 0 prior to a clock pulse, then the Q output will not change with clocking. If the T input is in state 1 then the output will change state at every clock pulse, regardless of its state prior to the arrival of the clock pulse. This is called **toggling** and the output completes one cycle for every two input cycles. T-type flip-flops are usually made up from other types of flip-flop and used in sequential counting networks on account of their inherent 'divide-by-2' capability.

JK flip-flop

The operation of the JK flip-flop differs from that of the RS flip-flop in one respect, i.e. the state J = K = 1 is allowed. In this case, the flip-flop toggles like a T flip-flop with T = 1. Also if J = 1 and Q = 0 then a clock transition from 0 to 1 will change Q to 1. Similarly, if K = 1 and Q = 1 then a clock transition of 0 to 1 will change Q to 0. This action is shown in Table 4.12, where Q_n is the state of the Q output after the *n*th clock pulse and Q_{n+1} its state after one more clock pulse, the 'X' indicating that a state may be 0 or 1. Figure 4.36 shows the implementation of a JK flip-flop with NAND gates, and Figs 4.37 and 4.38 show the C, J, K, Q waveforms.

The operation of this circuit will sometimes be faulty, because the output will tend to

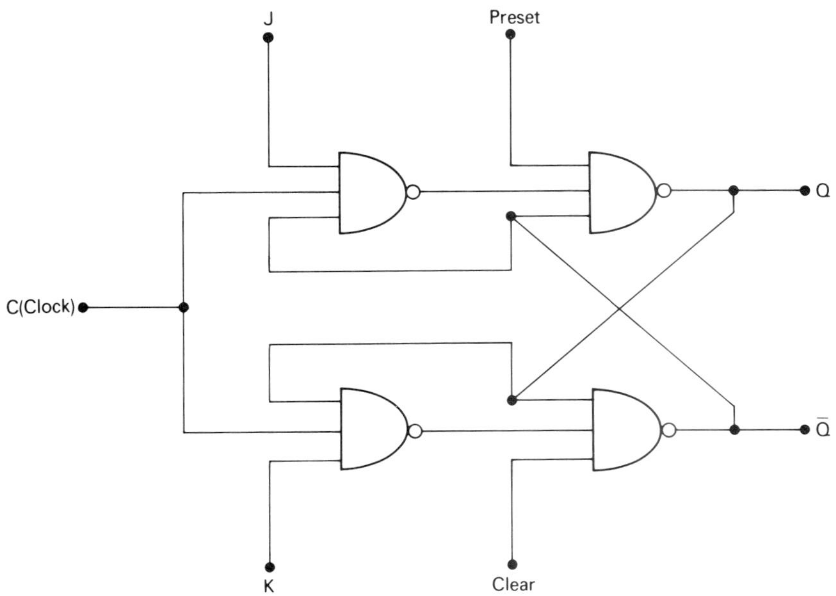

Fig. 4.36 JK flip-flop implemented with NAND gates

Table 4.12 Action of the J and K inputs

J	K	Q_n	Q_{n+1}
0	X	0	0
1	X	0	1
X	1	1	0
X	0	1	1

oscillate between 0 and 1 during the interval t_c when the clock is at logic 1. For example, when J = K = 1 and Q = 0, on the arrival of a clock pulse, Q will change to 1 after a propagation delay of Δt so that the conditions at this stage will be J = K = 1, C = 1 and Q = 1. From Fig. 4.36 it is obvious that Q will be forced to 0 after a further delay Δt. The output Q will thus oscillate between 0 and 1 for the duration t_c of the clock pulse, and might settle to either 0 or 1 when the clock returns to 0. The phenomenon is called the **race-round** condition and often occurs when $\Delta t \ll t_c$. The occurrence of this condition may be avoided by using the **master-slave** clocking mechanism shown in Fig. 4.39. This consists of two flip-flops in cascade, the leading one, or 'master', connected as a JK flip-flop and the second, or 'slave', connected as a RS flip-flop. Clock pulses C are used to enable the master, and inverted clock pulses $\overline{C}$ are used to enable the slave. Thus if J = 1 and $Q_m = 0$ initially, as soon as the clock C changes from 0 to 1, Q_m changes from 0 to 1 so that the S input of the slave is 1. The slave flip-flop is enabled when $\overline{C}$ changes from 0 to 1 or when C changes from 1 to 0. A change that occurred at the output of the master at the leading edge of the clock pulse is transferred to the output of the slave at the trailing edge of the same clock pulse. As the slave cannot change its state until the end of the clock pulse, the race-round condition is avoided.

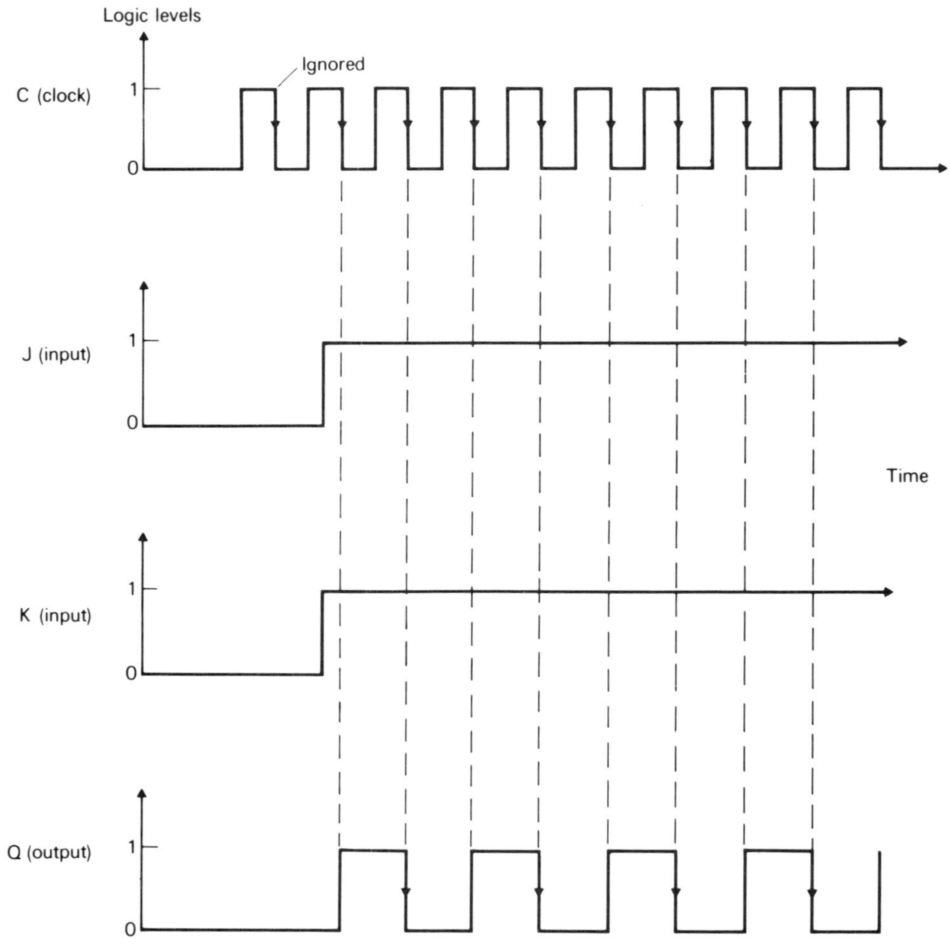

Fig. 4.37 JK flip-flop waveforms when both the J and K inputs are at logic 1

D-type flip-flop

This type is characterised by a single data input D and a clock input. The output Q assumes the logical value of the input at the time of the clock pulse. This action is shown in Table 4.13, where Q_n is the state of the Q output after the *n*th clock pulse and Q_{n+1} its state after one more clock pulse. The D-type flip-flop is used primarily in data storage and in register applications where temporary data storage is required.

Lastly, the JK flip-flop can be used to implement both D-type and T-type functions, as shown in Fig. 4.40. The D-type function is accomplished by using an inverter between the J and K inputs. The T-type function is implemented by connecting the J and K inputs together.

Counters

Counters are cyclic sequential circuits which **reset**, or return to their initial state, after a specified number of input pulses. At any instant before recycling, the output of the counter

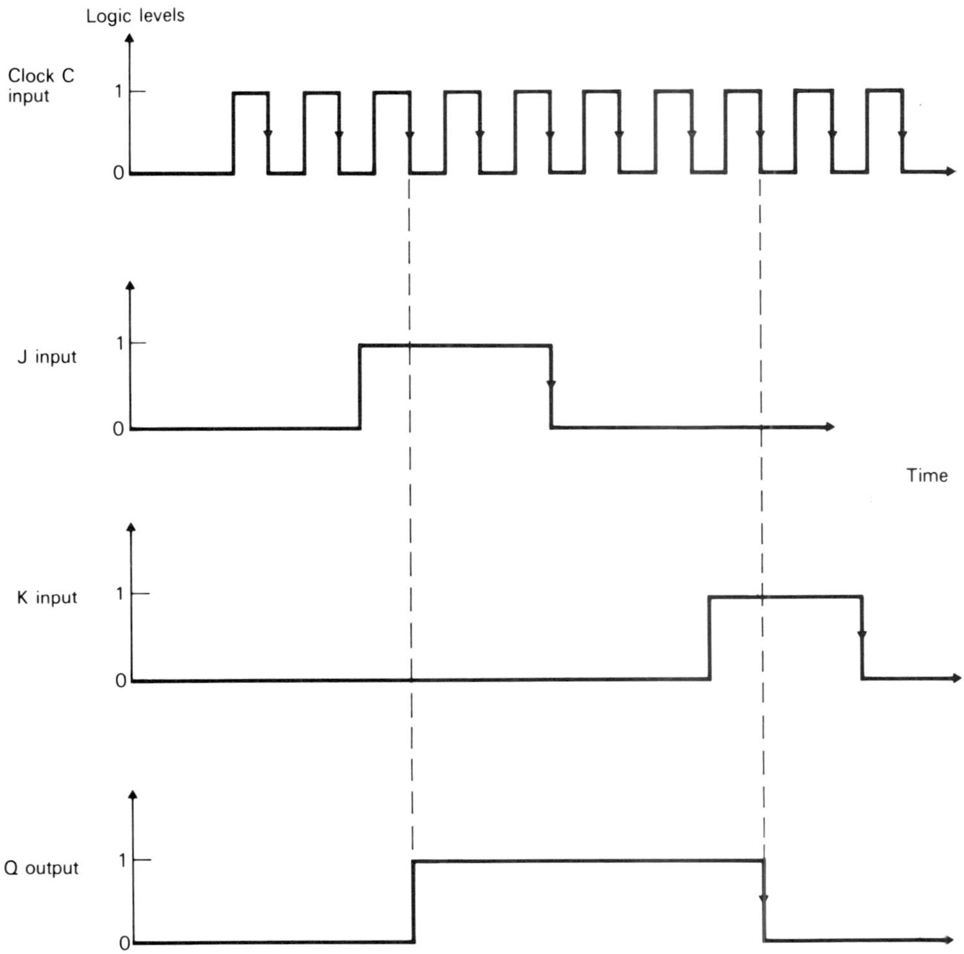

Fig. 4.38 The J and K input controls

indicates the number of input pulses, usually in a coded form. The commonly-used codes in IC counters are

a) binary;
b) binary coded decimal (BCD);
c) Gray codes.

Binary counting
This may be illustrated by connecting a number of JK flip-flops in cascade, as in Fig. 4.41 in which the Q output of one is connected to the C (clock) input of the next. In this case, if all the JK flip-flops are initially reset, then each flip-flop changes state when its clock input 'sees' a 1 to 0 transition. The resultant four-bit binary count is shown in Table 4.14. Obviously such a counter can be used as a frequency divider, each stage dividing the input pulse train by a factor of two, i.e. there are half as many trailing edges in Q as in C.

If there are n flip-flops, a frequency division by a factor of 2^n will result. When used as

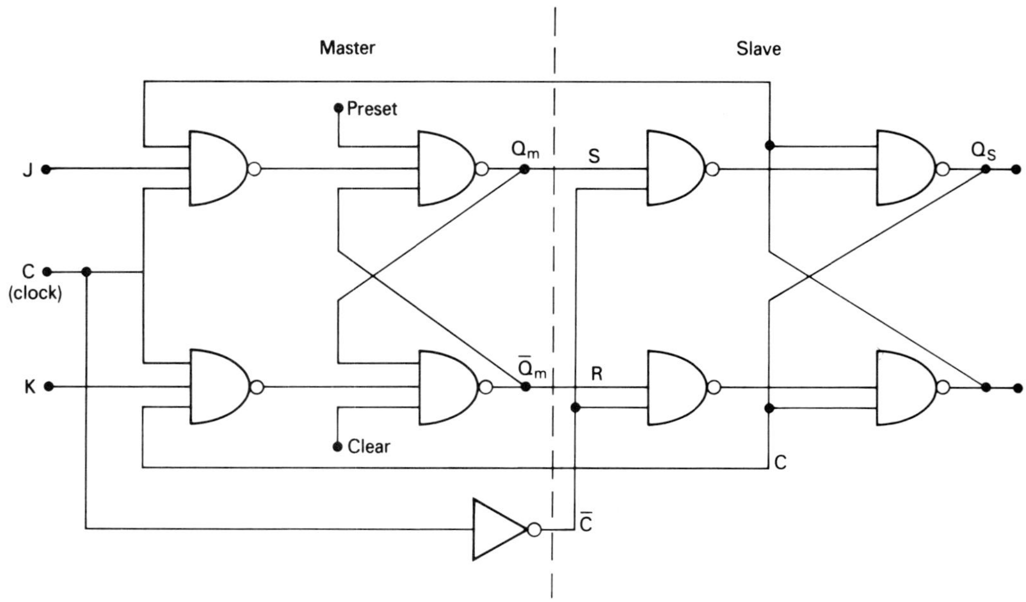

Fig. 4.39 Master-slave clocking mechanism

Table 4.13 Action of the D input

D	Q_n	Q_{n+1}
0	0	0
0	1	0
1	0	1
1	1	1

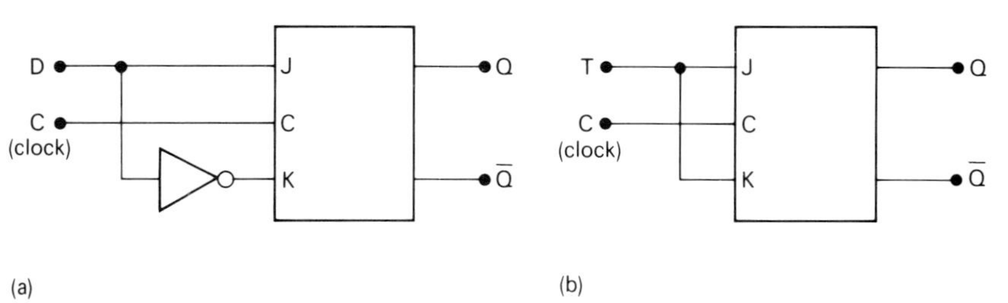

Fig. 4.40 (a) JK to D-type conversion; (b) JK to T-type conversion

binary storage, the maximum number of counts that can be contained before the counter recycles will be N, where N is given by

$$N = (2^n - 1)$$

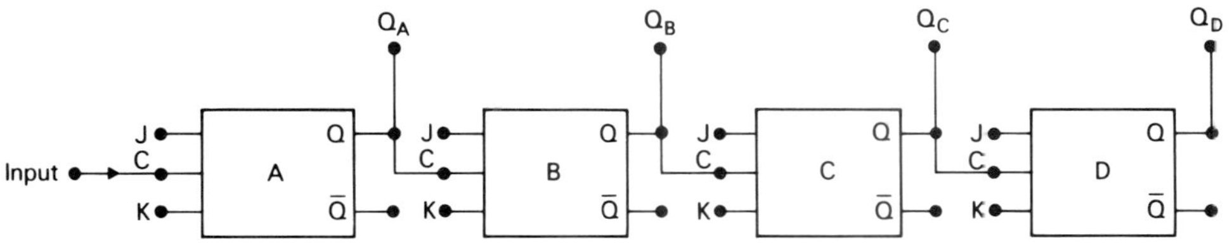

Fig. 4.41 Binary counter made up of JK flip-flops connected in cascade with Q outputs to C inputs

Table 4.14 Truth table of a four-bit binary counter made of JK flip flops

Clock pulse number	Q_D	Q_C	Q_B	Q_A
0	0	0	0	0
1	0	0	0	1
2	0	0	1	0
3	0	0	1	1
4	0	1	0	0
5	0	1	0	1

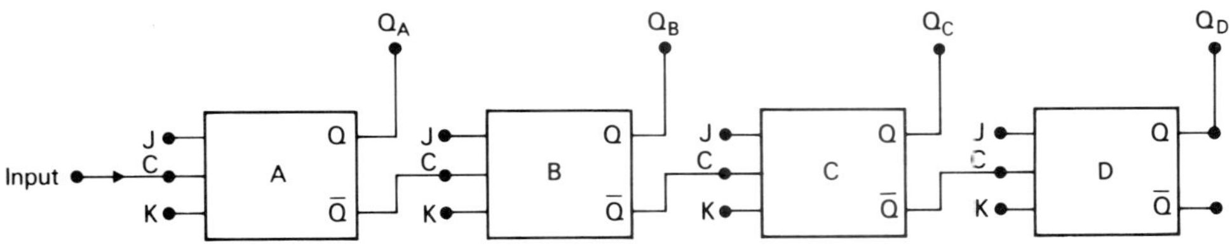

Fig. 4.42 A binary counter made up of JK flip-flops connected in cascade with $\bar{Q}$ outputs to C inputs

In this case, each 1 to 0 transition of the input clock causes the binary count to increase, and such counters are known as **count up** counters. If the $\bar{Q}$ outputs are connected to the clock inputs, Fig. 4.42, then a **count down** version results. The corresponding pulse counting sequence is shown in Table 4.15.

When the flip-flops are connected in cascade, the counters are known as as **ripple counters** or **asynchronous counters**. Their speed of counting is limited by the propagation delay, which is the sum of all the propagation delays associated with the chain of flip-flops. Apart from using nonsaturating logic circuits such as the ECL family which has a small propagation delay, high speeds of counting can be achieved by using synchronous counters. In this case, Fig. 4.43, the pulse to be counted is impressed on all the clock inputs simultaneously, but the transitions of the flip-flops are controlled by their J and K inputs. For example, the only time that the flip-flop B can change its logic state is when the normal output Q_A is at logic 1. The AND gates 1 and 2 are used to control the JK inputs of flip-flops C and D so that the binary counting sequence is the same as that of Table 4.14. As the flip-flops change simultaneously,

Table 4.15 Truth table of a 'count down' four-bit binary counter

Clock pulse number	Q_D	Q_C	Q_B	Q_A
15	1	1	1	1
14	1	1	1	0
13	1	1	0	1
12	1	1	0	0
11	1	0	1	1
10	1	0	1	0

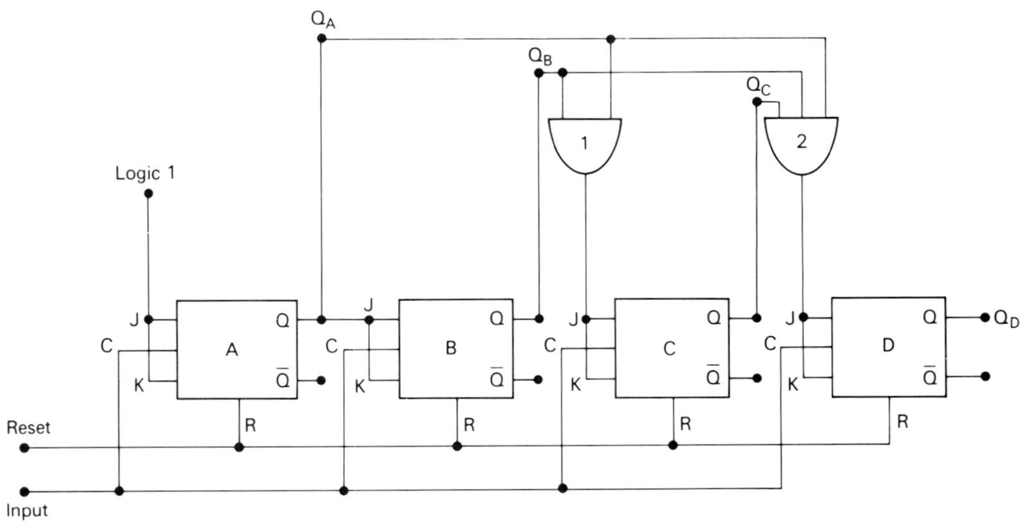

Fig. 4.43 A synchronous counter made up of JK flip-flops

the maximum delay between the input pulse and the change of state of the outputs is only as long as the longest propagation time of the flip-flops in the circuit.

In all cases, the initial states of the counters are defined by using the reset and preset lines. For example, if all the normal outputs of a chain of JK flip-flops are to be defined to logic 0, then their common reset line is held at logic 1 and momentarily brought to logic 0 with a pulse of about 100 ns duration.

In practice, it is seldom necessary to construct counters from individual flip-flop integrated circuits because versatile MSI circuits are already available. For example, the 74193 IC is a MSI four-bit synchronous binary counter, which can be used for either up or down counting, has a separate reset or clear input and has the facility to preset to any binary number applied by some external four-bit parallel source.

Binary coded decimal (BCD) counters

These are sequential circuits which count in tens. They have ten discrete states to represent 0 to 9, and such a counter is sometimes referred to as a **decade counter**.

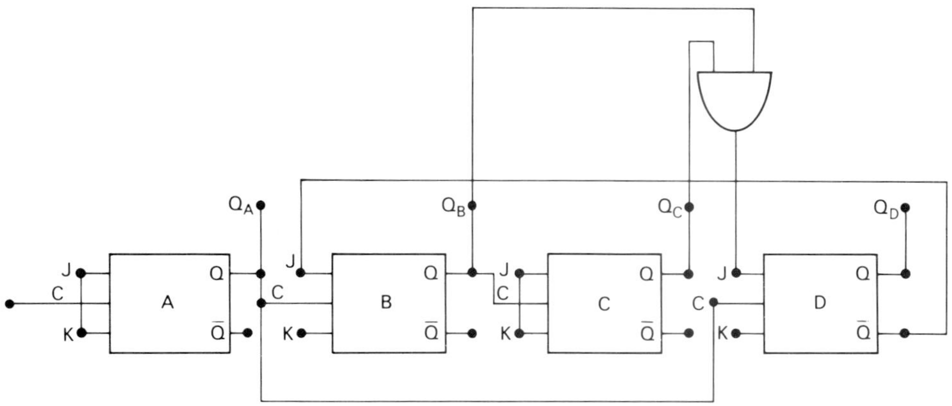

Fig. 4.44 An asynchronous 8-4-2-1 BCD counter made up of JK flip-flops

Figure 4.44 illustrates the mechanism of an asynchronous 8-4-2-1 BCD counter consisting of four JK flip-flops connected in a feedback loop, so that the standard four-bit counter recycles every ten input pulses as shown in Table 4.16. Comparing Figs 4.37 and 4.45, it is obvious that the count sequence for the first eight input pulses is identical to that of a four-bit pure binary counter. If all the flip-flops are initially reset, then the outputs of B and C will be at logic 0 and the AND gate will inhibit the flip-flop D via its J input. When the trailing edge of the eighth pulse occurs, the counter changes to 1000 and the complement of the D flip-flop Q_D inhibits the B flip-flop. The ninth pulse merely changes the A flip-flop so that the count is 1001. The tenth pulse changes both A and D, but B is not affected as its J input is at logic 0, so that the counter is reset to 0000. Numerous variations of this basic BCD counting mechanism are possible, including synchronous operation in order to increase the counting speed.

Table 4.16 Truth table of an asynchronous 8-4-2-1 BCD counter

Decimal number	Q_D 8	Q_C 4	Q_B 2	Q_A 1
0	0	0	0	0
1	0	0	0	1
2	0	0	1	0
.				
.				
.				
8	1	0	0	0
9	1	0	0	1

The most widely-used BCD decade counter is the MSI asynchronous counter 7490 that counts in the standard 8-4-2-1 code. More versatile synchronous BCD up-down counters, such as the 74191, are also available. As a 1 to 0 transition takes place at Q_D, Fig. 4.45, every tenth pulse, BCD counters can be cascaded as shown in Fig. 4.46. If all the counters are

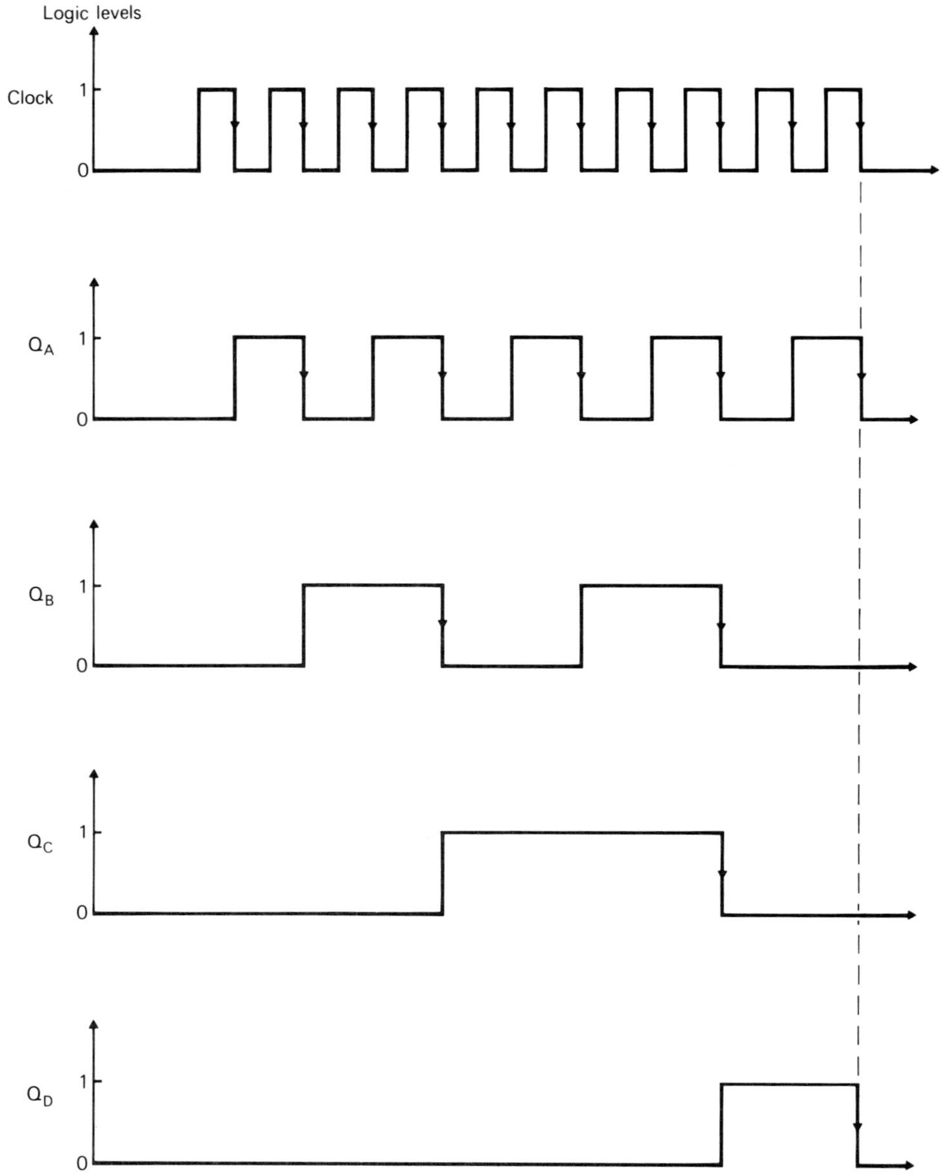

Fig. 4.45 Waveforms of a BCD decade counter

initially reset, then each BCD counter represents a decimal digit which can be determined by observing the outputs of the flip-flops $A_1B_1C_1D_1$, giving the least significant digit (LSD), and the outputs of the flip-flops $A_3B_3C_3D_3$, giving the most significant digit (MSD). When used for frequency division, the Q_D outputs do not give a square wave and the trailing edges of the output pulses are usually utilised.

Special-purpose counters such as those used for dividing the frequency of an input pulse train by numbers other than even powers of 2 or by 10 can be constructed from standard JK

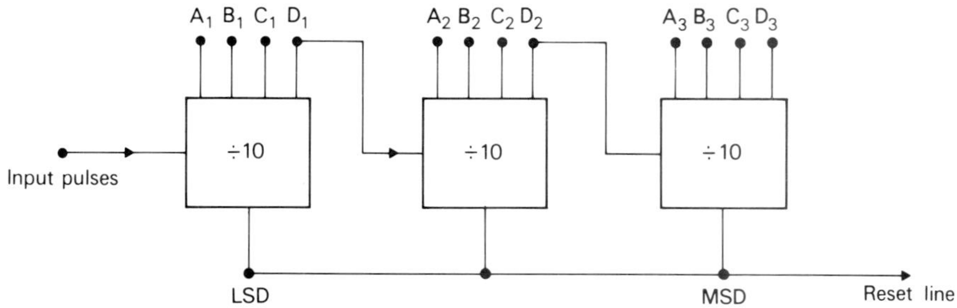

Fig. 4.46 Cascaded BCD counters

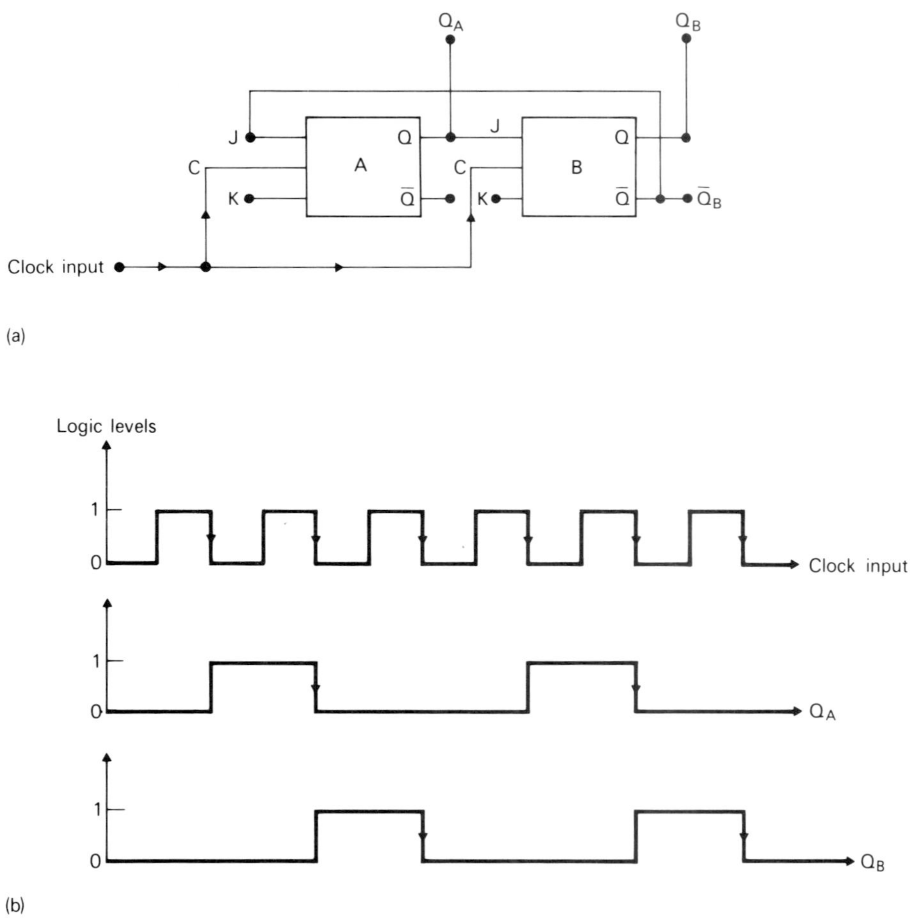

Fig. 4.47 (a) Modulo-3 counter made up of JK flip-flops; (b) waveforms at Q_A and Q_B

flip-flop elements by using feedback and appropriate gating via the JK inputs. For example, a modulo-3 counter can be constructed from two JK flip-flops as shown in Fig. 4.47(a). If both A and B flip-flops are initially reset to zero, then the trailing edge of the first clock pulse will change A but not B since the J input to B is at zero. The trailing edge of the second clock pulse will change Q_A to zero and Q_B to 1 so that $\overline{Q}_B$ will be at zero. Since $\overline{Q}_B$ is connected to the J input of A, the trailing edge of the third clock pulse will change Q_B to zero and leave Q_A unaltered, so that both Q_A and Q_B are at zero – the counter has recycled to its original reset condition. Figure 4.47(b) shows the waveforms at Q_A and Q_B. The design of synchronous and asynchronous modulo-N counters using Karnaugh maps is discussed in detail elsewhere.[3]

Lastly, in weighted codes a weight W_j is given to the jth binary digit so that

$$N_{\text{decimal}} = \sum_{j=1}^{4} W_j S_j$$

where S_j is the value of the jth binary digit. Two commonly-used code combinations are the 2-4-2-1 and 5-4-2-1 shown in Tables 4.17 and 4.18.

Table 4.17 The 2-4-2-1 code

Decimal	D 2	C 4	B 2	A 1
0	0	0	0	0
1	0	0	0	1
2	0	0	1	0
3	0	0	1	1
4	0	1	0	0
5	1	0	1	1
6	1	1	0	0
7	1	1	0	1
8	1	1	1	0
9	1	1	1	1

Table 4.18 The 5-4-2-1 code

Decimal	D 5	C 4	B 2	A 1
0	0	0	0	0
1	0	0	0	1
2	0	0	1	0
3	0	0	1	1
4	0	1	0	0
5	1	0	0	0
6	1	0	0	1
7	1	0	1	0
8	1	0	1	1
9	1	1	1	0

Gray codes

These are codes in which adjacent numbers differ in one bit only so that the race-round condition which often occurs when two or more bits change simultaneously, is avoided. One of the best known Gray codes is the reflected binary code shown in Table 4.19, where all combinations below the centre line have 1 as the most significant digit and those above the centre line have 0 as the most significant digit. If the MSB is ignored then the numbers 8 to 15 are the mirror images of the numbers 0 to 7. Plotted on a Karnaugh map, Fig. 4.48, it is obvious that only one cell boundary is crossed from one combination to the next.

Shift registers

A shift register is a sequential logic circuit using binary storage elements such as flip-flops. It has the ability to move the data, one bit at a time, from one storage element to another, at the command of a shift pulse. Figure 4.49 illustrates this basic principle by considering a shift register made up of four flip-flops, initially storing a binary number 1011 and having a number 0110 available at its input for serial loading. It will be noticed that at the end of four shift pulses, the old number is shifted out and lost, while the new number is shifted in and stored. Similarly, shift registers can be constructed to shift data from right to left and also to have parallel inputs and outputs. This ability to combine both serial and parallel operation

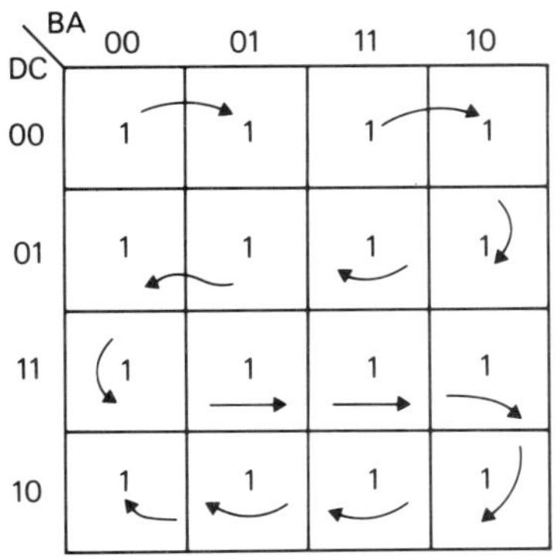

Fig. 4.48 Karnaugh map representation of the reflected binary code of Table 4.19

Table 4.19 The reflected binary code

Decimal	D	C	B	A
0	0	0	0	0
1	0	0	0	1
2	0	0	1	1
3	0	0	1	0
4	0	1	1	0
5	0	1	1	1
6	0	1	0	1
7	0	1	0	0
8	1	1	0	0
9	1	1	0	1
10	1	1	1	1
11	1	1	1	0
12	1	0	1	0
13	1	0	1	1
14	1	0	0	1
15	1	0	0	0

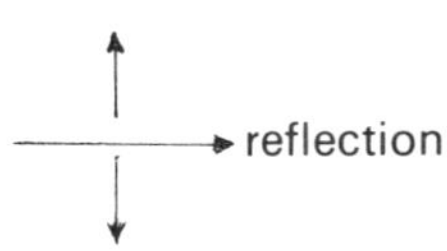

makes the shift register an ideal circuit for performing serial-to-parallel and parallel-to-serial data conversions.

A simple example to illustrate the implementation of the shift register principle using JK flip-flops is shown in Fig. 4.50(a). The waveforms of Fig. 4.50(b) show how a binary number 1010 is loaded serially into the register with four shift pulses when all the four flip-flops are initially reset from the common reset line. The initial binary 1 of the input is seen to be moving to the right at each clock pulse.

In most practical situations, shift registers consist of a number of bits that is some multiple of four, and are available in MSI form with other facilities such as parallel loading or shifting to the left. For example, the TTL MSI shift register type 7495, Fig. 4.51, is a versatile four-bit shift register which can perform the following functions.

i) When the mode control line is at zero, and pulses are applied to the serial input line, the

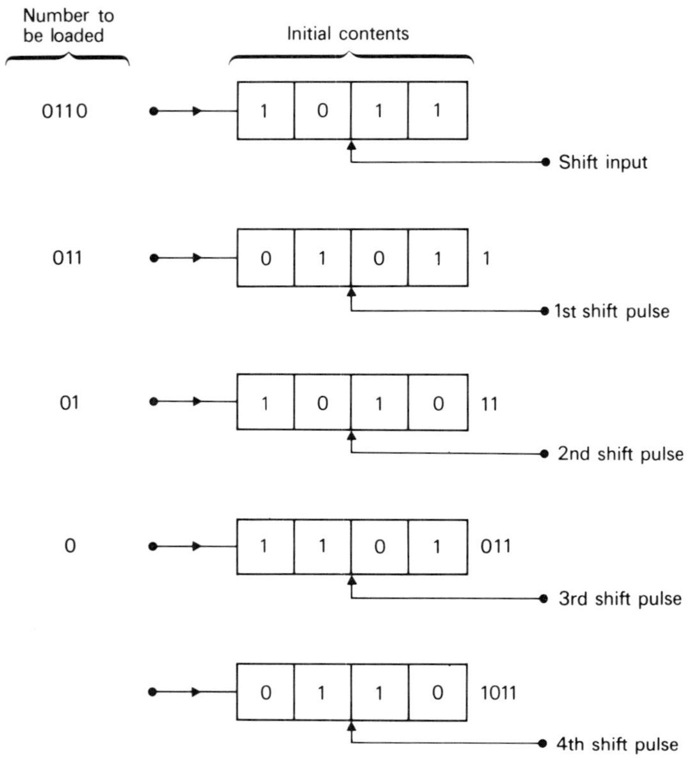

Fig. 4.49 Basic principles of a shift register

shift register performs the shift right operation at every 1 to 0 transition of clock (1).

ii) When the mode control line is at 1, the shift left operation is performed at every 1 to 0 transition of clock (2) provided that the D flip-flop output is connected to the C data input, the C output to the B data input, and the B output to the A data input. The D data input is used as a serial input line for external data which is shifted from D to C, C to B and B to A.

iii) When the mode control is at 1, clock (2) can actuate the flip-flops so that a four-bit parallel word can be loaded into the register.

The following are some of the most common applications of shift registers.

Parallel-to-serial and serial-to-parallel data conversion
In Fig. 4.52(a), the binary number 1101 is preset into the register and on applying four pulses to the shift input, the data is shifted out serially. In Fig. 4.52(b), the binary number 1001 is shifted into the register by applying four pulses to the shift input. The outputs of the four flip-flops can be monitored simultaneously to obtain the parallel data output.

Multiplication and division by factors of two
Shifting data one position to the left has the effect of *multiplying* the original number of a factor of two, and shifting it one position to the right has the effect of *dividing* it by two. For example, if a binary number 000110 (decimal 6) preset in a shift register, Fig. 4.53(a), is shifted

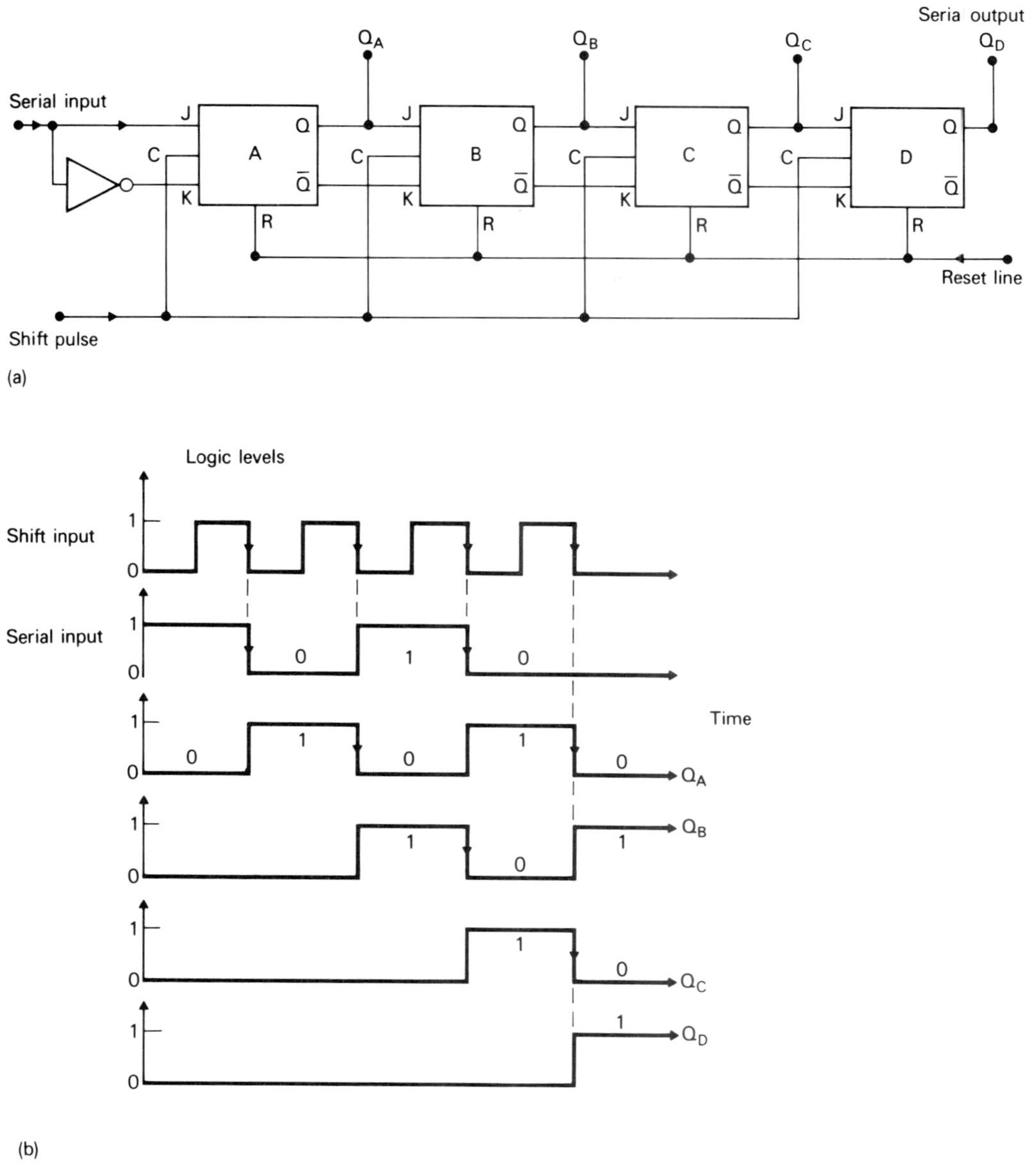

Fig. 4.50 (a) A simple shift register made up of JK flip-flops; (b) waveforms at Q_A, Q_B, Q_C and Q_D

to the left, the result will be 001100 or decimal 12, Fig. 4.53(b). Similarly, shifting the original data to the right, Fig. 4.53(c), will give 000011 or decimal 3. It is, however, important to note that for scaling operations, a shift register must be large enough so that data is not shifted out and lost.

Sequence generators or ring counters

Shift registers can conveniently be used for generating a sequence of pulses, each pulse being used to initiate or control a series of operations. Figure 4.54 shows one such system in which

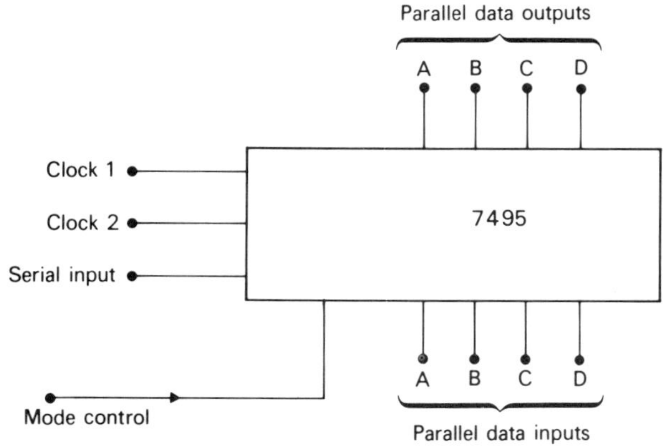

Fig. 4.51 A versatile four-bit shift register

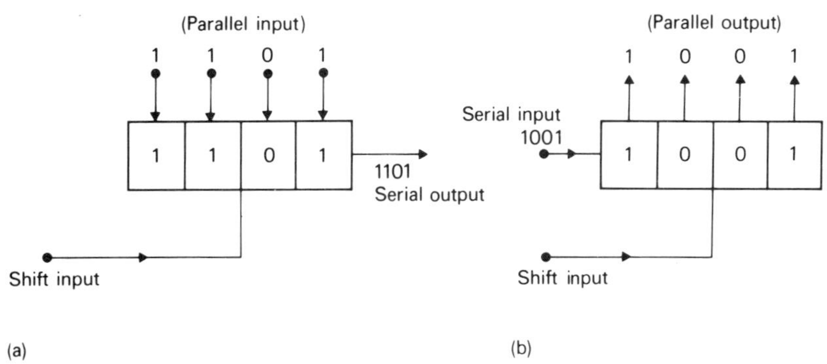

Fig. 4.52 (a) parallel input and serial output; (b) serial input and parallel output

four JK flip-flops are connected in a closed feedback loop or a ring configuration. A logic 0 applied to the reset line will cause the A flip-flop to set and the B, C and D flip-flops to reset. Pulses applied to the shift line will circulate the logic 1 in A to B, from B to C, from C to D and from D to A, Fig. 4.55. In this mode, the shift register can be used as a frequency divider because any flip-flop will recycle after a number of shift pulses numerically equal to the number of flip-flops in the ring. For example, when initially reset, the ring counter of Fig. 4.54 will divide the shift pulse train by a factor of four. Similarly, decade counters capable of driving a decimal display directly may be constructed using ten flip-flops. Such counters have the main disadvantage of not having a binary readout, but are useful in cases that require a decimal output.

If a shift register with a longer cycle length is required, then the simplest modification to a ring counter such as Fig. 4.54 is to invert the outputs of the final stage before feeding them back into the first stage. The resultant ring counter known as the **twisted ring** or **Johnson counter** has a cycle length of $2N$, where N is the number of stages. For example, Fig. 4.56 shows a four-stage Johnson counter and the truth table 4.20 shows the counting sequence starting from the reset condition. The outputs may be decoded into decimal using two input AND gates. However, if the counter is not initially reset, it may go into any of its allowed

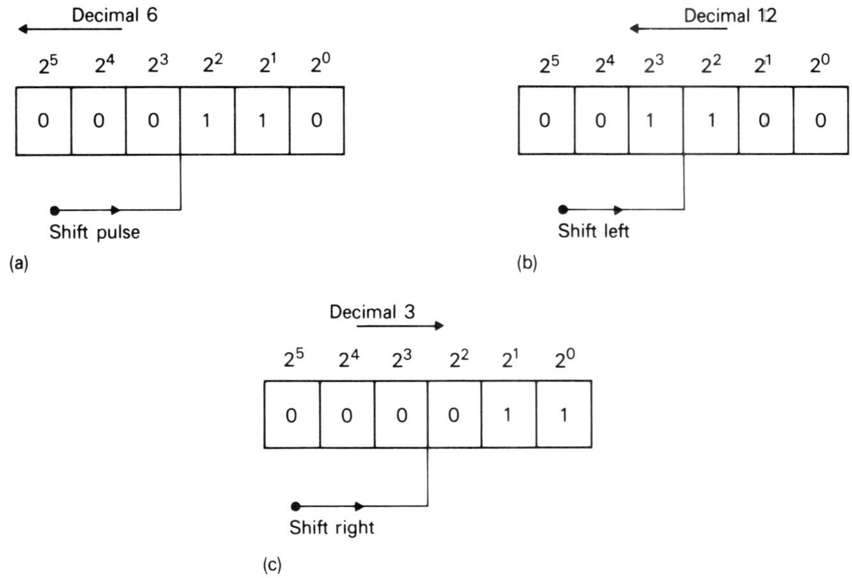

Fig. 4.53 (a) A binary number 000110 or decimal 6 preset in the register (b) a left shift giving 001100 or decimal 12; (c) a right shift giving 000011 or decimal 3

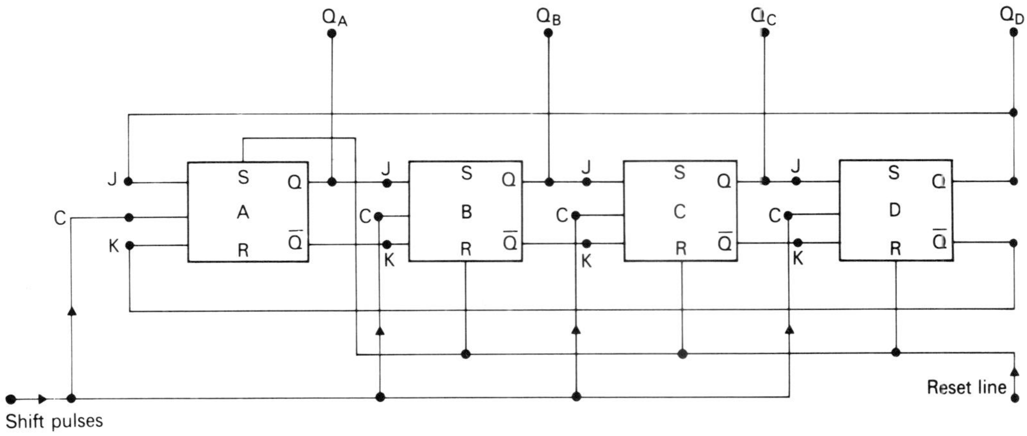

Fig. 4.54 A ring counter using JK flip-flops

sequences such as the one shown in truth table 4.21. To ensure that the counter reverts back to the desired sequence of truth table 4.20, one possible solution is to decode the 1001 state using a four-input AND gate with $Q_A\overline{Q}_B\overline{Q}_CQ_D$ as its input, Fig. 4.57, and to connect its output so as to override the normal input of the counter and generate the 1100 state of truth table 4.20.

4.9 Monostable circuits and clock pulse generators

Most **monostable** multivibrator circuits currently in use are in IC form, the duration of the

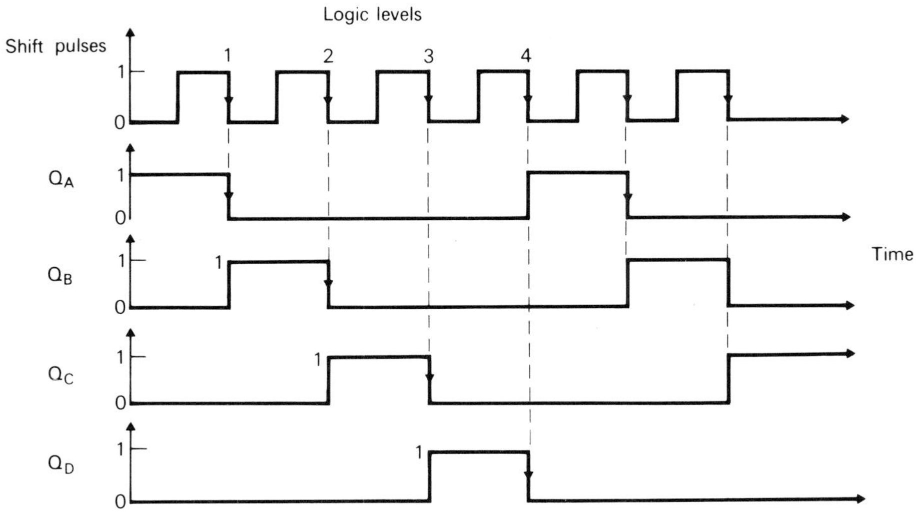

Fig. 4.55 Waveforms at Q_A, Q_B, Q_C and Q_D

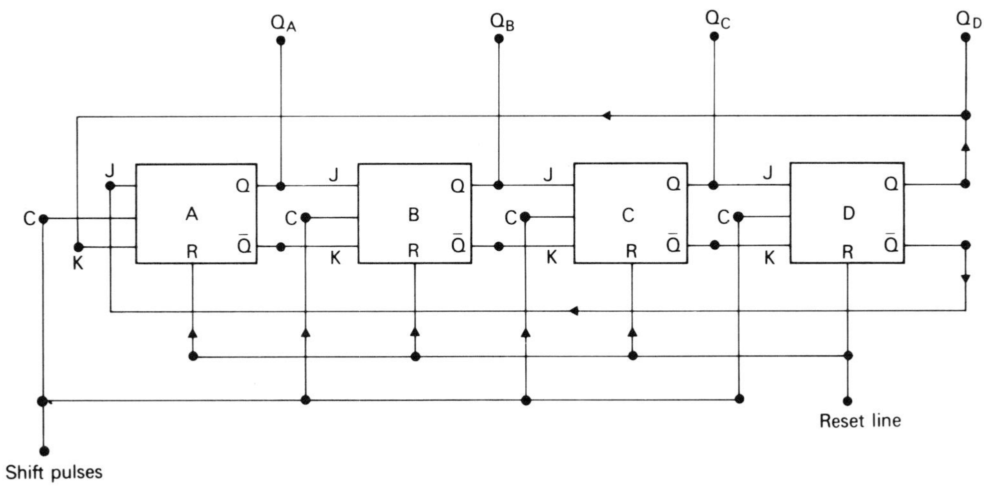

Fig. 4.56 Four-stage Johnson counter

output pulse being adjustable by varying the values of the externally connected timing capacitor and resistor. For example, the 74121 monostable IC, Fig. 4.58(a), can be triggered on the trailing edge of a pulse applied at inputs A_1 or A_2 if the input B is held at logic 1. Also, if A_1 and A_2 are held at logic 0 then the monostable will trigger on the rising edge of a pulse applied to B. The trigger input and the output waveforms at Q and $\overline{Q}$ are shown in Fig. 4.58(b); the duration of the pulse τ is $0.69RC$, where the limits on the values of R and C are specified by the manufacturer. In general, the value of R is limited to about 56 kΩ while any value of C between 10 pF and 100 μF can be used.

Another type of monostable integrated circuit currently available is the retriggerable one

Table 4.20 Truth table of the Johnson counter of Fig. 4.56

Pulse number	Q_A	Q_B	Q_C	Q_D	*Decoding to decimal*
0	0	0	0	0	$\bar{A}.\bar{D}$
1	1	0	0	0	$A.\bar{B}$
2	1	1	0	0	$B.\bar{C}$
3	1	1	1	0	$C.\bar{D}$
4	1	1	1	1	$A.D$
5	0	1	1	1	$\bar{A}.B$
6	0	0	1	1	$\bar{B}.C$
7	0	0	0	1	$\bar{C}.D$

Table 4.21 Another valid sequence of the Johnson counter of Fig. 4.56

Q_A	Q_B	Q_C	Q_D
0	0	1	0
1	0	0	1
0	1	0	0
1	0	1	0
1	1	0	1
0	1	1	0
1	0	1	1
0	1	0	1
0	0	1	0

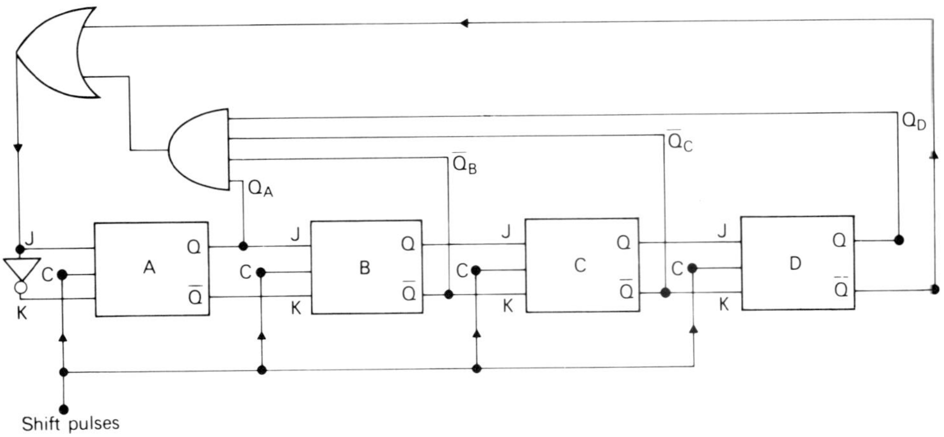

Fig. 4.57 Johnson counter with automatic reset

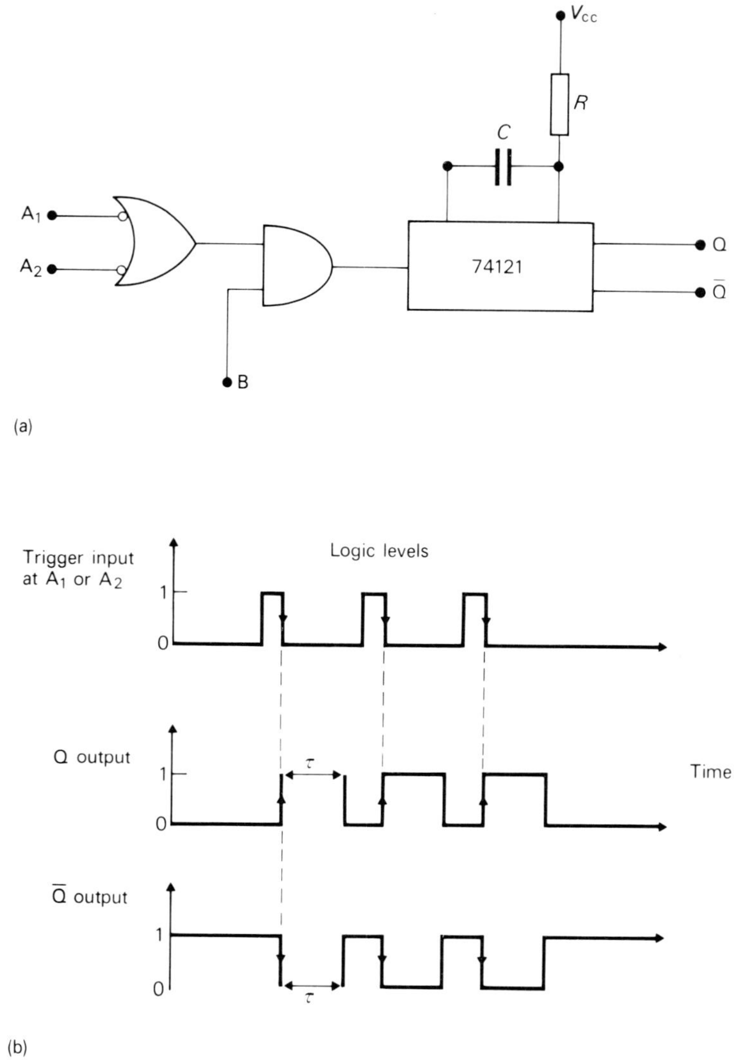

Fig. 4.58 (a) A typical IC monostable that can be triggered from the leading or trailing edge of an input pulse; (b) waveforms at A_1 or A_2 input and at Q and $\overline{Q}$ outputs

such as the 74123 whose output waveform is illustrated in Fig. 4.59. The monostable is triggered on the trailing edge of input pulse (1). However, before the output can complete its duration τ, which is defined by the external resistor and condenser, the input pulse (2) occurs. Its trailing edge automatically terminates the first timing interval and a new timing interval is initiated. This happens so quickly that the output remains in its triggered state and, in the absence of any further input pulses, resets to logic 0 after an interval τ. In addition to generating pulses of very long duration, the retriggerable monostable can be used as a **missing pulse detector**, simply by making the duration τ longer than the period of the input trigger pulses. Thus, if one of the input pulses should disappear due to malfunction or noise, the monostable will time out and reset to its stable state of logic 0.

Monostables are commonly used in the following applications.

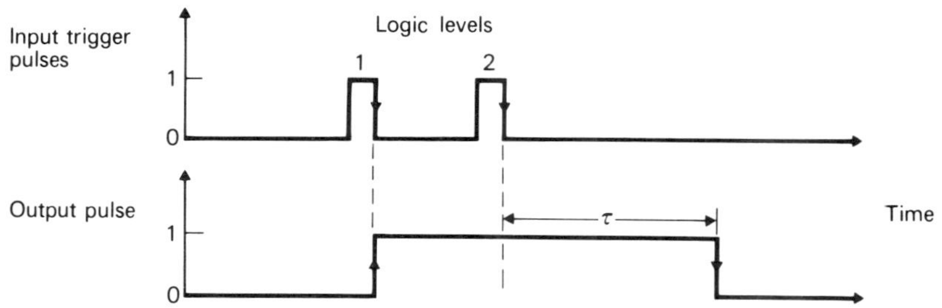

Fig. 4.59 Input and output waveforms of a retriggerable monostable IC

Pulse 'stretching

As the monostable triggers on the rising or trailing edge of an input pulse, it is a convenient circuit for adjusting the duration of a narrow pulse or 'spike' by simply choosing suitable values of R and C.

Time delay or phase shift

It is often required to generate a single pulse or a pulse train that is delayed or shifted in phase with respect to a given pulse. This can be achieved by using a dual monostable such as the 74123 in which the first monostable is triggered by the trailing edge of the input pulse, Fig. 4.60, and its output is made to have a duration t_1 governed by its external components R_1 and C_1. The trailing edge of this monostable triggers the second monostable, the duration τ of its output being governed by its external components R_2 and C_2 In this way the output pulse is shifted in phase or is delayed with respect to the input pulse by an interval t_1 which can be altered by changing R_1C_1. The mark-to-space ratio (τ/T) of the output pulse can be adjusted by changing R_2C_2 and will remain fixed as long as the period T of the input pulse train is constant.

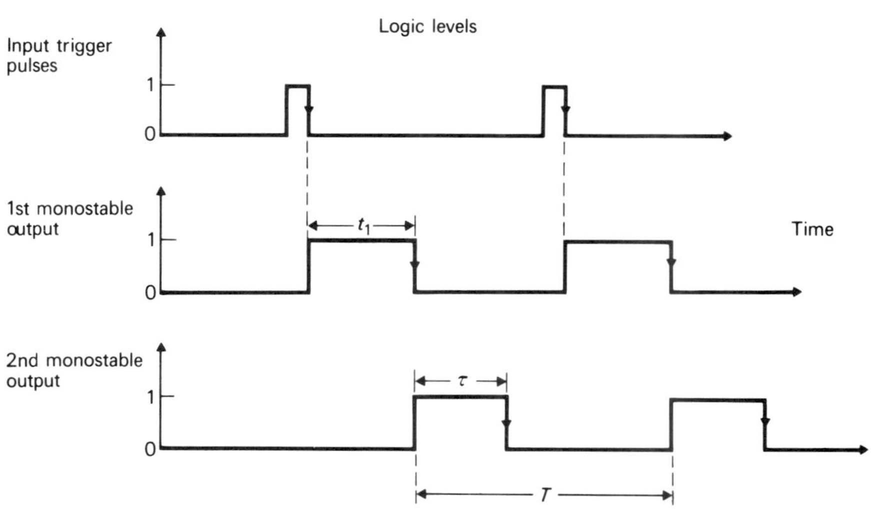

Fig. 4.60 Phase shift or time delay using a dual monostable IC

Sequence generation

A sequence of pulses of various durations can easily be generated by cascading several monostable units as shown in Fig. 4.61. Such operations are often used for synchronising sequencing and timing in digital circuitry.

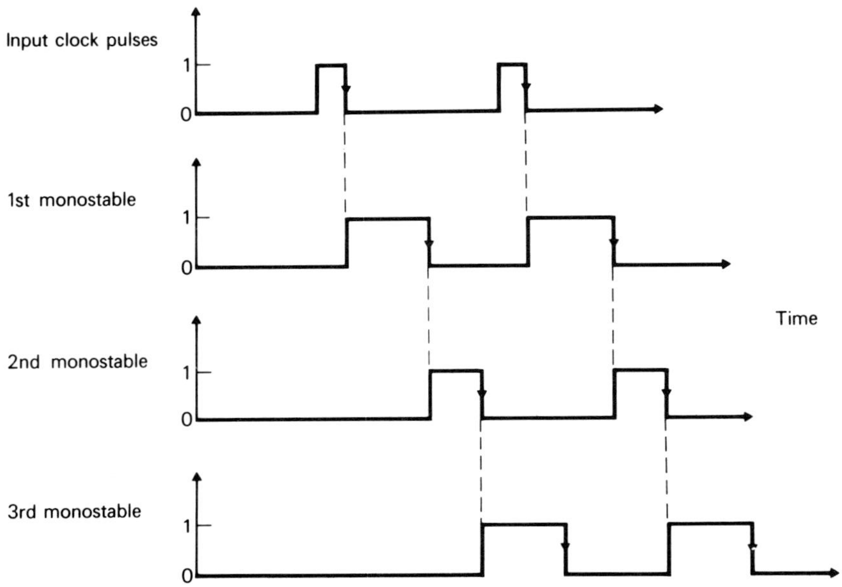

Fig. 4.61 Generation of pulse sequences using a monostable

Clock pulse generators used in digital circuitry are usually quartz crystal controlled pulse oscillators that give a frequency stability of about ±20 ppm over a wide temperature range. The fundamental resonant frequency of the quartz crystal is in the range of 1–5 MHz, and the drive circuits are a stable multivibrators made up of discrete transistors or ICs such as the 7404. The desired clocking frequency is obtained by frequency division and a specified pulse width from a monostable IC such as the 74121.

4.10 Digital memories

Digital memories may be classified into:

i) moving surface devices such as magnetic tape or disk memories used as 'file' memories in computer systems;
ii) memories with no moving parts, usually made up of microelectronic devices using LSI techniques.

The second type have virtually replaced ferrite core memories in small storage systems because they are physically smaller, have a lower power consumption and a significantly lower cost. The important features of memories are as follows:

a) storage capability
b) reliability and speed of operation;

c) cycle time;
d) data transfer rate;
e) cost of storage.

The smallest block of information accessible to a memory system can be a single binary digit or bit represented by 0 or 1. A group of eight or nine bits is called a **byte** or **character**. Depending on the digital system used, a group of 12 to 64 bits is called a **word**. Large magnetic disk memories can store 10^9 to 10^{10} bits on a disk surface, and a single LSI memory chip is capable of storing as many as 16 384 bits in an area less than 0.5 cm^2. A memory such as a magnetic disk, in which the information contents are retained after the power supply is switched off, is called **non-volatile**. The speed of operation is defined in terms of **access time**, i.e. the time required to read at any given storage location. The **cycle time** is the minimum time required to complete a read or write operation. The **data transfer rate** is the rate at which information is transferred to or from sequential storage positions. Finally, the cost per bit of storing information in moving surface memories is lower than that of microelectronic memories, but the access time for the former is longer.

LSI memories can be divided into two main categories, i.e. **read-only memories (ROM)** and **random access memories (RAM)**. The general organisation of both these types is basically the same, and both contain a number of memory locations where data can be stored and accessed in any order. In ROM the data is stored permanently and only reading out is allowed, whereas in RAM data may at any time be written into or read out of any portion of the memory.

The operation of a ROM is shown in Fig. 4.62. It consists of three major sections: the address decoder; the memory storage elements such as flip-flops; and output circuits. In this case, the address decoder input is a five-bit word, so that $2^5 = 32$ different states can be decoded. Thus, when the decoder input (generally known as the **address**) is 00100 or decimal

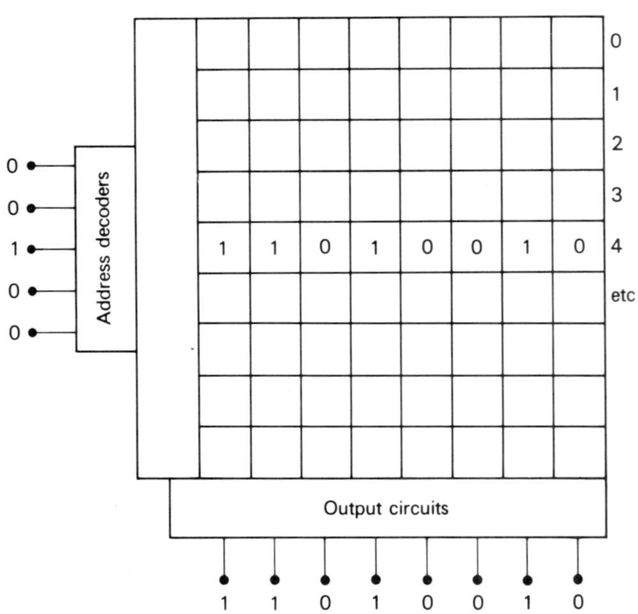

Fig. 4.62 Operation of a ROM

4, the contents of the fourth row will appear at the output via the output circuits which buffer the memory contents.

There are many ways of implementing the ROM function but MOS circuitry, although slow, is particularly suitable for IC memory chips because of its small size, low cost and low power consumption. For expanding the size of the memory, an extra address line known as the **chip enable** line is supplied. For example, two ROMs, each of which can store 32 words (each word comprising 8 bits, totalling 32 × 8 bits) can be put together to form a memory storing 64 eight-bit words. In Fig. 4.63, the five address lines are in parallel and the sixth chip enable (CE) line will enable one ROM at a time since the inverter (I) keeps these lines complementary.

RAM can take one of two basic forms:

a) dynamic RAM;
b) static RAM.

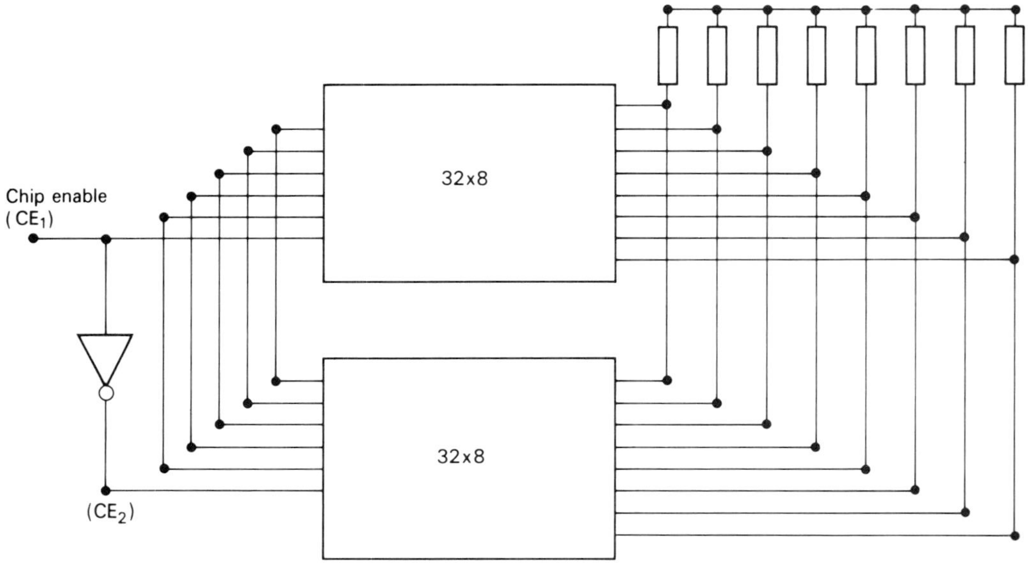

Fig. 4.63 A method of expanding the size of a ROM

Dynamic RAM

In most economical dynamic RAM, the information is stored as a minute charge ($\approx 500 \times 10^{-15}$ C) on a capacitor, a binary zero being represented by zero charge. Transistor switches are used to connect the capacitors to the data line, which is shared by a large number of capacitor-transistor cells, only one cell being connected to it at any one time. As the charge on the capacitors is likely to leak away, it is necessary to regenerate or **refresh** the stored charge periodically, usually every 2 ms. Continuous refreshing within each storage cell is usually achieved by using more than one transistor per cell. This, however, leads to a larger silicon chip area per cell and a higher cost and power consumption.

Static RAM

This type does not require refreshing because its memory cells consist of bistable circuits, one stable state representing a logic 1 and the other a logic 0. They are useful for small memory systems as they require a minimum of external support circuitry and have a very low power consumption. For example, CMOS memories, used in pocket calculators, operate satisfactorily for weeks on small batteries, but are slower than those using bipolar transistors.

Whenever random access memories are required to store information which is rarely altered, such as the control programs of pocket calculators, **programmable** read-only memory (PROM) is used. In one type of PROM the cells are made up of small fusible links from the transistors to ground instead of the storage capacitors. A pattern of stored information can then be placed in the array by applying a pattern of electrical signals capable of burning out the unwanted connections. Once a fusible link is burnt out, it cannot be reformed so that such memories are non-volatile.

Whenever the read operations are far more frequent than the write operations and a non-volatile memory is required, an **electrically alterable** read-only memory (EAROM) is used. These are metal-nitride-oxide-semiconductor (MNMOS) devices and basically consist of an interface between two dielectric materials, silicon dioxide and silicon nitride, which can be selectively charged and discharged by signals applied to a single overlying gate electrode. All the cells are erased simultaneously and an information pattern is then imposed. Although the potential usefulness of these devices is large, they do not yet have the combination of low cost, fast writing speed, long-term retention and reliability.

In random access memories, the time required for a read or write operation is independent of the physical location within the storage array of the cell being accessed. If random access is not required, then substantial economies in the design of memories can be achieved by using serial access organisation in which stored bits circulate as if they were in a closed pipe-line. For example, in charge-coupled device (CCD) memory organised in this way, each bit that is stored is transferred sequentially through 64 or more storage locations, between the time it is written into the memory and the first time it becomes available for reading. The time taken to shift the bits from one storage site to the next in a serially organised CCD memory is about the same as the cycle time for a random access memory. However, the total silicon chip area per bit of a serially accessed CCD memory is far smaller than that of a random access memory. Reset designs of serially accessed CCD memories have a storage of 65 536 bits on a 3.5 mm × 5 mm chip, and are adequate for applications such as video displays, in which the memory is scanned point-by-point in a repeating linear pattern.

Another type of compact non-volatile memory device, suitable for the storage of a few million bits, is the **magnetic bubble memory**. Magnetic bubbles or cylinders are magnetic domains that can be generated and moved around in a thin magnetic layer. Movement is then governed by propagation paths processed on the layer and magnetised by external fields. There are two main types of storage material capable of supporting bubbles. One is a magnetic garnet, grown epitaxially on a non-magnetic garnet substrate. The other is an amorphous metallic magnetic layer, sputtered on a substrate such as glass. Bubble memory chips are packaged with two wirewound solenoids, positioned at right angles to one another, to provide the rotating drive field. A permanent magnet is used to generate the bias fields which keep the bubbles stable. The absence or presence of a bubble is used to represent logic 0 and logic 1 respectively.

There are currently two forms of bubble memory organisation. One is a simple serial data loop with characteristics similar to those of a cassette; it gains simplicity of operation at the expense of a slow access time of about 100 ms. The second form is built with a series of minor and major loops for data storage. This requires a much more sophisticated controller, but is faster and more flexible, taking about 1 ms to access data blocks.

4.11 Microprocessors

A microprocessor is the control and processing part of a small computer or microcomputer. Moreover, the term microprocessor currently implies a type of processor that is built with LSI MOS circuitry, usually on one chip as in the case of the popular MC 6800 (Motorola Ltd) central processing unit (CPU) which is available in 40-pin, dual-in-line, packaging. This chip contains all the functions required for multi-instruction processing:

1) an arithmetic and logic unit (ALU);
2) instruction decode and address registers;
3) instruction register;
4) clock and logic circuits required for timing;
5) data bus input and output circuitry;
6) address bus circuitry.

The chip is equivalent to about 120 MSI TTL packages.

All microcomputers utilising microprocessors are made up of similar building blocks, although the internal organisation and design of these blocks differ between manufacturers. The basic blocks are as follows:

1) CPU;
2) ROM to hold the program of instructions for the CPU;
3) RAM to hold data;
4) Input/output (I/O) devices for connecting any given system to the microprocessing unit (MPU) which manipulates data in a manner defined by the program of instructions.

In the case of the MC 6800 central processor, random access memory for fast logic control, a read-only memory for storing the system program parameters and a set of input/output chips are available as a 'family' of circuits.

There are numerous types of microprocessors, some inherently faster than the single chip MOS devices. Details and applications are discussed in other texts.[2, 4, 5]

Finally, a microprocessor-based system is worth considering if an existing design uses 50 or more MSI packages. The main advantages offered by microprocessing systems are: lower cost; fewer components; increased reliability and versatility. However, program development can be time-consuming in situations where a large number of instructions is required, despite the aid of a large computer using a high level language such as FORTRAN or ALGOL.

References

(1) Zissos, D., and Duncan, F. G., Fan-in restrictions in logic circuits. *Proc. IEE*, **118**, 2 (February 1971)
(2) Altman, Laurence (ed.), *Microprocessors*. Electronics Book Series, McGraw-Hill (1975)
(3) ITT Semiconductors Application Note. *Modulo-*N *Counter Circuits*
(4) Altman, Laurence, New arrivals in the bulk storage industry. *Electronics* (April 13, 1978)
(5) *Electronics*, Special Issue on Microprocessors (April 15, 1976)

Further reading

Bozic, S. M., Cheng, R. M., and Parsons, J. D., *Electronic and Switching Circuits*. Edward Arnold (1968)
Morris, N. M., *Logic Circuits*. McGraw-Hill (1969)

Morris, N. M., and Miller, J. R., *Designing with TTL Integrated Circuits.* McGraw-Hill (1971)
Motorola Semiconductors. *MC MOS Handbook* (1974)
Poppelbaum, W. J., *Computer Hardware Theory.* Macmillan (1972)
Signetics Corporation. *Memory and Interface Handbook* (1972)
Zacharov, B., *Digital Systems and Logic Circuits.* Allen and Unwin (1968)
Zissos, D., *Logic Design Algorithms.* OUP (1972)

5

Principles of digital-to-analogue and analogue-to-digital conversion

5.1 Introduction

With the advent of inexpensive digital integrated circuits, digital data acquisition, processing and display have gained an ever-increasing popularity. Preamplification of an analogue signal, often required before converting it to digital form, introduces noise. However, once an analogue-to-digital conversion is accomplished, digital circuits with high noise immunity, no drift, high speed and low cost can perform mathematical operations with a high degree of accuracy. Moreover, several conventional analogue functions can be implemented using digital integrated circuits. For example, function generators can be constructed using ROMs and long time delays with high resolution can be generated using shift registers. However, there are several special-purpose analogue modules, such as log/antilog amplifiers and multipliers/dividers, for use in signal conditioning. Whenever their accuracy and cost compatibility is adequate, they can relieve a digital processor of expensive time-consuming software and of computational burdens.

Wherever an analogue system interfaces with a digital one, or vice versa, an analogue-to-digital converter (ADC) or a digital-to-analogue converter (DAC) is of prime importance. The analogue-to-digital signal conversion problem can be greatly eased by suitable choice of transducer. For example, the rotation of a mechanical shaft may be monitored by a digital shaft encoder or by a potentiometric device. In the latter case, a further A/D conversion is necessary, and the choice must therefore be a favourable balance between cost, accuracy, speed and complexity.

In the following sections, the specifications, limitations and features of a wide variety of D/A and A/D circuits are discussed along with some of their common applications. Some of the circuits are commercially available as economical building blocks.

5.2 Types of digital-to-analogue converters

There is a wide variety of digital (D/A) devices, some of which give a voltage output and some a current output. Also, some require an external reference source, while others have an internal reference device comprising a temperature-compensated reference diode and an operational amplifier. The outputs of such devices may be unipolar or bipolar.

The three commonly-used types are as follows:

1) VDAC, i.e. digital-to-analogue converters with a *voltage* output;
2) IDAC, i.e. digital-to-analogue converters with a *current* output;
3) multiplier-type D/A converters.

VDAC

The two techniques commonly-used in D/A converters with voltage outputs are illustrated in Fig. 5.1. A set of binary-weighted resistors are connected to a reference voltage V_{ref}, Fig. 5.1(a), via a set of electronic switches. An operational amplifier is used to hold one end of all the resistors at zero, or 'virtual earth', potential. The switches are operated by the digital input levels, a logic 0 representing an open switch and a logic 1 representing a closed switch. Each closed switch adds a binary-weighted increment of current given by $(V_{ref}/2^nR)$, where 2^nR is the least significant bit (LSB) and 2^0R is the most significant bit (MSB). The increment

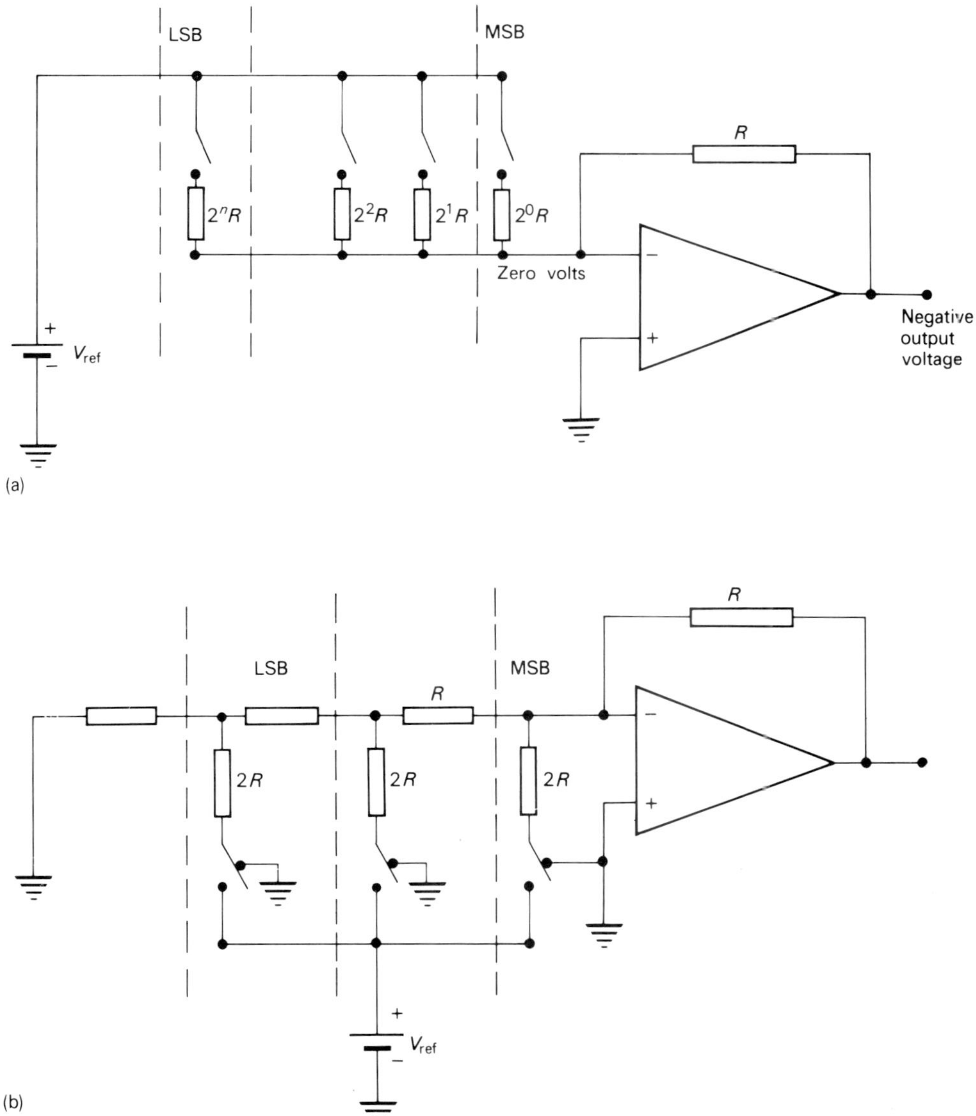

Fig. 5.1 (a) D/A converter using binary-weighted resistors with a voltage output; (b) D/A converter using R-$2R$ resistive ladder network with a voltage output

is added via the summing 'bus' connected to the INV(−) input of the operational amplifier. The negative output voltage is proportional to the total input current and therefore to the value of the binary number. For example, in practical application, a 12-bit D/A converter would need a range of 2^{12} resistor values so that if the MSB had a $10^4\,\Omega$ resistor, the LSB would need a $2^{12} \times 10^4 = 40.96 \times 10^6\,\Omega$ resistor. A significant disadvantage of this simple approach using weighted resistors is that the accuracy and stability of the DAC is dependent on the absolute accuracy of the resistors and their ability to track each other with temperature, especially when the range of values required for good resolution is very large. The conversion rate of these converters is relatively slow, mainly because of the high input impedances and the speed limitations of the voltage switches.

An alternative technique, illustrated in Fig. 5.1(b), uses an R-$2R$ resistive ladder network[1]. If all the branches are normally grounded and one branch is connected to the reference voltage V_{ref} via an electronic switch, then a current is produced in that branch. The current flows through the ladder network and is divided by a factor of two at each junction, so that the LSB is on the left-hand side. It can be shown that if the nth bit switch is connected to the reference then it will produce an increment of $(V_{ref}/2^n)$ volts in the output voltage. One of the advantages of the R-$2R$ ladder network, where R may have some convenient value between 5 kΩ and 50 kΩ, is that it lends itself to manufacture by thin-film deposition techniques; this improves its temperature tracking capabilities. For example, thin-film nickel-chromium R-$2R$ ladder networks, featuring tracking with temperature to better than 1 part per million per °C, are commercially available.

IDAC

In this type of conversion, binary-weighted currents are generated using active sources, and the output is obtained by summing them on a common output 'bus'. This principle is illustrated in Fig. 5.2.

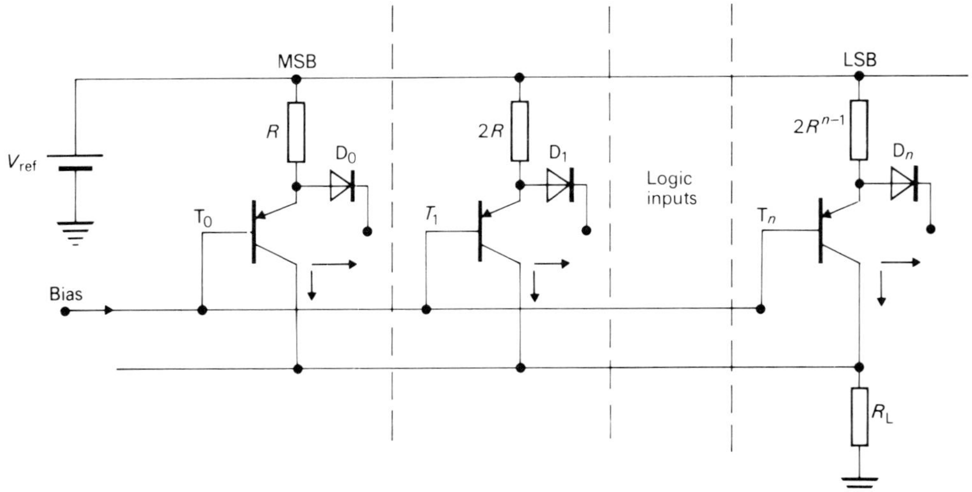

Fig. 5.2 D A converter with a current output

Grounded base transistors, $T_0, T_1, \ldots, T_n$, with their binary-weighted emitter resistances, represent the active current sources. The digital input controls the total current flowing through the terminating resistance R_L via the diodes D_0, D_1, . . . , D_n. For example, if the cathodes of all the diodes are at logic 1 then the current in each branch is proportional to the binary-weighted emitter resistance. The total current through R_L will be a maximum, i.e. the sum of all the currents in the individual branches. However, if any of the cathodes is at logic 0 then the current in that branch will flow through that diode to earth, the corresponding transistor will turn off and the contribution of that branch to the total current will be eliminated. As in the case of the VDAC, the use of an *R*-2*R* ladder network with equal currents would have several advantages. However, the errors involved in this type of conversion usually limit its accuracy to about 12 bits. A full-scale output current of 10 mA is ideal, since a 1 V output would develop across a load resistance of 100 Ω, an approximate impedance level for high-speed coaxial cable operation.

Multiplier-type DAC

A DAC which is specially designed to accept a varying reference voltage is known as the **multiplying DAC (MDAC)**. The magnitude of its output, which may be a voltage or a current, is directly proportional to the product of the reference voltage and the digital input. The reference voltage to this type of converter can be a bipolar or AC source and its performance depends mainly on the type of electronic switches used. For example, FET switches are most suitable for the bipolar reference mode.

MDACs are extensively used in programmable test equipment where the amplitude of a varying or DC signal can be controlled by a digital computer. They are also used for graphic displays since they can perform direct multiplications such as $r \sin \theta$ which are required for conversion of polar to rectangular coordinates in microseconds.

All the D/A converters discussed above are currently available in modular form. A wide variety of circuits also exist for implementing the functions described with discrete components. These circuits give satisfactory results but are bulky and inconvenient to construct.

5.3 D/A converter terminology

In practical applications, the following terms are used to describe the performance and characteristics of D/A converters.

Resolution

As applied to a DAC, the term exclusively refers to the maximum number of input bits that the converter is capable of handling or the corresponding discrete output levels. For example, a ten-bit DAC has $2^{10} = 1024$ levels, or its resolution is approximately one part in 10^3. A three-decade BCD signal will require a twelve-bit DAC, since each decade consists of four bits. The resolution will still be one part in 999, reflecting the fact that BCD coding, although convenient, is wasteful (see Chapter 4).

Linearity

This is generally defined as the maximum deviation from the best straight line drawn through the staircase output waveform, Fig. 5.3, obtained as a result of incrementing the DAC sequentially through all its successive states. The maximum allowable deviation from the best straight line must ideally be less than $\pm\frac{1}{2}$ LSB, or the full-scale deflection (FSD) divided by

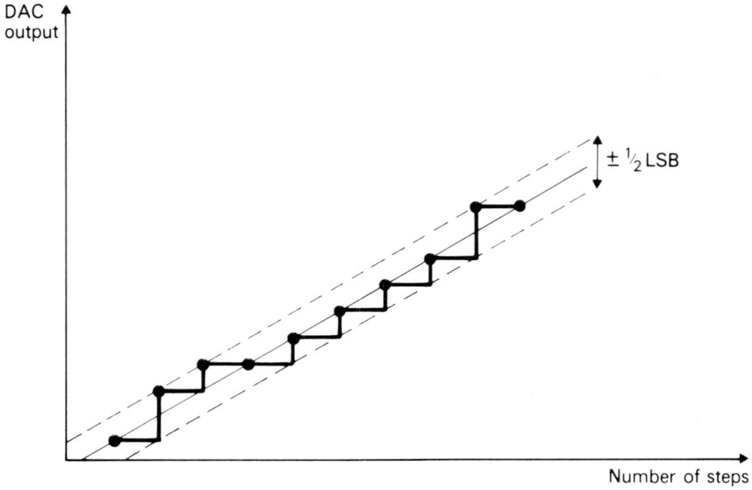

Fig. 5.3 Linearity of a DAC

the number of states. In practice, the degree of linearity or the variation in the individual step sizes depends on the degree of sophistication of the DAC circuitry and hence its cost.

Offset

This is the output of a DAC when the appropriate code for zero output, say (000 . . . 00 in binary) is applied to the input. This undesired output is specified (in millivolts or microvolts) as a fraction of the FSD output or as a fraction of an LSB. An external control for adjusting the offset at room temperature is usually available in most DAC modules, and in some cases it is utilised to produce a bidirectional output. The presence of an offset, whether it is deliberately introduced or present as a circuit imperfection, means that the value of the output for any given input must be corrected to allow for the offset.

Settling time

This is defined as the interval of time required for all transients to be sufficiently reduced while the output lies within some specified amount of its final value. The difference, Fig. 5.4, is generally of the order of the linearity error of $\pm\frac{1}{2}$ LSB. Settling time is primarily a function of the types of switches, resistors and amplifiers used, the current switching mode being the fastest.

Stability

An important feature of a DAC is its stability, governed mainly by its reference voltage and by the temperature stability of its active and passive components. The stability of a DAC without an internal reference is obviously higher than that of one with a cheap 'built-in' source, because the contribution of the reference source towards the overall stability is significant. Thus a considerable improvement can be achieved by using a more elaborate external reference source. The absolute value of the output of a DAC often depends on the magnitude and stability of the external DC power supply, which must be well-regulated and free from noise, as it often gets coupled into the DAC output.

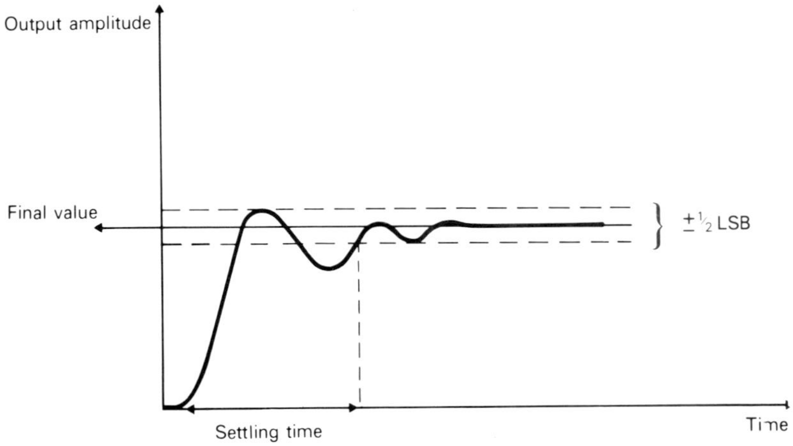

Fig. 5.4 Settling time of a DAC

5.4 D/A conversion with magnitude and sign

In some experimental situations the polarity of the DAC output must reverse under the command of a 'sign bit' included in the digital input. There are several ways of implementing this requirement, but a straightforward and relatively inexpensive approach is to incorporate in series with the output an amplifier whose gain can be switched between +1 and −1. The switching is done by a two-pole two-way electronic switch S, Fig. 5.5, activated by the sign bit signal. When S is in position (1) the amplifier works in the **invert mode**, while in position (2) it works in the **non-invert mode**. One of the advantages of this technique is that the output of the DAC is used, so that the magnitude of the input signal can be in any convenient digital code.

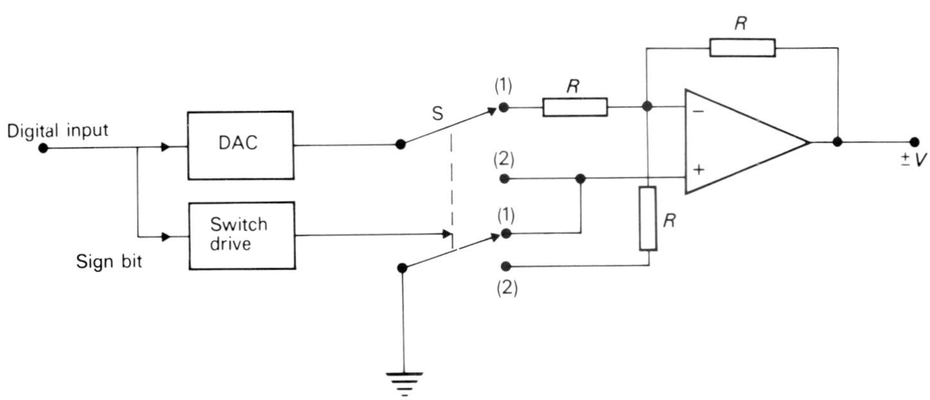

Fig. 5.5 D/A conversion with magnitude and sign

5.5 Types of analogue-to-digital converters

Most modern digital instruments, such as digital voltmeters, have a built-in analogue-to-digital (A/D) converter whose primary requirement is the compatibility of its output with the rest of the digital circuitry in the instrument. There is a large number of circuits for converting an analogue signal into digital form[2], but only a few types are commercially available in the modular form which is suitable for interfacing with digital systems such as microcomputers and data loggers.

The most popular types use one of the following techniques:

1) successive approximation;
2) single and dual slope integration;
3) voltage comparator – counter loop;
4) multiple comparator type;
5) voltage-to-frequency conversion.

Each approach has characteristics that make it most useful for a specific class of applications. The choice is based on linearity and speed of conversion, accuracy, stability and cost.

Successive approximation

In this technique, Fig. 5.6, a 'start conversion' pulse clears the output register and turns ON (logic 1) its MSB so that the DAC gives half of a full-scale deflection, say 5 V for a 10 V full-scale system. This output is compared with the analogue input, and if the latter is less than 5 V then the comparator turns OFF (logic 0) the MSB in the output register, via the control logic unit. However, if the analogue input is greater than 5 V, then the MSB in the output register is left ON. The next clock pulse turns ON the next significant bit in the output register via the shift register and its contribution of 2.5 V adds to the previous bit provided that it was left ON, so that a comparison is made to determine whether the analogue signal is greater than or less than the new total of 7.5 V. If it is greater, then the comparator leaves the bit ON but if it is less, then the bit is turned OFF. This process is repeated n times for a n-bit converter, until the LSB has been compared. Then the clock stops, to be restarted only by the

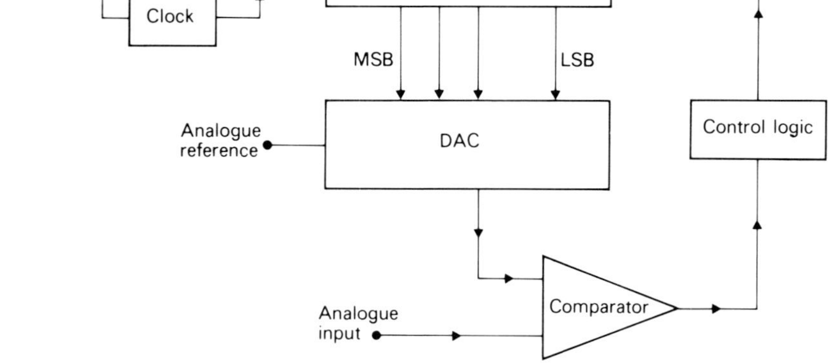

Fig. 5.6 A successive approximation type A/D converter

next conversion pulse. The 'status' line changes state, to indicate that the contents of the output register now constitute a valid conversion. During the conversion interval, typically of a few microseconds, in slow converters the input analogue signal must not change appreciably. If it does, a 'sample hold' device (see Section 5.5, p. 144) is used to hold the input signal value constant during the conversion interval, and the 'status' signal is used to release the 'hold' mode at the end of the conversion.

The accuracy, linearity and speed of conversion in successive approximation A/D converters is primarily dictated by the properties of its D/A converter, its reference and the comparator. In general, the settling time of the D/A converter and the response time of the comparator are considerably longer than the switching time of the digital elements.

Successive approximation type converters are widely used for interfacing analogue signals to digital computers because they are capable of both high resolution (up to 16 bits) and high speed (conversion interval of less than 1 μs). The conversion time is fixed and independent of the input voltage. Each conversion is unique and independent of the result of previous conversions because the internal logic is cleared at the start of a conversion.

Single and dual slope integration

Figure 5.7 illustrates the principle of a single slope integration type A/D converter. It consists of an analogue integrator, a clock pulse generator, a voltage comparator and a digital counter (see Chapter 4). The integrator and the counter are initially reset via the logic control unit which initiates the integrator and the count simultaneously. The integrator generates a precise ramp while the counter is incremented, one LSB at a time, by the fixed-frequency clock. When the ramp reaches a level equal to that of the input analogue signal, the comparator changes state and stops the count. The contents of the counter are then the digital representation of the analogue input, indicated by the change of state of the status

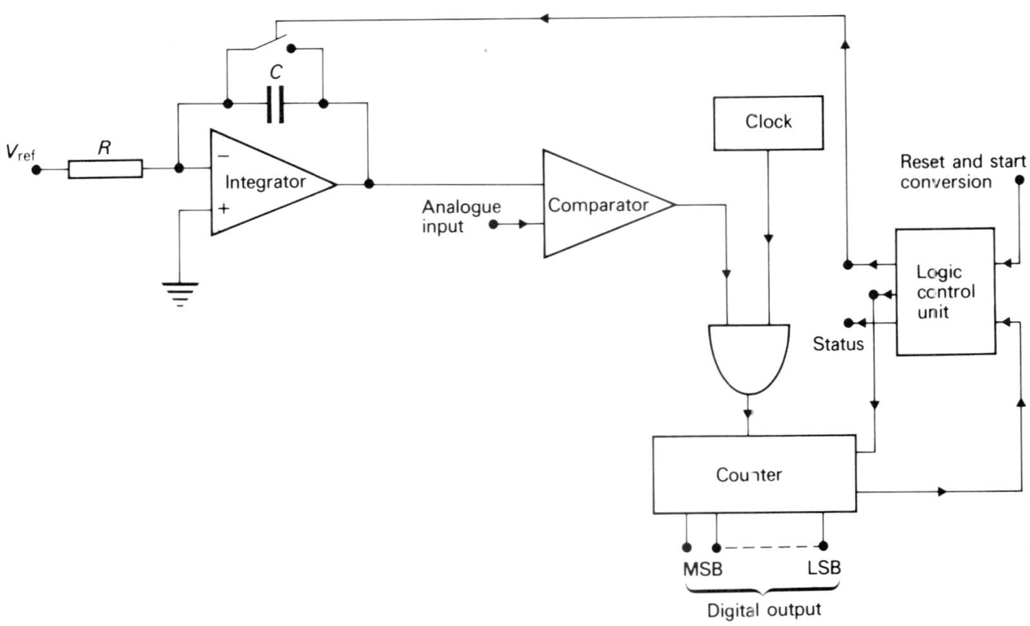

Fig. 5.7 A single-slope integration type A/D converter

terminal. The method is inherently slow and has several problems such as accurate resetting, high linearity ramp generation, spurious triggering of the comparator and the variation of the clock frequency with temperature; these problems give rise to errors in the conversion process.

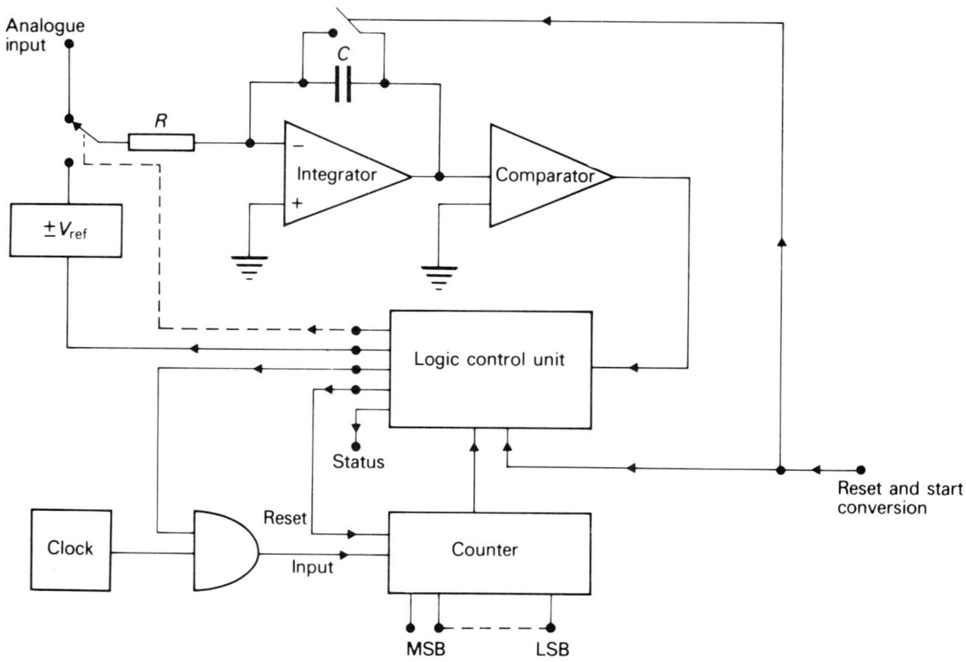

Fig. 5.8 A dual-slope integration type A/D converter

Some of the limitations can be overcome by using the dual slope integration technique illustrated in Fig. 5.8. After an initial reset, the 'start conversion' pulse connects the analogue input to the integrator via the logic control unit, which simultaneously starts the counter. As soon as a preset count defining a precise integration time T_1 is reached, the logic control unit switches the integrator input to a reference voltage of appropriate polarity, so as to cause the integrator output to decrease at a constant rate. Also, at the instant of the switching, the counter is reset and starts a fresh count which stops when the comparator changes state, indicating that the integrator ramp has reached zero. If the time taken by the ramp to reach zero is t_1, then

$$V_{ref} t = e_{in} T_1$$

The count in the counter is proportional to t_1 and is, therefore, a digital representation of the analogue voltage e_{in}, indicated by a change in state of the status terminal when the conversion is complete.

The operation of the converter is thus independent of the properties of the integrator and of the frequency stability of the clock, as long as they remain constant during the conversion interval, because both T_1 and t_1 are affected to the same extent. This technique is used in situations where the speed of conversion is not important; for example in digital voltmeters (DVM) and digital panel meters (DPM). However, a high degree of rejection of hum (50 Hz), or any other fixed-frequency interference, can be achieved by a suitable choice of the sampling time T_1. For example, if 50 Hz and its harmonics are to be rejected, then the minimum integrating time will be 20 ms. In practice, the maximum number of conversions per second is limited to about 30.

Voltage comparator–counter loop

In this type of converter, Fig. 5.9, a D/A converter, a voltage comparator, two three-input gates and an up-down counter driven by a fixed-frequency clock are used in a closed feedback loop. If the D/A output is less than the analogue input, then the counter counts *up*. If the D/A output is greater than the analogue input, then the counter counts *down*. In short, the feedback loop tries to 'lock in' on the analogue input value continuously, and can give a valid 'conversion complete' report only during the clock period immediately following the change of state of the comparator. By stopping the count via the external control input following a complete conversion, this type of converter can be used as a 'sample hold' device (see Section 5.5, p. 144) with an arbitrarily long 'hold' time and no 'droop'.

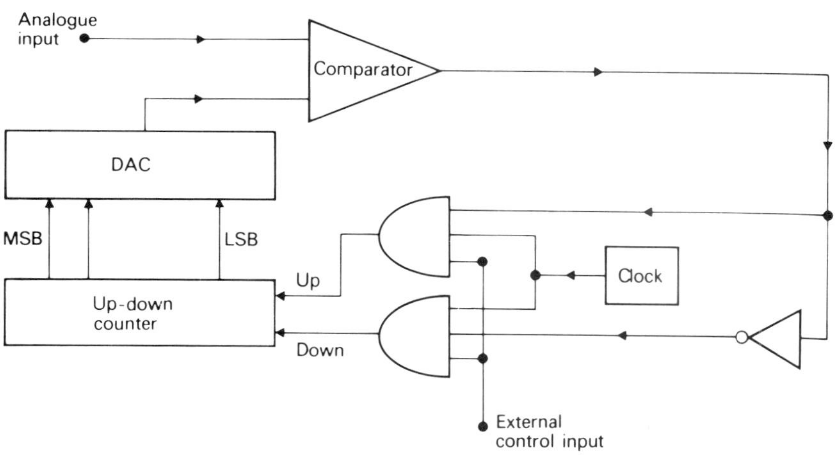

Fig. 5.9 Voltage comparator–counter loop type A/D converter

Multiple comparator type

The fastest and simplest approach to A/D conversion is illustrated in Fig. 5.10. A string of n comparators $C_1, C_2, \ldots, C_n$, each with a successively smaller reference, one LSB apart and starting with $\frac{1}{2}$ LSB, sorts the analogue input into n thresholds. However, since each comparator can only indicate whether its input is above or below its reference, the n comparator outputs must be encoded into a more efficient and convenient code. The obvious advantage of this approach is that the conversion occurs in parallel with speeds limited by the switching time of the comparators and the encoder gates. However, since each state requires one comparator, $2^n - 1$ comparators are needed to digitise to n bit resolution. For example, an eight-bit converter would require $2^8 - 1 = 255$ comparators, so that their cost and that of the associated circuitry would be prohibitive. This technique is, however, useful in cases where a high conversion speed, with up to four-bit resolution (or 15 converters), is required.

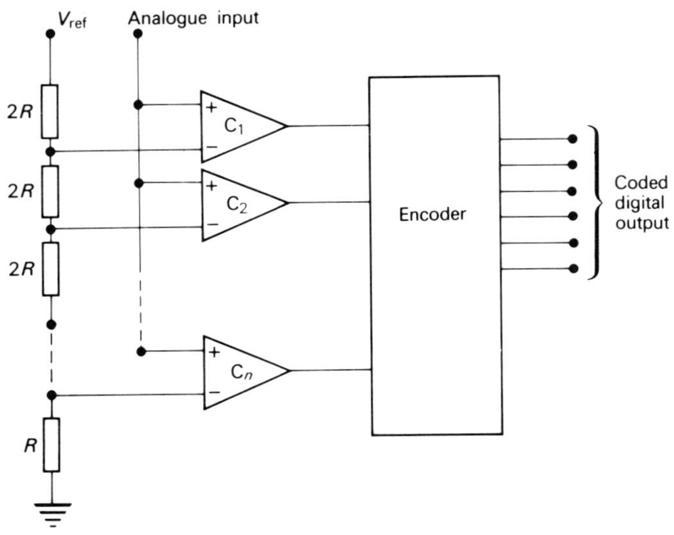

Fig. 5.10 Comparators-encoder type A/D converter

Voltage-to-frequency converters

In voltage-to-frequency (v/f) converters, a train of pulses is generated with a repetition rate proportional to the input analogue voltage. By counting the pulses over a fixed interval of time, an A/D conversion is effected.

There are several ways of achieving a voltage-to-frequency conversion but the most common method is illustrated in Fig. 5.11. The analogue input is connected to an integrator which resets when the amplitude of its output ramp reaches that of the fixed reference voltage V_{ref} of the comparator. Each time the comparator changes state and the integrator is reset, the pulse thereby generated is shaped and fed to the counter via a gate which stays open for a fixed interval T. The repetition rate of these pulses is obviously proportional to the magnitude of the analogue input voltage, since the latter controls the charging rate of the capacitor of the integrator.

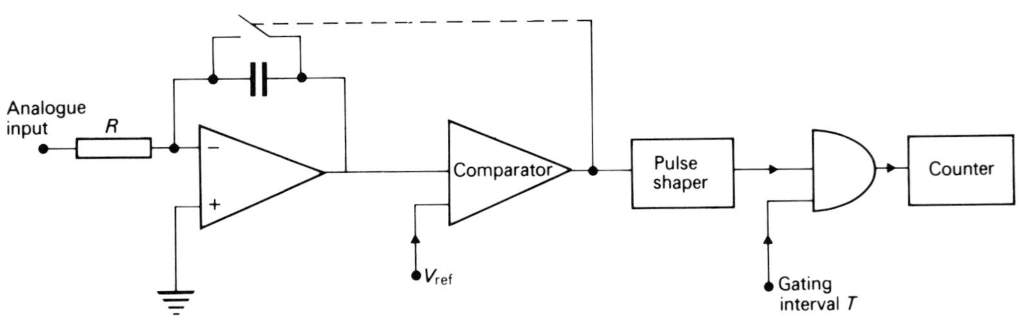

Fig. 5.11 Principle of a simple voltage-to-frequency converter

There are several circuits commercially available in modular form for implementing this principle, but two that merit description are as follows:

a) a bidirectional voltage-to-frequency converter;
b) a ratiometric voltage-to-frequency converter.

Bidirectional v/f converter
This device produces positive pulses when the analogue input voltage is positive with respect to earth and negative pulses when the input voltage is negative. If a very low-frequency sinusoidal wave is applied to the input, Fig. 5.12, the number of pulses produced is proportional to the amplitude of the input. Also, an axis-crossing pulse is generated when the input voltage changes sign. This pulse can be used to measure the phase difference between the input signal and a reference of the same frequency. The device has several applications.[3, 4, 5]

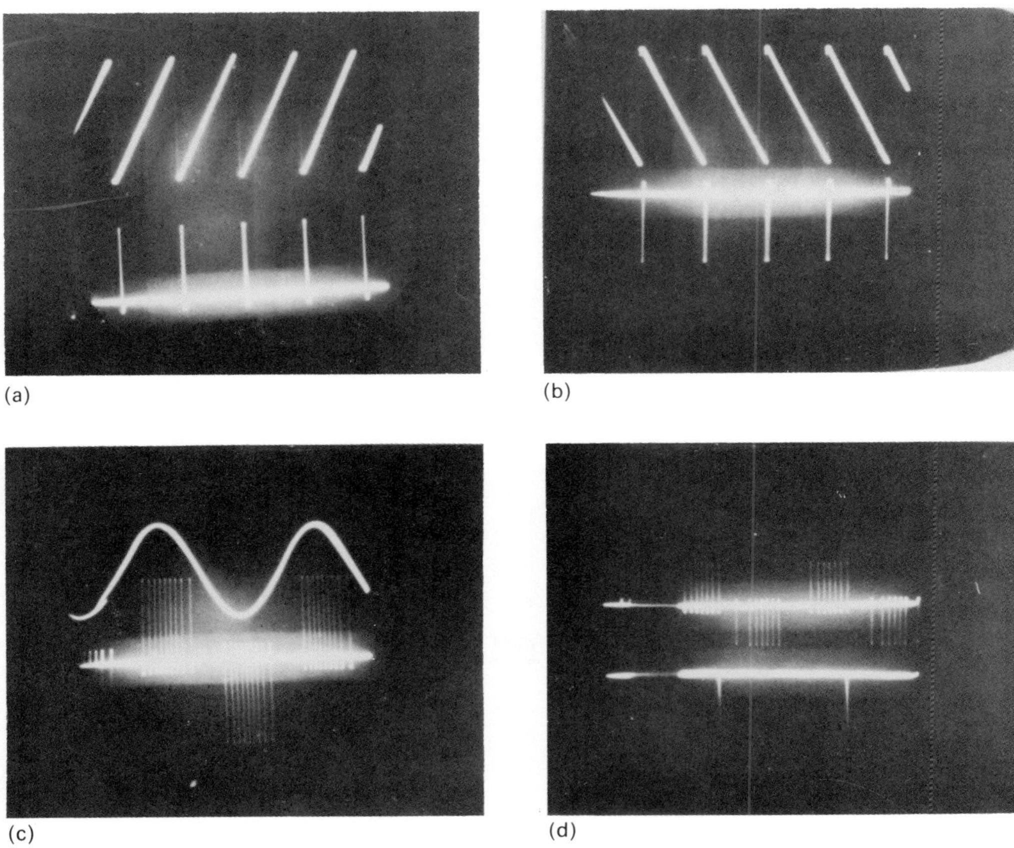

Fig. 5.12 (a) v/f converter output when the analogue input voltage is positive; (b) v/f converter output when the analogue input voltage is negative; (c) v/f converter output when the analogue input is a 5 Hz sinusoidal signal; (d) v/f converter output and axis-crossing pulses when the analogue input is a 5 Hz sinusoidal signal

Ratiometric v/f converter
This device produces an output frequency which is proportional to the ratio of two analogue input voltages, i.e.

$$f_{out} = K\left(\frac{V_x}{-V_y}\right) \quad \text{Hz}$$

where K is a constant, V_x and V_y are the two analogue input voltages and f_{out} is the output pulse frequency. This type of ratiometric v/f converter has been successfully used in optical absorption spectroscopy, to compensate for the random fluctuations in the intensity of light sources.[6]

Sample and hold devices

These devices 'sample' or 'track' an input signal faithfully in the **sample** mode, and retain the last value of the signal at the instant of arrival of the **hold** command. Figures 5.13(a) and 5.13(b) illustrate the **sample and hold** (S/H) principle. The control signals are logic 1 for sample and logic 0 for hold. There are four distinct states:

a) sample;
b) transition from sample to hold;
c) hold;
d) transition from hold to sample.

Ideal S/H devices have the following characteristics:

1) tracking that is error free;
2) instantaneous signal acquisition and release;
3) zero settling time;
4) infinite hold times.

Commercially-available modules are usually specified in terms of the extent to which they deviate from this ideal.

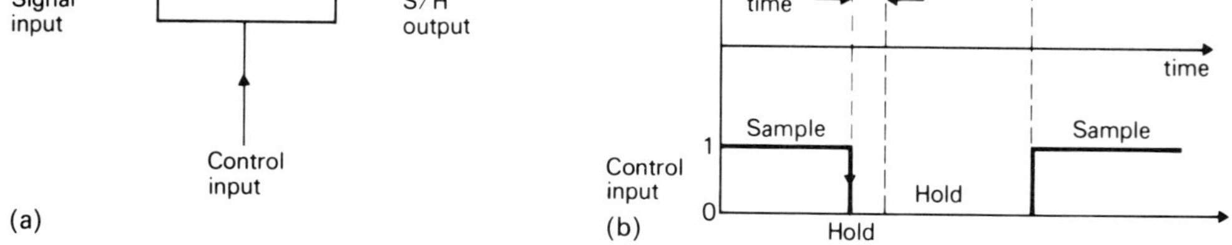

Fig. 5.13 (a) Sample-hold (S/H) block diagram; (b) S/H principle

The S/H principle can be implemented digitally as illustrated on p. 141, or by analogue means using a condenser as storage. The latter technique is shown in Fig. 5.14 in an open loop configuration using buffer amplifiers, A_i for input and A_o for output. When the electronic switch S is connected, the condenser C charges exponentially to the input voltage

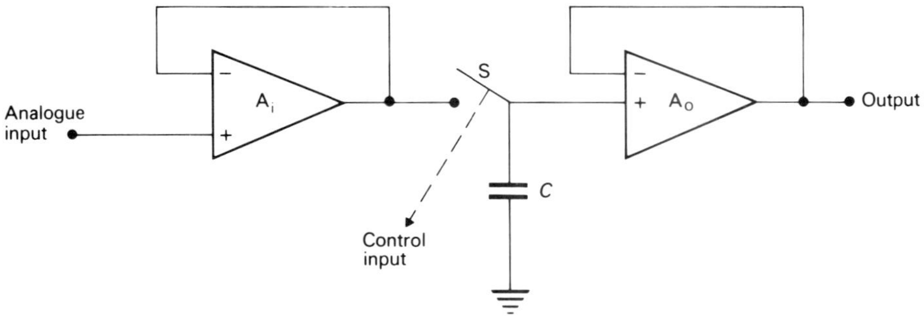

Fig. 5.14 Sample-hold open loop configuration

via the buffer A_i and the output amplifier A_o follows the condenser voltage. This is the **sample mode** of Fig. 5.13(a). When the switch S is disconnected at the 'hold' control input, the charge remains on the condenser and represents the value of the input signal at the instant of the arrival of the command. This is the **hold mode** of Fig. 5.13(b). Open loop circuits have the essential advantage of fast acquisition and settling time. If low-frequency tracking accuracy is more important than speed, then the closed loop configuration of Fig. 5.15 will yield better results. Most analogue circuits have several sources of error. For example, **aperture time**, Fig. 5.13(b), is the time elapsed between the command to hold and the actual disconnecting of the switch S. Thus if a signal is changing at 1 V μs^{-1}, the aperture uncertainty must be less than 10 ns if a resolution of 0.1 % of a full-scale deflection of 10 V is required. Secondly, the output of all S/H devices is subject to **droop**, Fig. 5.13(b), which a undirectional drift mainly due to the leakage of the charge on the condenser, when long hold times are required. More detailed descriptions of S/H modules can be found in other texts.[1,7]

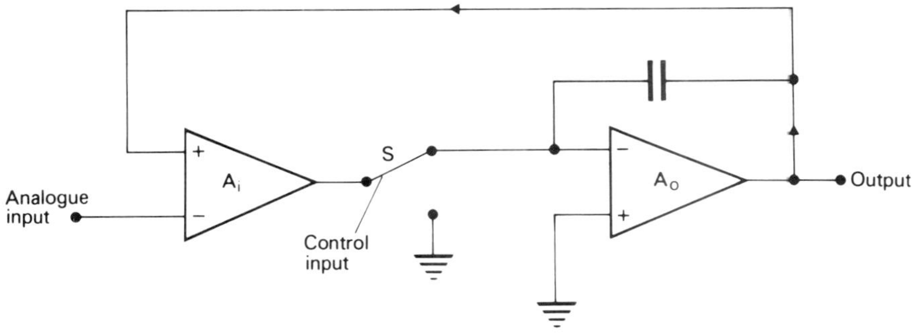

Fig. 5.15 Sample-hold closed loop configuration

5.6 A/D converter terminology

In practical applications the following terms are used to describe the performance and characteristics of A/D converters.

Resolution

The analogue signal is divided into 2^n discrete ranges for n-bit conversion. All analogue values within a given quantum are represented by the same digital code. There is, therefore, an inherent quantisation uncertainty of $\pm\frac{1}{2}$ LSB associated with the resolution, in addition to the actual conversion errors. Nominal resolution for an n-bit binary converter is 2^{-n}, often expressed in parts per million or simply as 'n bits'. Useful resolution is the smallest uniquely distinguishable bit under all operating conditions. For example, a twelve-bit converter may have a useful resolution of only ten bits over its temperature range.

Conversion time

This is the time required for a complete measurement by an A/D converter. In successive approximation converters, its value ranges from 0.8 μs to 400 μs, and popular twelve-bit general purpose A/D converters have a conversion time of about 25 μs.

Linearity and noise

Linearity determines the relative accuracy of converters. Once a converter is adjusted and calibrated, deviation from linearity gives rise to absolute errors. For an A/D converter, noise in the input signal, the input circuitry or the conversion device effectively increases the size of the quantisation band. For all except the integration type converters, peak noise must be considered carefully, especially in cases where the number of readings is small and the processing capacity is limited.

Output signals and codes

The output of A/D converters may be in pure binary, BCD or any other specified code. It may be parallel or serialised but the output levels corresponding to logic 0 and 1 define its compatibility with the interface circuitry. For example, in TTL compatible devices, logic 0 is less than +0.4 V and logic 1 is greater than +2.4 V.

'Convert command' and 'status'

The polarity and minimum width of the trigger pulse that initiates an A/D conversion is specified and must be observed. The completion of a conversion is indicated by a change in level from 1 to 0 at the status terminal.

5.7 Appropriate choice of converters for certain applications

The cost of converters is dictated by the performance requirements of their accuracy and speed of conversion. Accuracy involves resolution, linearity, noise, temperature dependence, etc., and speed of conversion involves the conversion time, bandwidth, settling time of the input circuitry, etc. In certain experimental situations, the converter specification can be relaxed and the cost of implementation reduced if suitable building blocks are added. Modest accuracy is substituted for extreme accuracy, at a considerably lower cost. For example, in data acquisition systems, if data compression is achieved by using logarithmic amplifiers prior to A/D conversion, then a wide dynamic range of input amplitudes can be handled with an inexpensive eight-bit converter. The technique is adequate except for applications in which extremely small errors are required at all points in the range. However, the same dynamic range may be covered with a high degree of accuracy by using an expensive 20-bit converter.

Similarly, a sample and hold device may be used to raise the upper frequency limit of a converter for constant amplitude signals that lie within the limits of its resolution. For example, a 10 bit, 10 μs, 10 V full-scale converter, in order to have an output error of $\pm\frac{1}{2}$LSB, must have the rate of change of its input less than $\pm\frac{1}{2}(10/2^{10}) \approx 5$ mV in 10 μs (500 V per sec). If the input is a constant amplitude sinusoidal wave $E_0 \sin 2\pi ft$, where the amplitude $E_0 = 5$ V, then the maximum rate of change of the signal is given by

$$\frac{d}{dt}[E_0 \sin 2\pi ft]_{t=0} = 2\pi f E_0$$

If the output error is to be $\pm\frac{1}{2}$LSB then the highest signal frequency f_{max} that can be applied to the converter is given by

$$2\pi f_{max} E_0 = 500$$

therefore

$$f_{max} = 15 \text{ Hz}$$

Using an S/H device, f_{max} may be improved by the ratio of the conversion time of the converter (10 μs) to the aperture time of the S/H device (20 ns). Assuming that the S/H device has an adequate bandwidth, an improvement (10 μs/20 ns) or 500:1 is obtained.

References

(1) Brook, D. B., *Data Conversion Handbook*. Hybrid Systems Corporation (1974)
(2) Schmid, H., *Electronic Analog/Digital Conversions*. Van Nostrand Reinhold (1970)
(3) De Sa, A., Optical trace analyser. *J. Sci. Instrum.* **43**, pp. 946–7 (December 1966)
(4) De Sa, A., Digital integration and recording of low frequency signals. *J. Sci. Instrum.*, **43**, p. 614 (August 1966)
(5) De Sa, A., A digital method of measurement of amplitude and phase at low frequencies. *J. Sci. Instrum.*, **42**, pp. 265–7 (April 1965)
(6) De Sa, A., and McCartan, D. G., Ratiometric measurements in optical absorption spectroscopy. *J. Phys. E. Sci. Instrum.*, **9**, pp. 725–7 (1976)
(7) Sheingold, D. H. (ed.), *Analog-Digital Conversion Handbook*. Analog Devices Inc. (1972)

Further reading

Connor, F. R., *Networks*. Edward Arnold (1972)

6

Feedback theory and its applications

6.1 Introduction

The history of the development of feedback theory to its present state can be found in several papers.[1] Its mathematical basis is well-defined and its practical applications are numerous, especially in the field of automatic control.

In the following sections, the general theory underlying the behaviour of linear feedback systems is outlined along with several applications in electronic instrumentation. Particular reference is made to methods of assessing the stability of such systems, and several commonly-used stabilisation techniques are listed. For a given application, the choice of system obviously depends on the nature of the feedback mechanism in question.

A brief review of the basic mathematical concepts used in feedback theory is included at the beginning of the chapter.

6.2 Basic mathematical concepts in feedback theory

The Fourier and Laplace transforms

A repetitive signal $f(t)$ having a complex waveform can be represented in terms of a Fourier series as

$$f(t) = \frac{a_0}{2} + \sum_{n=1}^{\infty} (a_n \cos n\omega_0 t + b_n \sin n\omega_0 t) \tag{6.1}$$

where $\omega_0 = 2\pi/T$ is the lowest angular frequency component and the repetition period is T seconds. The Fourier coefficients

$$a_n = \frac{2}{T}\int_{-T/2}^{+T/2} f(t) \cos n\omega_0 t \, \mathrm{d}t \tag{6.2}$$

$$b_n = \frac{2}{T}\int_{-T/2}^{T/2} f(t) \sin n\omega_0 t \, \mathrm{d}t \tag{6.3}$$

and

$$a_0 = \frac{2}{T}\int_{-T/2}^{T/2} f(t) \, \mathrm{d}t \tag{6.4}$$

Equation 6.4 represents the DC component or the average value of $f(t)$ over one period.

The series can be expressed in complex form as

$$f(t) = \sum_{n=-\infty}^{\infty} C_n \, \mathrm{e}^{\mathrm{j}n\omega_0 t} \tag{6.5}$$

$$\text{where} \quad C_n = \frac{1}{T}\int_{-T/2}^{T/2} f(t)\,\mathrm{e}^{-\mathrm{j}n\omega_0 t} \tag{6.6}$$

Substituting for C_n gives

$$f(t) = \sum_{n=-\infty}^{\infty}\left(\frac{1}{T}\int_{-T/2}^{T/2} f(x)\,\mathrm{e}^{-\mathrm{j}n\omega_0 x}\,\mathrm{d}x\right)\mathrm{e}^{\mathrm{j}n\omega_0 t}$$

where x is a dummy integration variable. This gives

$$f(t) = \sum_{n=-\infty}^{\infty}\left(\frac{1}{2\pi}\int_{-T/2}^{T/2} f(x)\,\mathrm{e}^{-\mathrm{j}n\omega_0 x}\,\mathrm{d}x\right)\omega_0\,\mathrm{e}^{\mathrm{j}n\omega_0 t} \tag{6.7}$$

A non-repetitive signal may be considered as a limiting case of a repetitive signal in which the period $T \to \infty$ so that ω_0 becomes small. Let $\omega_0 = \Delta\omega$. Then the frequency of any harmonic $n\omega_0$ must correspond to the general frequency variable which describes the continuous spectrum. In other words, $n \to \infty$ as $\omega_0 = \Delta\omega \to 0$ such that the product is finite, i.e.

$$n\omega_0 = n\Delta\omega \to \omega$$

Therefore

$$f(t) = \sum_{-\infty}^{\infty}\left(\frac{1}{2\pi}\int_{-T/2}^{T/2} f(x)\,\mathrm{e}^{-\mathrm{j}\omega x}\,\mathrm{d}x\right)\mathrm{e}^{\mathrm{j}\omega t}\Delta\omega \tag{6.8}$$

In the limit as $T \to \infty$, $\Delta\omega \to \mathrm{d}\omega$ and the summation becomes an integration over ω.

The non-repetitive function $f(t)$ can be represented by

$$f(t) = \frac{1}{2\pi}\int_{-\infty}^{\infty} F(\omega)\,\mathrm{e}^{\mathrm{j}\omega t}\,\mathrm{d}\omega \tag{6.9}$$

$$\text{where} \quad F(\omega) = \int_{-\infty}^{\infty} f(t)\,\mathrm{e}^{-\mathrm{j}\omega t}\,\mathrm{d}t \tag{6.10}$$

Equations 6.9 and 6.10 are called a **Fourier transform pair** because they transform a time function, $f(t)$, into a function of angular frequency, $F(\omega)$, and vice versa. However, equation 6.10 will not converge if $f(t)$ remains finite or increases indefinitely as $t \to \infty$. If $\mathrm{j}\omega$ is replaced by $(\sigma + \mathrm{j}\omega) = s$, then a new function is obtained, i.e.

$$F(\sigma + \mathrm{j}\omega) = \int_{-\infty}^{\infty} f(t)\,\mathrm{e}^{-\sigma t}\,\mathrm{e}^{-\mathrm{j}\omega t}\,\mathrm{d}t \tag{6.11}$$

This integral converges for a much wider range of functions $f(t)$, particularly when $\int_{-\infty}^{\infty} |f(t)|\mathrm{e}^{-\sigma t}\,\mathrm{d}t$ is finite.

Equation 6.11 expresses the time function as an infinite number of infinitesimally small functions, each of the form $v = A\mathrm{e}^{\sigma t}\mathrm{e}^{\mathrm{j}\omega t}$, Fig. 6.1(a) when σ is negative and Fig. 6.1(b) when σ is positive. For the particular case when $\sigma = 0$, the equation reduces to the Fourier integral of equation 6.10. Writing $s = (\sigma + \mathrm{j}\omega)$ in equation 6.11, the definition of a **dual-sided** or **bilateral** Laplace transform is obtained as

$$L[f(t)] = \int_{-\infty}^{\infty} f(t)\,\mathrm{e}^{-st}\,\mathrm{d}t = \overline{f}(s) \tag{6.12}$$

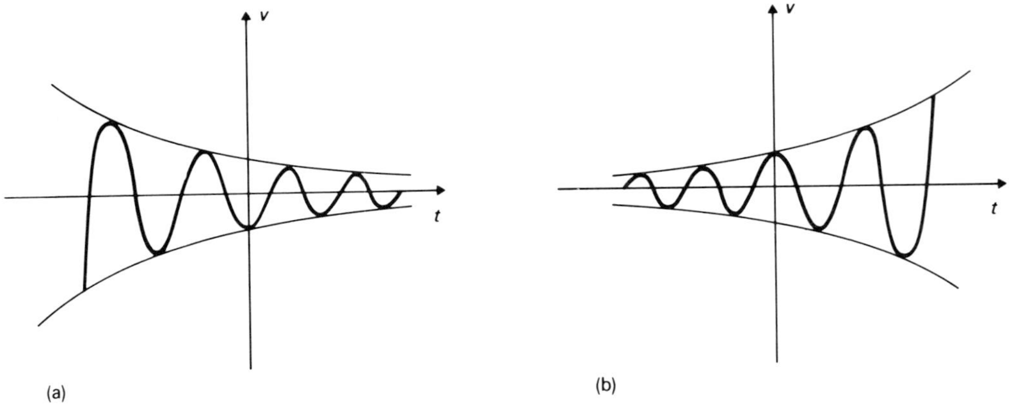

Fig. 6.1 (a) σ negative; (b) σ positive

If $f(t)$ is a known function of time t for values of $t > 0$ then the Laplace transform $\bar{f}(s)$ of $f(t)$ may be written as

$$\bar{f}(s) = \int_0^\infty e^{-st} f(t)\,dt \tag{6.13}$$

where s is a real positive number sufficiently large to make the integral of equation 6.13 convergent. The inverse transform is denoted by

$$\begin{aligned} L^{-1}[\bar{f}(s)] &= f(t) \\ &= \frac{1}{2\pi j}\int_{\sigma-j\infty}^{\sigma+j\infty} \bar{f}(s)\,e^{st}\,ds \end{aligned} \tag{6.14}$$

In practice it is not necessary to evaluate the integral of equation 6.14 because the inverse transforms can be found from tables. Details of the properties of the Laplace transform and its inverse, as well as a short table of Laplace transforms of frequently encountered time functions, are given in other texts.[2] The most important properties are listed below for ready reference.

Properties of Laplace transforms

Scaling
If

$$\bar{f}(s) = \int_0^\infty f(t)\,e^{-st}dt$$

and the scale in the time domain is changed by a factor a then

$$\bar{f}_1(s) = \int_0^\infty f(at)\,e^{-st}dt = \frac{1}{a}\bar{f}\left(\frac{s}{a}\right)$$

Time shifting
If $f(t)$ is delayed by τ then

$$\bar{f}_2(s) = \int_0^\infty f(t-\tau)e^{-st}dt = e^{-s\tau}\bar{f}(s)$$

Frequency shifting
If $f(t)$ is multiplied by $e^{-\alpha t}$ then

$$\bar{f}_3(s) = \int_0^\infty e^{-\alpha t}f(t)e^{-st}dt = \bar{f}(s+\alpha)$$

Time differentiation
The transform of the nth derivative of $f(t)$ is given by

$$\bar{f}_4(s) = s^n\bar{f}(s) - s^{n-1}f(0) - s^{n-2}\frac{df(0)}{dt} - \ldots - \frac{df^{n-1}(0)}{dt^{n-1}}$$

Time integration
If $f(t)$ is integrated between 0 and t, its transform becomes

$$\bar{f}_5(s) = \int_0^\infty \left(\int_0^t f(t)\,dt\right)e^{-st}dt = \frac{1}{s}\bar{f}(s)$$

Most dynamical or electrical systems are started, in a given way, at some instant which may be taken as the zero of time, i.e. $t = 0$. In such initial value problems, the Laplace transform technique is a convenient way of obtaining the solution of the differential equation describing the system, because negative values of time do not arise at all. This technique is illustrated in the following sections.

The unit impulse and unit step responses of a linear system

In order to make a mathematical study of the response of linear systems to arbitrary disturbances, it is useful to use a time function known as the unit impulse or **delta function** (δ-function) postulated by Dirac. It is an intense disturbance of duration ε seconds and an amplitude $1/\varepsilon$ where $\varepsilon \to 0$, as shown in Fig. 6.2.

The δ-function is defined by the following basic properties.

a) $$\int_{-\infty}^{\infty} \delta(t-t_1)\,dt = 1 \qquad (6.15)$$

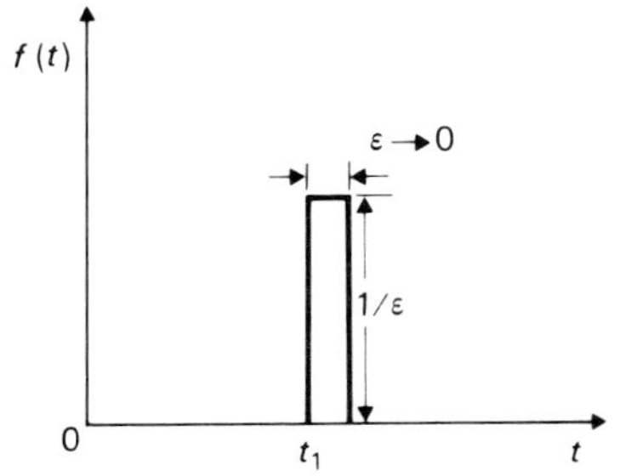

Fig. 6.2 Dirac delta function (δ)

This means that although the function has negligible duration, it nevertheless has a definite area or strength.

b) $$\int_{-\infty}^{\infty} \phi(t)\,\delta(t-t_1)\,\mathrm{d}t = \phi(t_1) \tag{6.16}$$

where $\phi(t)$ is an arbitrary function of time. This means that the product of an arbitrary function of time and a δ-function at t_1 only exists at t_1. It is an impulse function of strength $\phi(t_1)$.

c) $$L[\delta(t-t_1)] = \int_{0}^{\infty} \delta(t-t_1)\mathrm{e}^{-st}\,\mathrm{d}t = \mathrm{e}^{-st_1} \tag{6.17}$$

If the function occurs at the origin then $t_1 = 0$ so that

$$L[\delta(t)] = 1 \tag{6.18}$$

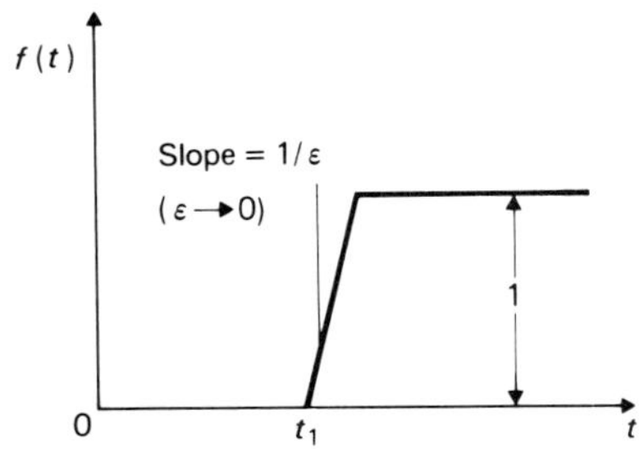

Fig. 6.3 Heaviside step function

Another useful concept is that of the **Heaviside step function** (Fig. 6.3) occurring at $t = t_1$ and defined as

$$\left.\begin{aligned} H(t-t_1) &= 0 \qquad \text{for } t < t_1 \\ H(t-t_1) &= 1 \qquad \text{for } t > t_1 \end{aligned}\right\} \tag{6.19}$$

It follows that

$$\frac{\mathrm{d}}{\mathrm{d}t}[H(t-t_1)] = \delta(t-t_1) \tag{6.20}$$

The δ-function is thus the result of differentiating a Heaviside step function. The Laplace transform of the unit step function is

$$L[H(t-t_1)] = \frac{1}{s}\mathrm{e}^{-st_1}$$

At the origin $t_1 = 0$, therefore

$$L[H(t)] = 1/s \tag{6.21}$$

The response of a system, initially at rest, to a unit impulse is the unit impulse response $y(t)$. The response to a step function is the unit step response $U(t)$, so that

$$U(t) = \int_{0}^{t} y(t)\mathrm{d}t \tag{6.22}$$

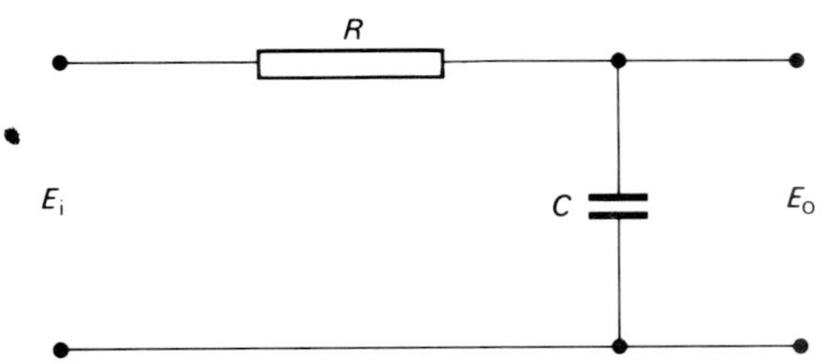

Fig. 6.4 Passive RC circuit

For example, consider the passive circuit in Fig. 6.4, which may be represented by the differential equation

$$E_i = E_o + RC\frac{d(E_o)}{dt} \tag{6.23}$$

where E_i is the input voltage. The output voltage E_o is given by q/C, where q is the charge on the condenser C; E_o is zero at $t = 0$. The impulse response can easily be determined by putting $E_i = \delta(t)$ and using the Laplace transforms to convert the differential equation 6.23 into an algebraic form, i.e.

$$s\bar{E}_o(s) + \alpha\bar{E}_o(s) = \alpha\,.\,1 \tag{6.24}$$

where $\alpha = 1/RC$ and $\bar{E}_o(s) = L[E_o(t)]$. Therefore,

$$\bar{E}_o(s) = \alpha/(s+\alpha) \tag{6.25}$$

Taking the inverse transform,

$$E_o(t) = L^{-1}[\bar{E}_o(s)] = \alpha e^{-\alpha t} = y(t) \tag{6.26}$$

which is the impulse response.

In practice, any pulse whose duration is short compared to the shortest time constant in a system can be approximated to a delta function whose strength is the product of its duration and amplitude. Any arbitrary disturbance, Fig. 6.5, can be represented by a large number of closely spaced pulses of short duration, the amplitude of each pulse being equal to the

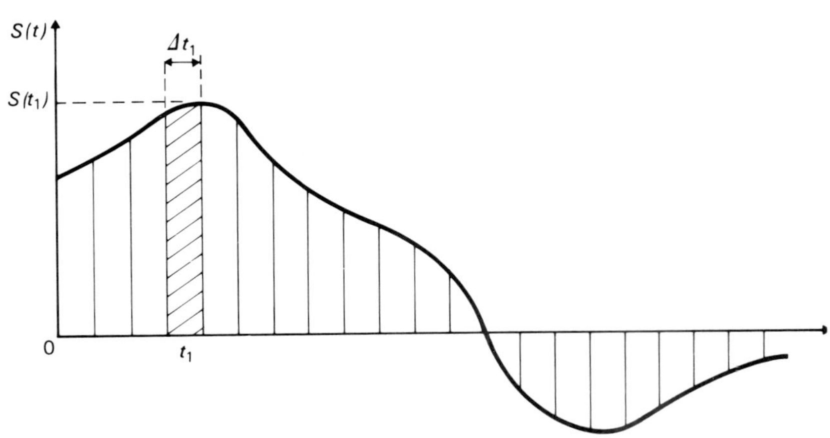

Fig. 6.5 Response of a linear system to an arbitrary disturbance

amplitude of the time function at the instant of its occurrence. The response of each individual pulse is determined at a subsequent time t and the response of all the pulses is summed. Because the principle of superposition applies, the sum of the responses is the response of the linear system to the arbitrary disturbance. At a time t the response to a delta function which occurred at a previous time t_1 will be $y(t-t_1)$. If the pulse at t_1 has a duration Δt_1 then it is equivalent to a delta function of strength $S(t_1)\Delta t_1$ and the response at time t is $S(t_1)\Delta t_1\, y(t-t_1)$ (see equation 6.16). The summed response of all the pulses contained in $S(t)$ up to time t is given by

$$R(t) = \sum_{t_1=0}^{t_1=t} S(t_1)\,y(t-t_1)\Delta t_1 \tag{6.27}$$

As Δt_1 becomes indefinitely small, the sum becomes an integral:

$$R(t) = \int_0^t S(t_1)\,y(t-t_1)\,\mathrm{d}t_1 \tag{6.28}$$

Changing the variables so that $(t-t_1)=\tau$ gives

$$R(t) = \int_0^t S(t-\tau)\,y(\tau)\,\mathrm{d}\tau \tag{6.29}$$

This is the **convolution integral**, and $S(t)$ and $y(t)$ are said to have convolved to form $R(t)$.

Transfer functions

If $y_1(t)$ and $y_2(t)$ are two functions of time then the product of their Laplace transforms is the Laplace transform of their convolution integral, i.e.

$$L[y_1(t)]L[y_2(t)] = L\left[\int_0^t y_1(t_1)y_2(t-t_1)\mathrm{d}t\right] \tag{6.30}$$

Putting $y_1(t)=S(t)$, the disturbance applied to a linear system, and $y_2(t)=y(t)$, the unit impulse response of the system, equation 6.30 gives

$$L[S(t)]\,L[y(t)] = L\left[\int_0^t S(t_1)\,y(t-t_1)\,\mathrm{d}t_1\right]$$

Hence, from equation 6.28,

$$L[S(t)]L[y(t)] = L[R(t)] \tag{6.31}$$

i.e. $$\overline{S}(s)\overline{y}(s) = \overline{R}(s)$$

so $$\overline{y}(s) = \overline{R}(s)/\,\overline{S}(s) \tag{6.32}$$

The Laplace transform of the unit impulse response of the system, $\overline{y}(s)$, is known as its **transfer function** because it relates its response to its stimulus. In other words, the transfer function of any device is the ratio of the Laplace transform of the output of the device to the Laplace transform of its input when the *system is initially at rest*. Sometimes the transfer function of a device is defined as the Laplace transform of the output of the device in response to a unit impulse input.

If the output of a device with a transfer function $\overline{y}_1(s)$ is connected to the input of another device with a transfer function $\overline{y}_2(s)$ then the overall transfer function, from equations 6.30

and 6.31, will be the product of the individual transfer functions, i.e.

$$\bar{y}_{12}(s) = \bar{y}_1(s)\,\bar{y}_2(s) \tag{6.33}$$

The transfer function of a network can be obtained if the differential equation relating its input and output is known. For example, for the network of Fig. 6.4 in the steady state, equation 6.23 can be written as

$$E_i = R\frac{dq}{dt} + \frac{q}{C} \quad \text{and} \quad E_o = \frac{q}{C}$$

These will transform to

$$\bar{E}_i(s) = \left(sR + \frac{1}{C}\right)\bar{q}(s) \quad \text{and} \quad \bar{E}_o(s) = \frac{1}{C}\bar{q}(s)$$

The transfer function is therefore given by

$$\frac{\bar{E}_o(s)}{\bar{E}_i(s)} = \frac{1}{(sRC+1)} = \frac{1}{(s\tau+1)}$$

This expresses the complex ratio of the output to input for *any type* of input waveform. In practice, the transfer functions of complicated networks may be found from tables.[3]

The *s*-plane

In any operational expression such as a transfer function, certain values of s make the expression zero and others make it tend to infinity. Such singular values of s are known as **zeros** and **poles** respectively. Thus $(s+a) = 0$ when $s = -a$, and $1/(s-b) \to \infty$ when $s = b$. The quantity $-a$ is therefore a zero of the expression $(s+a)$, while b is a pole of $1/(s-b)$. The Laplace operator s can be represented as a complex number $(\sigma + j\omega)$, and since poles and zeros can be real, imaginary or complex numbers they can be represented as points on an Argand diagram with σ and ω as coordinates. For example, if a closed loop transfer function is written as

$$\bar{y}_0(s) = \frac{A}{s^3 + 2s^2 + s + 2} \tag{6.34}$$

then to locate the poles of the function, it must be expressed as

$$\bar{y}_0(s) = \frac{A}{(s+6)(s-j)(s+j)}$$

The poles are $s = -2$, $s = j$ and $s = -j$. Their s-plane plot is shown in Fig. 6.6. If the operational expression has repeated roots then the corresponding singularities are known as nth-order poles or zeros.

The *y*-plane

In practical situations, when observing the steady-state behaviour of sinusoidally disturbed feedback systems, s is set to $j\omega$ and the resulting transfer function is known as the **frequency response function**. The modulus of the frequency response function is called the **gain modulus** or **gain**. For example, if $\bar{y}_1(j\omega)$ is the frequency response function of the forward

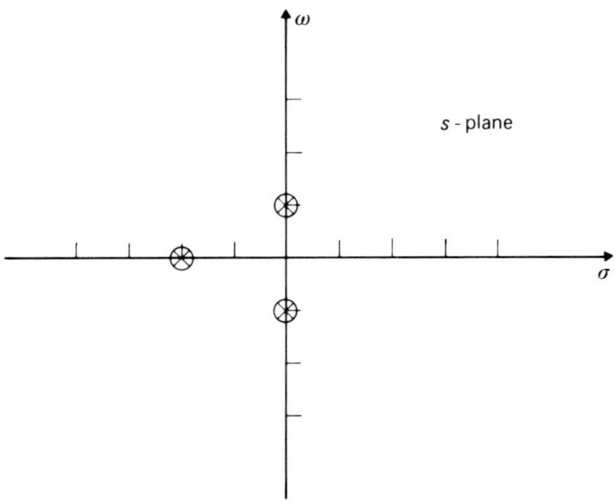

Fig. 6.6 s-plane plot with poles $s = -2$, $s = \mathrm{j}$ and $s = -\mathrm{j}$

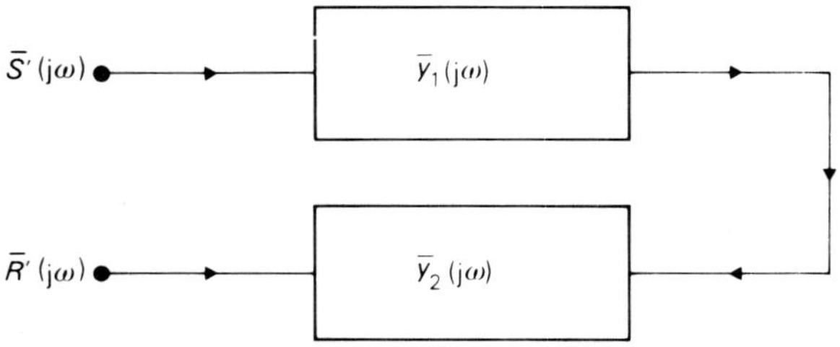

Fig. 6.7 An open loop frequency response function $y_L = y_1(\mathrm{j}\omega)\,y_2(\mathrm{j}\omega)$

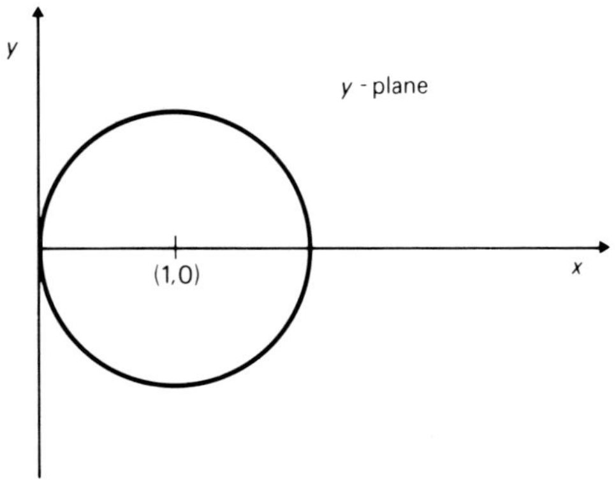

Fig. 6.8 y-plane plot of equation 6.37, the 'feedback cut-off boundary'

sequence and $\bar{y}_2(j\omega)$ is the frequency response function of the feedback sequence, then the quantity

$$\frac{\overline{R}'(j\omega)}{\overline{S}'(j\omega)} = \bar{y}_1(j\omega)\bar{y}_2(j\omega) = \bar{y}_L(j\omega)$$

is related to the close loop frequency response function by the expression

$$\bar{y}_0(j\omega) = \frac{\bar{y}_1(j\omega)}{1-\bar{y}_L(j\omega)} \tag{6.35}$$

(see p. 162). The open loop frequency response function can be expressed as a complex number $\bar{y}_L(j\omega) = x + jy$ where x and y are functions of the angular frequency ω. Therefore,

$$\frac{\bar{y}_0(j\omega)}{\bar{y}_1(j\omega)} = \frac{1}{1-x-jy} \tag{6.36}$$

or $$\left|\frac{\bar{y}_0}{\bar{y}_1}\right| = \frac{1}{(1-x)^2+y^2}$$

When the feedback is neutral $|\bar{y}_0| = |\bar{y}_1|$ so that

$$(1-x)^2 + y^2 = 1 \tag{6.37}$$

The y-plane plot, Fig. 6.8, is a circle with centre (1, 0) and radius unity. If at some frequency the vector representing $\bar{y}_L(j\omega)$ lies within this circle then positive feedback exists since $|\bar{y}_0| > |\bar{y}_1|$. The circle is called the **feedback cut-off boundary**.

When the vector $\bar{y}_L(j\omega)$ is plotted at varying angular frequencies, the tips of the resulting families of vectors form a locus called the **open loop frequency response locus** or the **Nyquist diagram**. For example, if the open loop transfer function of a feedback system is given by

$$\bar{y}_L(s) = \frac{K\beta}{(1+s\tau)^2} \tag{6.38}$$

then the open loop frequency response function will be

$$\begin{aligned}\bar{y}_L(j\omega) &= \frac{K\beta}{1-\omega^2\tau^2+2j\omega\tau}\\ &= \left(\frac{K\beta(1-\omega^2\tau^2)}{(1-\omega^2\tau^2)^2+4\omega^2\tau^2}\right) - j\left(\frac{2K\beta\omega\tau}{(1-\omega^2\tau^2)^2+4\omega^2\tau^2}\right)\end{aligned} \tag{6.39}$$

For $K\beta$ negative the locus crosses the feedback boundary at a frequency ω_c after which the feedback becomes positive, Fig. 6.9.

Contours in the *s*- and *y*-planes

Based on complex variable theory[4] for feedback applications, the representations of the s- and y-planes are made using the forms $\bar{y}(s) = x + jy$ and $s = \sigma + j\omega$. Hence for each value of s there is a unique value of $\bar{y}(s)$ so that any contour in the s-plane has its counterpart in the y-plane, although the transformation of the contour from one plane of coordinates to the other introduces distortion in the contour shape. If, for example, a contour in the s-plane traced in the anticlockwise direction does not encircle a zero, then the corresponding contour in the y-plane will exclude the y-plane origin. If, however, the contour in the s-plane encircles

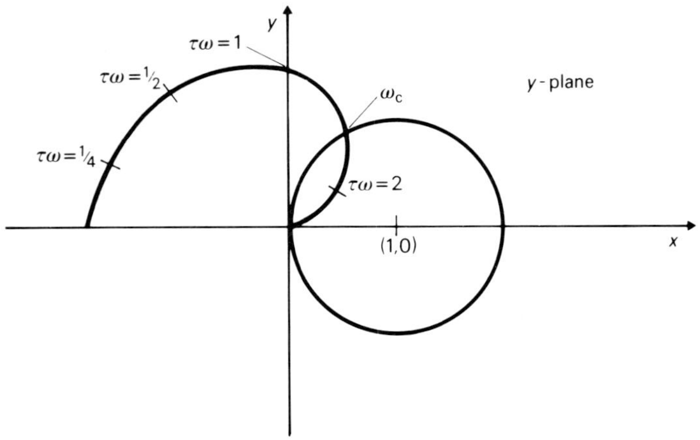

Fig. 6.9 y-plane plot of equation 6.39 and the feedback cut-off boundary

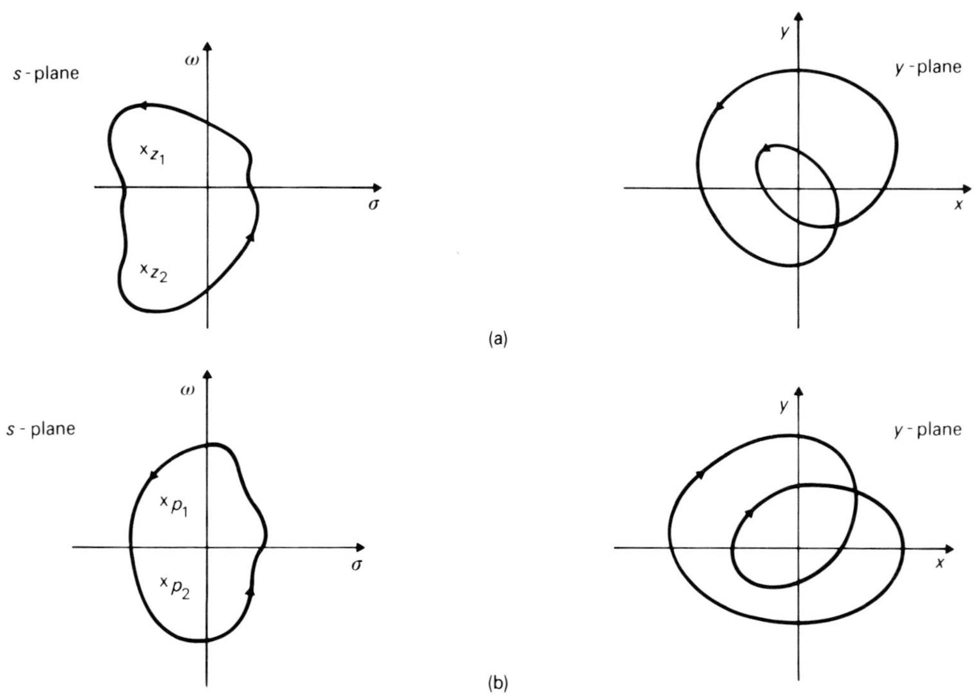

Fig. 6.10 (a) s-plane and y-plane contours with two zeros in the s-plane; (b) s-plane and y-plane contours with two poles in the s-plane

two zeros in the anticlockwise direction, Fig. 6.10(a), the corresponding contour in the y-plane will encircle the origin twice in the *anticlockwise* direction. Similarly, if the contour in the s-plane encircles two poles in the anticlockwise direction, Fig. 6.10(b), then the corresponding contour in the y-plane will encircle the origin twice in the *clockwise* direction.

When poles and zeros are present together, a transfer function can be written as

$$\bar{y}(s) = AB\,\frac{(s-z_1)(s-z_2)\ldots(s-z_n)}{(s-p_1)(s-p_2)\ldots(s-p_n)} \tag{6.40}$$

In general, if a contour in the s-plane encircles z zeros and p poles, Fig. 6.11, of the function $\bar{y}(s)$ in an anticlockwise direction, then the transformed contour in the y-plane encircles the origin $(z-p)$ times in the same direction.

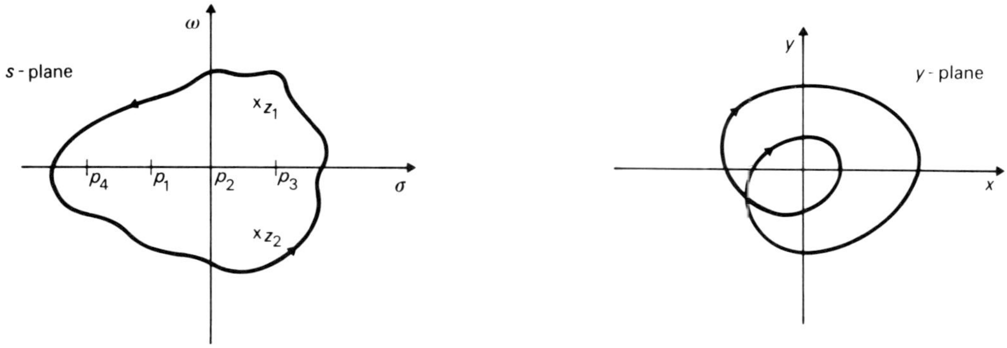

Fig. 6.11 s-plane and y-plane contours with z zeros and p poles

Lastly, in interpreting contours it must be noted that a point may be encircled once in a clockwise direction and once in the anticlockwise direction, thereby giving no net encirclement. In other words, a point may be *surrounded* but not *encircled* by a contour; point P in Fig. 6.12 is encircled once in the clockwise and once in the anticlockwise direction.

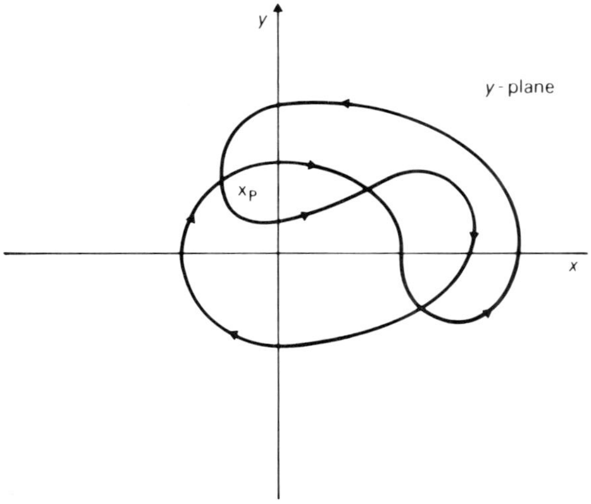

Fig. 6.12 The point P in the y-plane surrounded but not encircled

Minimum phase networks

A minimum phase network is a network whose transfer function has no zeros or poles appearing in the right-hand half of the complex s-plane. For example, consider the transfer function

$$\bar{y}(s)=\frac{K(1+sT_1)(1+sT_2)\ldots(1+sT_n)}{(1+sT_a)(1+sT_b)\ldots(1+sT_x)} \tag{6.41}$$

where K is a constant and $T_1,\ldots,T_n$ are the network time constants. This transfer function has zeros and poles in the left-hand half of the s-plane; the network is therefore a minimum phase structure. On the other hand, if the transfer function is

$$\bar{y}(s)=\frac{K(1-sT_1)(1+sT_2)\ldots(1+sT_n)}{(1+sT_a)(1+sT_b)\ldots(1+sT_x)} \tag{6.42}$$

then there is one zero in the right-hand half of the s-plane and the network is therefore a *non-minimum phase structure*. In general, the transfer functions of such structures have a factor either of type $(1-sT)$ or of type $(1-2sT+s^2T^2)$ in the numerator or denominator.

Minimum phase networks have the following properties.

1) A minimum phase network will have at every frequency a smaller phase shift than a non-minimum phase shift network having the same gain characteristics.
2) All minimum phase networks, irrespective of circuit configuration or component values, have the same mathematical relationship between gain versus frequency and phase versus frequency.

Hence if one of these characteristics is given for a network which is known to be a minimum phase type, the other characteristic is completely defined. The nature of this unique relationship is discussed in detail by Bode.[5, 6] Briefly, the phase shift at any frequency is proportional to a weighted average of the gain slope over the entire logarithmic frequency scale. The equation has the form

$$\theta(\omega_c)=\frac{1}{\pi}\int_{-\infty}^{\infty}\frac{\mathrm{d}A}{\mathrm{d}u}\ln\coth\frac{|u|}{2}\,\mathrm{d}u \tag{6.43}$$

where θ is the angle of lag at a frequency ω_c; $u=\ln(\omega/\omega_c)$; ω_c is a constant; $A(\omega)=\ln|\bar{y}(j\omega)|$. The quantity $\bar{y}(j\omega)$ is the frequency response function of the network. The lag at a particular frequency ω_c is proportional to the integral over the whole logarithmic frequency range from $u=-\infty$ to $u=+\infty$ of the product of $\mathrm{d}A/\mathrm{d}u$ and a weighting factor $\ln\coth(|u|/2)$. The term $\mathrm{d}A/\mathrm{d}u$ is the slope of the gain characteristic plotted on a logarithmic scale. The weighting factor determines the relative importance of the derivative term $\mathrm{d}A/\mathrm{d}u$ in the various parts of the frequency spectrum and is large in the vicinity of ω/ω_c.

A wide variety of circuit configurations satisfy the definition of a minimum phase network. Some satisfy the definition irrespective of the component values, whereas others prove to be minimum phase networks only if certain limitations are imposed on the components. A detailed discussion of various circuits that satisfy the minimum phase definition is given elsewhere.[7]

Second-order control systems

Many control systems can be approximately described by a general second-order differential equation of the type

$$\frac{d^2y}{dt^2} + 2\xi\omega_n\frac{dy}{dt} + \omega_n{}^2 y = \omega_n{}^2 x \tag{6.44}$$

The positive coefficient ω_n is the **undamped natural frequency** and the coefficient ξ is the **damping ratio** of the system.

When the initial conditions are zero, i.e. $\bar{y}(s) = L[y(t)]$ and $\bar{x}(s) = L[x(t)]$, then the transfer function is given by

$$\frac{\bar{y}(s)}{\bar{x}(s)} = \frac{\omega_n{}^2}{s^2 + 2\xi\omega_n s + \omega_n{}^2} \tag{6.45}$$

The poles of this function will be

$$s = -\xi\omega_n \pm \omega_n\sqrt{\xi^2 - 1} \tag{6.46}$$

If $\xi > 1$, both poles are negative and real and if $\xi = 1$, both poles are equal and negative. If $0 < \xi < 1$, the poles are complex conjugates with negative real parts, i.e.

$$s = -\xi\omega_n \pm j\omega_n\sqrt{1 - \xi^2}$$

or $$s = -\alpha \pm j\omega_d$$

where $1/\alpha = 1/\xi\omega_n$ is called the **time constant** of the system and $\omega_d = \omega_n\sqrt{1-\xi^2}$ is called the **damped natural frequency** of the system.

6.3 Flow diagrams and linear feedback systems

The overall transfer function of elements in cascade is the product of the transfer functions of the individual elements, and feedback systems can therefore be conveniently represented by a series of interconnected boxes, Fig. 6.13. Each box can then contain the transfer function of an element; the direction of information flow is indicated by arrows on interconnecting lines. The output variable is obtained by multiplying the input by the transfer function in the box. Two or more lines, each corresponding to a variable, may be joined and eventually connected to the input of a box to be operated upon by the transfer function in it. In this way, once the

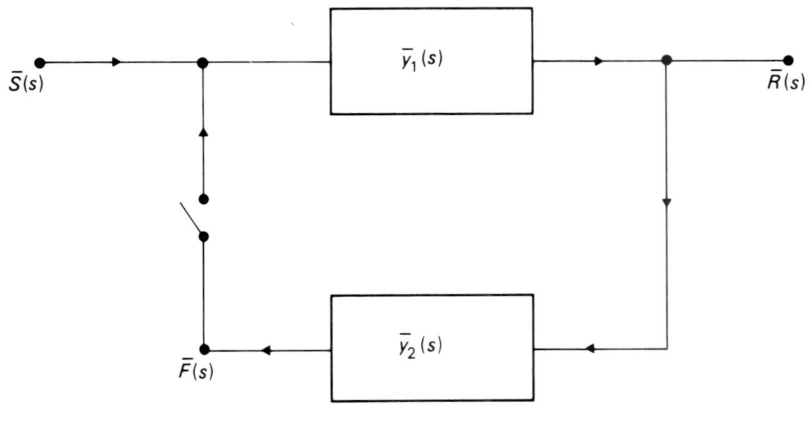

Fig. 6.13 A single-loop feedback system

mathematical picture or **flow diagram** of the feedback system is constructed, the physical nature of the components is of no importance to the analysis.

In Fig. 6.13, let $\bar{y}_1(s)$ be the transfer function of the forward path and $\bar{y}_2(s)$ the transfer function of the feedback path. When the switch T is in OFF position, the response $R(t)$ to a stimulus $S(t)$ can be expressed as

$$\bar{R}_0(s) = \bar{y}_1(s)\bar{S}(s) \tag{6.47}$$

The response is given by

$$\bar{F}(s) = \bar{y}_2(s)\bar{R}_0(s) \tag{6.48}$$

From equations 6.47 and 6.48, the open loop transfer function is given by

$$\frac{\bar{F}(s)}{\bar{S}(s)} = \bar{y}_1(s)\,\bar{y}_2(s) = \bar{y}_L(s) \tag{6.49}$$

When the switch T is in ON position, the feedback loop is closed, and

$$\bar{R}_c(s) = \bar{y}_1(s)\,[\bar{S}(s) + \bar{F}(s)]$$

Therefore

$$\bar{R}_c(s) = \bar{y}_1(s)\,[\bar{S}(s) + \bar{y}_2(s)\bar{R}_c(s)]$$

and

$$\frac{\bar{R}_c(s)}{\bar{S}(s)} = \frac{\bar{y}_1(s)}{1 - \bar{y}_1(s)\bar{y}_2(s)} = \bar{y}_0(s) \tag{6.50}$$

which is the closed loop transfer function.

Combining equations 6.49 and 6.50 gives

$$\bar{y}_0(s) = \frac{\bar{y}_1(s)}{1 - \bar{y}_L(s)} \tag{6.51}$$

which is an equation fundamental to single-loop feedback systems because it relates their open and closed loop transfer functions, $\bar{y}_L(s)$ and $\bar{y}_0(s)$ respectively. At this stage, it is worthwhile noting the following points about equation 6.51 and Fig. 6.13.

a) For a given frequency range, if $|\bar{y}_0(s)| < |\bar{y}_1(s)|$ then $\bar{y}_2(s)$ is said to provide **negative feedback**, i.e. the closed loop gain will be less than the gain of the forward path.
b) If, however, $|\bar{y}_0(s)| > |\bar{y}_1(s)|$ then $\bar{y}_2(s)$ is said to provide **positive feedback**, i.e. the closed loop gain will be greater than the gain of the forward path.
c) A given system may have negative feedback in one frequency range and positive feedback in another frequency range.
d) Stable individual elements can give rise to an unstable feedback loop.

6.4 Feedback and its consequences

In any physical system of the type outlined in Section 6.3, feedback in general gives rise to the following consequences.

Speed of response

Consider a single exponential lag in the forward path of the system in Fig. 6.13. Let the transfer functions be represented by $\bar{y}_1(s) = K/(s\tau + 1)$ and $\bar{y}_2(s) = \beta$ where K and β are frequency independent. The quantity K is positive but β may be positive or negative and τ is the time constant of the lag. The open loop transfer function, equation 6.49, will be

$$\bar{y}_L(s) = \frac{K\beta}{s\tau + 1}$$

and the closed loop transfer function, equation 6.50, will be

$$\bar{y}_0(s) = \frac{K}{(1 + s\tau - K\beta)} = \frac{K}{\tau}\left(\frac{1}{s + \alpha}\right)$$

where $\alpha = (1 - K\beta)/\tau$). The impulse response of the closed loop system is the inverse transform of $\bar{y}_0(s)$, i.e.

$$y_0(t) = (K/\tau)e^{-\alpha t} \tag{6.52}$$

The unit step response is given by

$$U(t) = \int_0^t y_0(t)dt$$

$$= \frac{K}{\tau}\int_0^t e^{-\alpha t}dt = \frac{K}{\alpha\tau}(1 - e^{-\alpha t})$$

The steady-state closed loop gain is given by

$$G = \frac{K}{\alpha\tau} = \left(\frac{K}{1 - K\beta}\right) \tag{6.53}$$

The time constant is given by

$$t_f = \frac{1}{\alpha} = \frac{\tau}{1 - K\beta} \tag{6.54}$$

therefore

$$t_f/\tau = G/K \tag{6.55}$$

In the case of negative feedback, β will be negative and the closed loop system will have a decreased gain but a faster response. Similarly, in the case of positive feedback, β will be positive and the closed loop system will have an increased gain but a slower response.

The speed of response of a more complex feedback system must always be compatible with its overall stability (see Section 6.5, p. 167), and the step response technique is a quick way of assessing both in most experimental situations. Figures 6.14(a) and (b) show typical step responses of a second-order system photographed from a double-beam oscilloscope, the lower trace being a fixed-frequency calibration signal.

Effect on the sensitivity to variations in the parameters of a given system

The main parameters in question are those involved in the transfer functions of (i) the

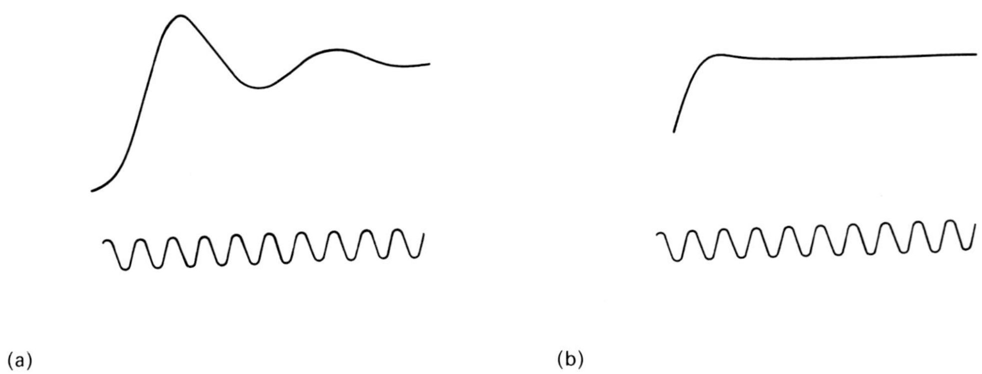

Fig. 6.14 (a) Step response of an underdamped feedback system (the lower trace is a 200 Hz calibration signal); (b) step response of the same system when critically damped

forward path, and (ii) the feedback path. For example, in the simplest but physically unrealisable case, when the forward and feedback paths are frequency-independent, the transfer functions $\bar{y}_1(s)$ and $\bar{y}_2(s)$ of equation 6.51 degenerate into pure numbers, say $\bar{y}_1(s) = K$ and $\bar{y}_2(s) = \beta$, and $\bar{y}_0(s) = G$ so that equation 6.51 can be written as

$$G = \frac{K}{1 - K\beta} \tag{6.56}$$

Differentiating equation 6.56 with respect to K and rearranging gives

$$\frac{\mathrm{d}G}{G} = \frac{\mathrm{d}K}{K}\left(\frac{1}{1 - K\beta}\right) \tag{6.57}$$

When β is increasingly negative, G, the closed loop transfer function, becomes increasingly insensitive to changes in K, the transfer function of the forward path. Similarly, differentiating equation 6.56 with respect to β and rearranging gives

$$\frac{\mathrm{d}G}{G} = \frac{\mathrm{d}\beta}{\beta}\frac{K}{(1/\beta) - K}$$

When $K \gg 1/\beta$,

$$\frac{\mathrm{d}G}{G} = -\left(\frac{\mathrm{d}\beta}{\beta}\right) \tag{6.58}$$

In other words, the sensitivity of the closed loop transfer function of a system with negative feedback and a high open loop gain is governed mainly by the transfer function of the feedback sequence.

In the general case, when all the transfer functions are frequency dependent, the description of the effect of feedback must consider the percentage variations of both amplitude and phase. For instance, it can be shown[8] that in amplifiers, over a narrow frequency range, it is possible to sacrifice phase stability for gain stability and vice versa.

Linearisation and distortion

Negative feedback is often used to reduce the effect of nonlinear elements in the forward path and thereby to linearise the overall performance of a system. The feedback path is designed to have a negligibly small deviation from linearity, and the open loop gain used has as high a value as is compatible with the stability conditions.

One of the ways of investigating the effect of nonlinearities in a system is to inject a pure sinusoidal waveform at the input and to examine the distortion of the output waveform caused by the generation of harmonics. For example, if the input to a nonlinear open loop system with a frequency response function $y_1(\mathrm{j}\omega)$ is given by

$$x_{\mathrm{i}}(t) = a_1 \mathrm{e}^{\mathrm{j}\omega_1 t} \tag{6.59}$$

where ω_1 is the angular frequency of the input, then the output can be represented by

$$\begin{aligned} x_{\mathrm{o}}(t) &= (a_1 \mathrm{e}^{\mathrm{j}\omega_1 t}) y_1(\mathrm{j}\omega_1) + D_{\mathrm{o}1} a_1 \mathrm{e}^{\mathrm{j}\omega_1 t} y_1(\mathrm{j}\omega_1) \\ &\quad + D_{\mathrm{o}2} a_1 \, \mathrm{e}^{2\mathrm{j}\omega_1 t} y_1(\mathrm{j}\omega_1) + D_{\mathrm{o}3} a_1 \mathrm{e}^{3\mathrm{j}\omega_1 t} y_1(\mathrm{j}\omega_1) + \;.\;\;. \\ &\quad + D_{\mathrm{o}n} \times D_{\mathrm{o}n} a_1 \mathrm{e}^{n\mathrm{j}\omega t} y_1(\mathrm{j}\omega_1) \\ &= a_1 y_1(\mathrm{j}\omega_1)\,(\mathrm{e}^{\mathrm{j}\omega_1 t} + D_{\mathrm{o}1} \mathrm{e}^{\mathrm{j}\omega_1 t} + D_{\mathrm{o}2} \mathrm{e}^{2\mathrm{j}\omega_1 t} + \ldots \\ &\quad + D_{\mathrm{o}3} \mathrm{e}^{3\mathrm{j}\omega_1 t} \ldots D_{\mathrm{o}n} \mathrm{e}^{n\mathrm{j}\omega t}) \ldots \end{aligned} \tag{6.60}$$

where $D_{\mathrm{o}1}, D_{\mathrm{o}2}, \ldots, D_{\mathrm{o}n}$ are the distortion coefficients and are functions of a_1 and ω_1. If negative feedback is applied via a path having a frequency response function $y_2(\mathrm{j}\omega)$ and negligible nonlinearities, then its effect on the distortion produced by the overall system may be expressed approximately in terms of the distortion coefficients as

$$x_{\mathrm{oc}}(t) = a_1 y_1(\mathrm{j}\omega_1) D_{\mathrm{o}n} \mathrm{e}^{n\mathrm{j}\omega t} + y_1(n\mathrm{j}\omega_1) y_2(n\mathrm{j}\omega_1) x_{\mathrm{oc}}(t) \tag{6.61}$$

in which it is assumed that (i) the sinusoidal input to the system is increased after introducing negative feedback, so as to keep the first term of equation 6.60 unaltered and (ii) the nonlinearity of y_1 itself in the second term of equation 6.61 is ignored. This gives

$$x_{\mathrm{oc}}(t) = \frac{a_1 y_1(\mathrm{j}\omega_1) D_{\mathrm{o}n} \mathrm{e}^{n\mathrm{j}\omega t}}{1 - y_1(n\mathrm{j}\omega_1) y_2(n\mathrm{j}\omega_1)} \tag{6.62}$$

Comparing with the nth term of the open loop output equation 5.60, it is obvious that the effect of negative feedback is to reduce the harmonic content by a factor determined by the feedback ratio calculated at the frequency of the harmonic in question.

In general, for a given output level, the effect of negative feedback is to reduce the distortion components in the region of any frequency ω by a factor $1/[1 - y_1(\omega) y_2(\omega)]$.

Input impedance, output impedance and noise

The effect of negative feedback on the input and output impedances of a system depends on the mode in which the feedback is applied. There are four basic modes shown in Fig. 6.15, the most general feedback cricuit being a linear combination of these types of circuits.

For example, in the case of amplifiers, if R_{i} represents the input resistance, R_{o} the output resistance, A_{v} the open loop voltage gain and A_{i} the open loop circuit gain, Table 6.1 shows

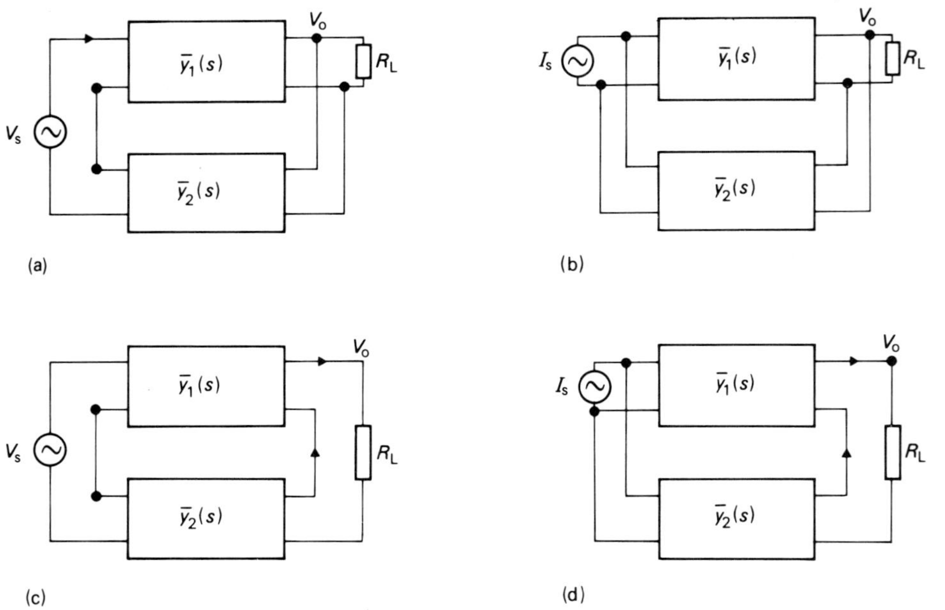

Fig. 6.15 (a) Feedback network connected in parallel with the output and in series with the input; (b) feedback network connected in parallel with the output and in parallel with the input; (c) feedback network connected in series with the output and in series with the input; (d) feedback network connected in series with the output and in parallel with the input

Table 6.1 Variation of parameters in negative feedback circuits

Feedback network connection	R_i	R_o	A_v	A_i
In *parallel* with the output and in *series* with the input (see Fig. 6.3(a))	increase	decrease	decrease	no change
In *parallel* with the output and *parallel* with the input (see Fig. 6.3(b))	decrease	decrease	no change	decrease
In *series* with the output and in *series* with the input (see Fig. 6.3(c))	increase	increase	decrease	no change
In *series* with the output and *parallel* with the input (see Fig. 6.4(d))	decrease	increase	no change	decrease

the variations in these parameters when negative feedback is used in the four different configurations of Fig. 6.15.

As far as noise generated within a system is concerned, benefit from negative feedback is obtained only if additional stages of amplification are not used to make up for the loss due to negative feedback. Thus, noise generated in preamplifiers is seldom combatted by negative feedback whereas the hum introduced by the power supply may be decreased appreciably. The changes in the noise figure (see Chapter 9) are governed by the nature of the noise generating mechanism within the system as well as the feedback configuration used. A more generalised discussion of noise in feedback circuits is given by van der Ziel.[9]

It is interesting to note, in general, the effect on a feedback loop of random fluctuations entering a system of the type described in Fig. 6.13, in the forward or feedback path. For example, in Fig. 6.16, if a random fluctuation $N(t)$ enters a feedback loop in the forward path then the response to a stimulus $S(t)$ will be

$$\{[\overline{S}(s)+\overline{R}(s)\overline{y}_3(s)]\,\overline{y}_1(s)+\overline{N}(s)\}\,\overline{y}_2(s)=\overline{R}(s) \tag{6.63}$$

$$\overline{R}(s)[1-\overline{y}_1(s)\,\overline{y}_2(s)\,\overline{y}_3(s)]=\overline{S}(s)\overline{y}_1(s)\,\overline{y}_2(s)+\overline{N}(s)\,\overline{y}_2(s)$$

$$\overline{R}(s)=\frac{\overline{S}(s)\overline{y}_1(s)\,\overline{y}_2(s)}{1-\overline{y}_1(s)\,\overline{y}_2(s)\,\overline{y}_3(s)}+\frac{\overline{N}(s)\,\overline{y}_2(s)}{1-\overline{y}_1(s)\,\overline{y}_2(s)\,\overline{y}_3(s)} \tag{6.64}$$

The first term in equation 6.64 is the effect of the stimulus and the second term is that of the noise. A closer look at Fig. 6.16 and equation 6.64 leads to the following observations.

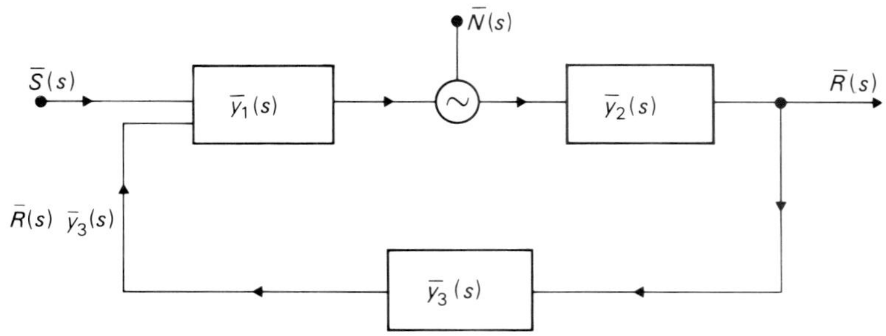

Fig. 6.16 Random fluctuations $N(t)$ entering a feedback system in the forward path

The effect of the fluctuations is minimal when they occur towards the output point, and is large when they occur towards the input point.

Equation 6.64 implies that a large amount of negative feedback would make the system insensitive to imposed fluctuations.

In order to combat random fluctuations in the environment, the system must have negative feedback characteristics for a range of frequencies at least as wide as the spectrum of the expected fluctuations.

Lastly, if the unwanted fluctuations enter the system via the input terminal itself, then they can be combatted only if the characteristics of the wanted stimulus and unwanted noise are distinctly defined, by selecting the parameters of the feedback loop.

6.5 Instability of linear feedback systems

Types of instabilities

Instabilities in linear systems resulting from feedback loops that are introduced deliberately, or loops due to inevitable internal and stray coupling between components, manifest themselves in three distinct forms when a small disturbance is applied to the input:

a) the output is oscillating with ever-increasing amplitude which is eventually limited by saturation, Fig. 6.17(a);

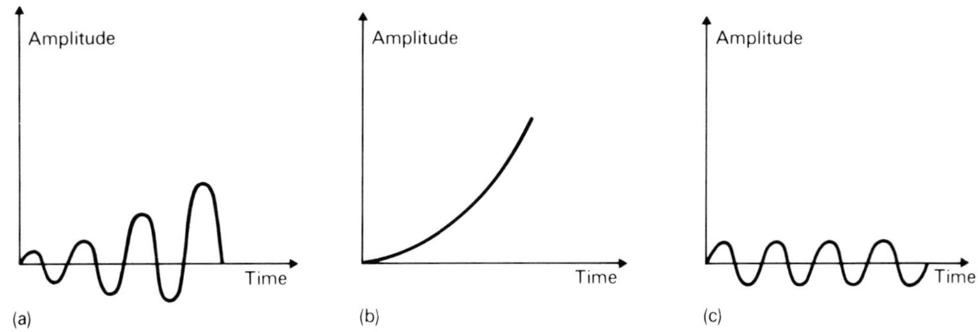

Fig. 6.17 (a) Output oscillating with ever-increasing amplitude; (b) output increasing undirectionally; (c) output oscillating with amplitude defined by the circuit parameters

b) the output increases undirectionally until it is limited by saturation, Fig. 6.17(b);
c) the output is oscillating with its amplitude defined by the circuit parameters, Fig. 6.17(c).

Quantitatively, these instabilities can be described in terms of the impulse response of the close loop transfer function as follows.

Consider a closed loop transfer function of the type

$$\bar{y}_0(s) = \frac{1}{s^2 - 2\alpha s + (\alpha^2 + \omega^2)} \tag{6.65}$$

Expressing in partial fractions and obtaining the inverse transform which is the impulse response, we get

$$\bar{y}_0(s) = \frac{A}{s-(\alpha+j\omega)} + \frac{B}{s-(\alpha-j\omega)} \tag{6.66}$$

therefore

$$\bar{y}_0(t) = Ae^{\alpha t}e^{j\omega t} + Be^{\alpha t}e^{-j\omega t} \tag{6.67}$$

This is an oscillatory response of ever-increasing amplitude, Fig. 6.17(a), obviously due to the positive real indices in the exponentials. In general, if the closed loop transfer function contains factors of the type $1/(s-\alpha)$, where α is positive, then the impulse response will be of the form $e^{\alpha t}$, a undirectionally increasing output like Fig. 6.17(b). In other words, the necessary condition for stability is that the denominator of the closed loop transfer function must not contain positive real roots or complex roots with positive real parts. This statement is the basis of several methods of predicting instabilities in linear feedback systems (see Section 6.5, p. 169).

Lastly, a system for which the denominator of the closed loop transfer function has some roots with their real parts equal to zero, but none with positive real parts, is said to be **marginally stable**. The impulse response will not, in general, decay to zero, Fig. 6.17(c), but in practice it might do so because of changes in the circuit characteristics with time.

Methods of predicting instabilities in linear feedback systems and their relative merits

Routh's method

If the closed loop transfer function of a linear feedback system is expressed as

$$\bar{y}_0(s) = \frac{1}{(a_0 s^n + a_1 s^{n-1} + a_2 s^{n-2} + \ldots + a_n)} \tag{6.68}$$

where a_0 is positive and all the coefficients are real, then the system is stable only if the equation

$$a_0 s^n + a_1 s^{n-1} + a_2 s^{n-2} + \ldots + a_n = 0 \tag{6.69}$$

has no positive real parts. Routh's method indicates the presence of such roots by inspection of the coefficients a_0 to a_n. The coefficients are written down in two rows:

Row 1 → $a_0 \quad a_2 \quad a_4 \quad a_6 \quad a_8$ → (even)

Row 2 → $a_1 \quad a_3 \quad a_5 \quad a_7 \quad a_9$ → (odd)

The first column is cross-multiplied with each succeeding column in turn so as to form $n-1$ rows, i.e.

Row 3 → $(a_1 a_2 - a_0 a_3) = b_1 \quad (a_1 a_4 - a_0 a_5) = b_2 \quad (a_1 a_6 - a_0 a_7) = b_3 \quad (a_1 a_8 - a_0 a_9) = b_4$

Row 4 → $(b_1 a_3 - a_1 b_2) = c_1 \quad (b_1 a_5 - a_1 b_3) = c_2 \quad (b_1 a_7 - a_1 b_4) = c_3$

Routh's criterion states that the equation contains no roots with positive real parts if, and only if, all the terms in the first column, i.e. a_1, b_1, c_1, . . . , etc. are positive.

The method has three main disadvantages:

1) each coefficient of equation 6.69 must be known – a difficult, tedious or impossible task in most experimental situations;
2) the method does not indicate how to stabilise an unstable system;
3) in higher order systems, the array of rows and columns is somewhat cumbersome to formulate.

The root locus method

Routh's stability criterion can be restated in terms of poles and zeros in the s-plane as follows. For stable response there must be no poles of the closed loop transfer function $\bar{y}_0(s)$, i.e. no zeros of $[1 - \bar{y}_L(s)]$ in the right-hand half of the s-plane. The locus obtained for the values of s satisfying the characteristic equation as a system parameter is varied, and is called the **root locus diagram**. The closed loop transient response and parameters such as the open loop gain, the damping ratio and the undamped natural frequency can be found from it. For example, the second-order feedback system of Fig. 6.18 has a closed loop transfer function givn by

$$\bar{y}_0(s) = \frac{k}{s^2 + 4s + k} \tag{6.70}$$

The open loop transfer function is

$$\bar{y}_L(s) = \frac{k}{s(s+4)} \tag{6.71}$$

The poles of $\bar{y}_L(s)$ are $s_1 = 0$, $s_2 = -4$. The poles of $\bar{y}_0(s)$ are the roots of the equation

$$s^2 + 4s + k = 0 \tag{6.72}$$

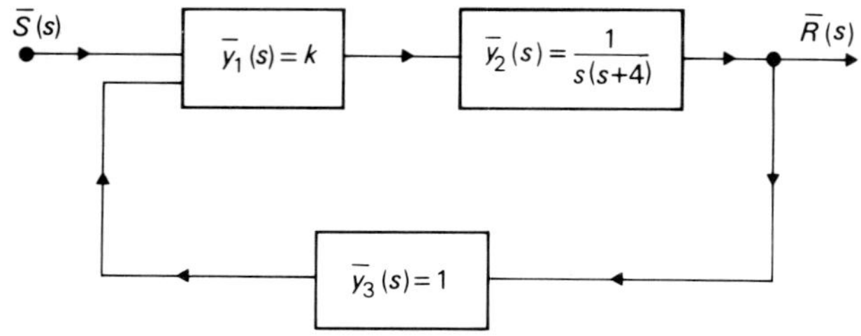

Fig. 6.18 A second-order feedback system

given by

$$s = -2 \pm \sqrt{2^2 - k} \tag{6.73}$$

The poles of $\bar{y}_L(s)$ are plotted in the complex s-plane of Fig. 6.19 along with the locus of s from equation 6.73 for $0 < k < 4$. Thus at $k = 4$, $s = -2$, the two loci coincide and further increases in k will result in complex conjugate roots such as

$$s = -2 \pm j\sqrt{k - 4} \tag{6.74}$$

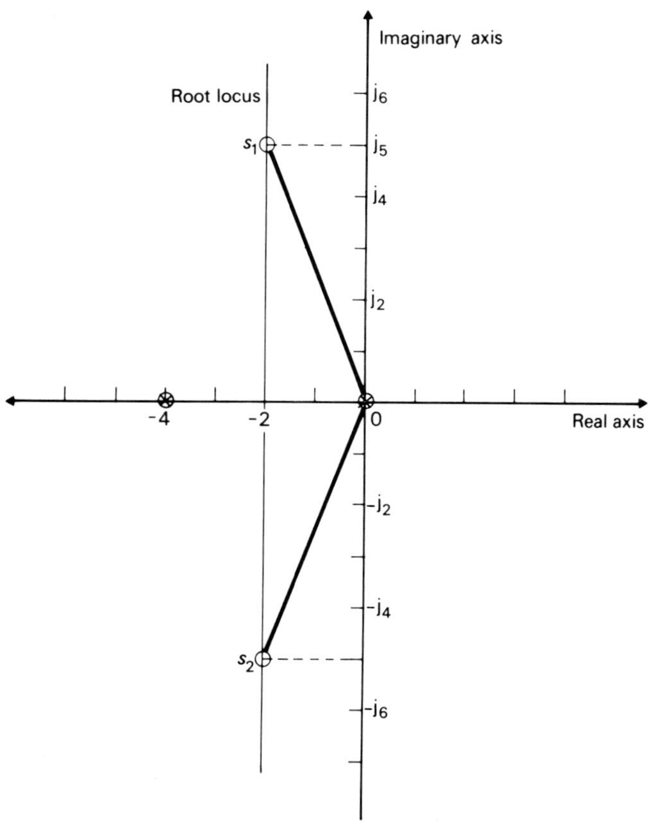

Fig. 6.19 Root locus of Fig. 6.18

At $s = -2$ the root locus breaks away from the negative real axis. Arbitrarily selecting a root $s_1(-2, j5)$ in Fig. 6.19, the value of k is given by $|s_1 - 0| \times |s_1 - (-4)|$ or $|-2+5j| \times |+2+5j|$, i.e. $k = 2^2 + 5^2 = 29$.

The undamped natural frequency ω_n is given by $|s_1|$, i.e.

$$\omega_n = \sqrt{25+4} = 5.4 \text{ rad s}^{-1}$$

As in Routh's method, the transfer function of the system must be known. It is often difficult to find the roots of higher order equations, and the analysis is performed graphically according to certain rules.[2]

Nyquist's method

This is a graphical method, originally devised for predicting the closed loop behaviour of feedback amplifiers from their open loop frequency response function[10].

In particular, in cases where the closed loop frequency response function $\bar{y}_0(j\omega)$ is related to the open loop frequency response function $\bar{y}_L(j\omega)$ by equation 6.51, the Nyquist criterion for stability may be stated as follows: a feedback system is stable if the locus of the open loop frequency response function $y_L(j\omega)$ in the y-plane does not encircle the point (1, 0) when ω is varied over all positive and negative values. The open loop frequency response of a system may be obtained experimentally by measuring the amplitudes of the input and output and the phase difference between them, at various frequencies. The Nyquist method provides a means of predicting whether a system is going to be stable when its feedback loop is closed, by merely examining the y-plane locus of its open loop frequency response function. Fig. 6.20(a) shows the y-plane locus (Nyquist diagram) of an unstable system, Fig. 6.20(b) that of a stable one and Fig. 6.20(c) that of a conditionally stable one.

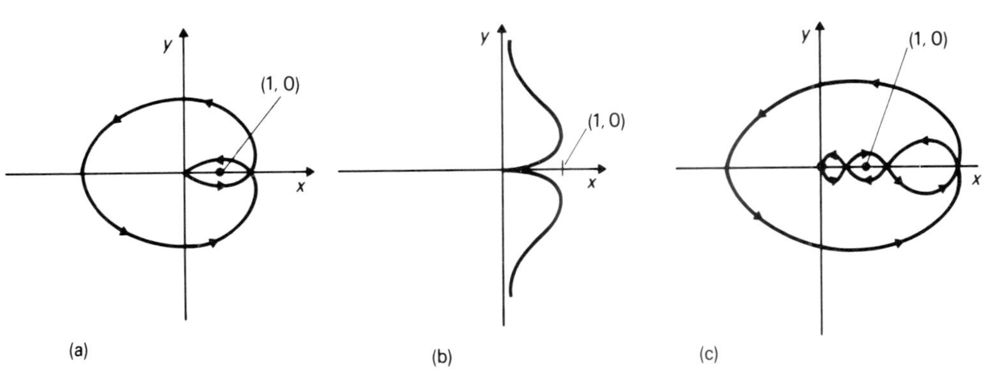

Fig. 6.20 (a) Unstable system; (b) stable system; (c) conditionally stable system

In terms of the feedback cut-off boundary, Fig. 6.8, assessment of stability may be made as follows. If the $\bar{y}_L(j\omega)$ locus in the y-plane makes an intercept X on the x-axis, then if the gain is increased by $1/X$, the locus will pass through (1, 0) giving rise to a closed loop response that is on the verge of instability. The quantity $1/X$ is called the **gain margin** of the system. Instability can also be induced in a stable system by altering the phase round the loop. The amount of additional phase shift that may be introduced in the open loop, before instability sets in, is equal to the angle θ, Fig. 6.21, for which $|y_L(j\omega)| = 1$. This value is known as the **phase margin**. Thus the gain margin and the phase margin give an indication of the form of

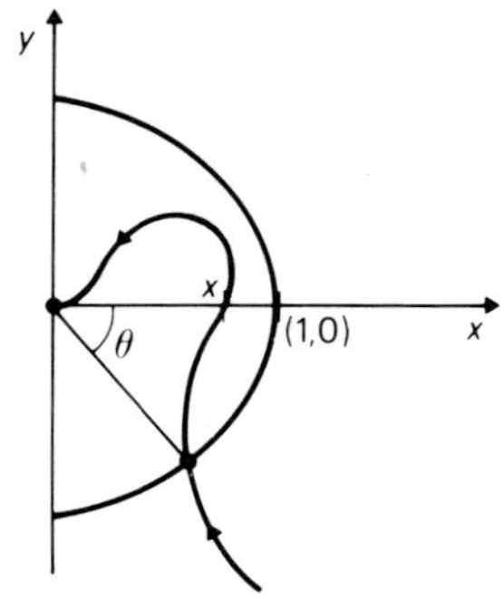

Fig. 6.21 Gain margin 1/x and phase margin θ

the closed loop response. To allow for changes in component values, it is generally recommended that the gain margin should be at least 2 and the phase margin not less than 30°.

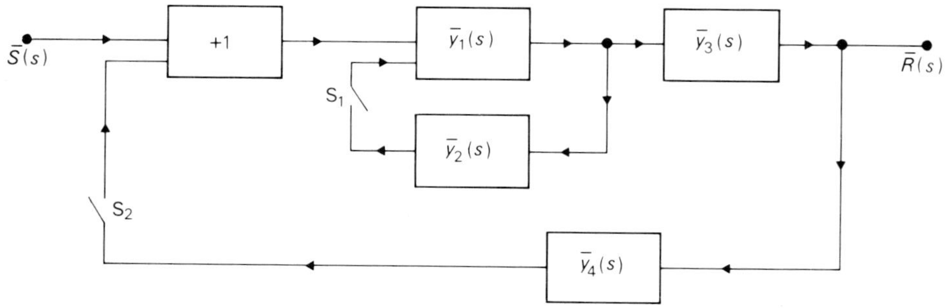

Fig. 6.22 A multiple-loop feedback system

Feedback systems often comprise multiple feedback loops, Fig. 6.22. A peculiarity of such systems is that although the inner loop comprising $\bar{y}_1(s)\bar{y}_2(s)$ may be unstable, the system as a whole, when both the switches S_1 and S_2 are closed, can be stable. To apply the Nyquist technique for predicting the stability of such systems, the locus of $\bar{y}_1(j\omega)\bar{y}_2(j\omega)$ must first be plotted in order to locate the number of poles in the right-hand side of the s-plane of the function

$$\bar{y}_0(s) = \frac{1}{1 - \bar{y}_1(s)\bar{y}_2(s)} \tag{6.75}$$

Let P be the number of such poles. The locus of $\bar{y}_0(j\omega)\bar{y}_3(j\omega)\bar{y}_4(j\omega)$ is then examined for the number of encirclements of the point (1, 0). If the locus encircles (1, 0) P times in the clockwise direction to account for the P positive poles of $\bar{y}_0(s)$ then the system is stable. If, however, the locus encircles (1, 0) less than P times in the clockwise direction, or any number of times in the anticlockwise direction, then the system is unstable because this implies the presence of poles with positive real parts in the overall transfer function

$$\bar{y}_{01}(s) = \frac{1}{1 - \bar{y}_0(s)\bar{y}_3(s)\bar{y}_4(s)} \tag{6.76}$$

Nyquist's graphical method enables prediction of the behaviour of a system when all its feedback loops are closed, by simply plotting the locus of the open loop frequency response function $\bar{y}_L(j\omega)$ for the relevant values of the frequency ω. The prediction may be carried out by calculation if all other constants are known, or by measurement. Experimental determination of the frequency response locus, however, requires the measurement of the input and open loop output amplitudes and also their phase difference; a difficult procedure at low frequencies. Unlike Routh's method, the Nyquist method gives a definite indication of the degree of instability to be expected of a system when its feedback loop is closed, in terms of its gain and phase margins.

Bode plots

Instead of representing the open loop frequency response function $\bar{y}_L(j\omega)$ by a vector locus (Nyquist diagram), its modulus $|\bar{y}(j\omega)|$, often called the **gain modulus** or **gain**, and its argument can be plotted separately as functions of frequency. For a large number of frequency response functions, pertaining to minimum phase elements (see Section 6.2, p. 160), Bode[5] has shown how to determine the minimum possible phase shift associated with a given gain characteristic. In such cases the measurement of phase shift with frequency is unnecessary.

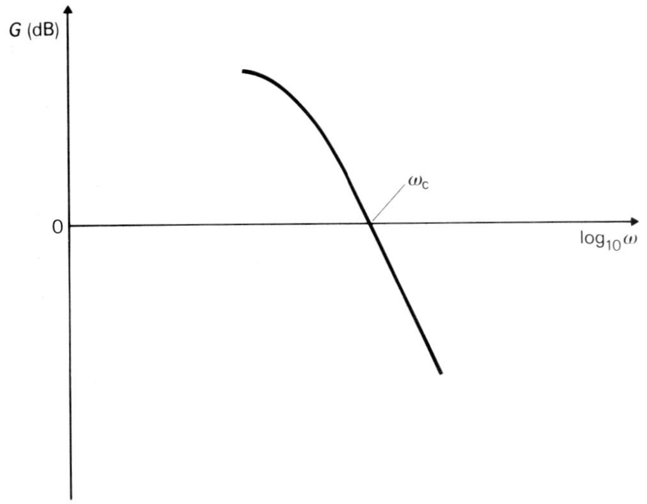

Fig. 6.23 A typical Bode plot

For practical purposes, the approximate form of Bode's relationship may be summarised as follows. If the slope of the Bode plot of $\log_{10}|\bar{y}_L(j\omega)| = G$ (dB) versus $\log_{10}(\omega)$, Fig. 6.23, is m dB per octave at some frequency then the phase shift at that frequency is $\pi m/12$ radians $= 15m$ degrees. Obviously, at a frequency ω_c where $|\bar{y}_L(j\omega)| = 1$ or G (dB) $= 0$, the slope of the Bode plot must be less than 12dB per octave (i.e. $m = 180/15$), or the phase shift must be less than 180°, for the closed loop to be stable as required by the Nyquist criterion. If a phase margin of 30° is required, then the slope at unity gain (G (dB) $= 0$) must be 10 dB per octave ($m = 150/15$). Figure 6.24(a) shows the Bode plot of a first-order stable system and Fig. 6.24(b) that of an unstable third-order system.

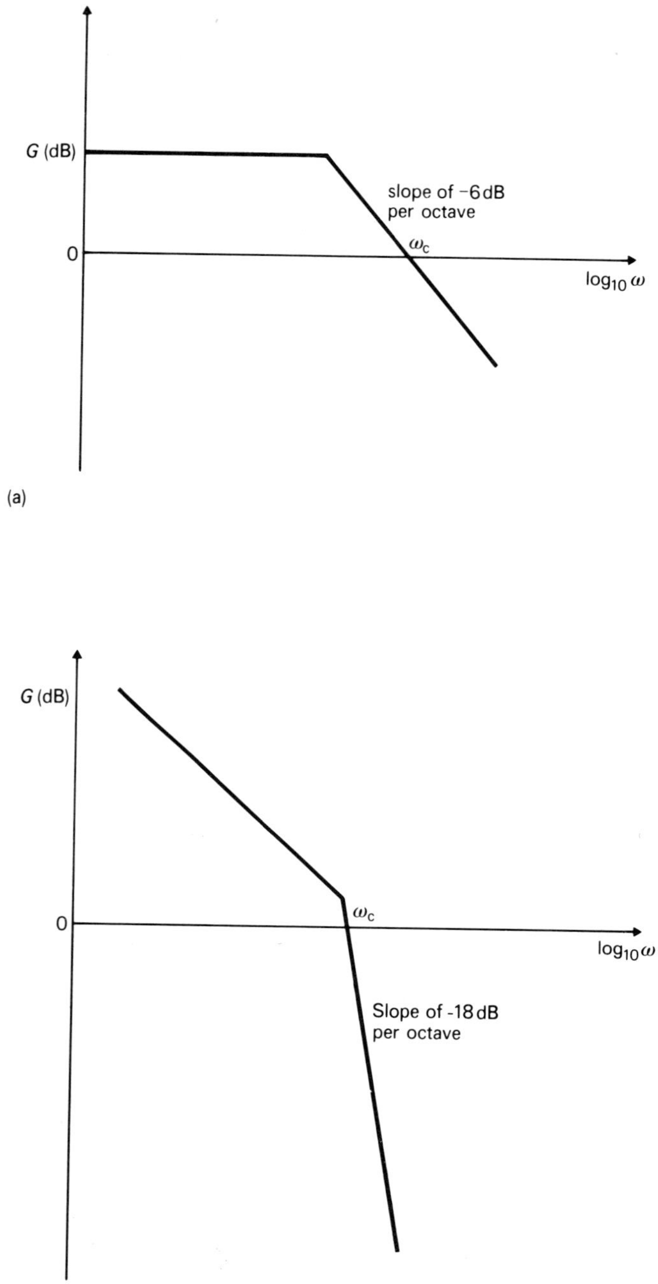

Fig. 6.24 (a) Bode plot of first-order stable sytem with a slope of − 6 dB per octave; (b) Bode plot of a third-order unstable system with a slope of − 18 dB per octave

One of the advantages of Bode's logarithmic plots is in the analysis of feedback systems in which large open loop gains are achieved by cascading elements with open loop frequency response functions $\bar{y}_{1L}(j\omega)$, $\bar{y}_{2L}(j\omega)$, $\bar{y}_{3L}(j\omega)$, etc. The corresponding gain moduli G_1(dB), G_2(dB), G_3(dB) etc. and the slopes of the Bode plots are additive and hence easier to handle.

Lastly, it must be emphasised that although the Bode plot technique is convenient to use (i.e. it needs no phase shift measurements, it gives an indication of the gain and phase margins, it shows the effect of additional elements over frequency range of interest by changes in the slope) it is only applicable to minimum phase systems.

Miscellaneous techniques
In addition to the four classical methods described above, there are several other techniques most commonly used in the design of linear control mechanisms, although they are equally applicable to any other linear feedback system. One approximate technique is that described by West and Potts.[11] The technique utilises the fact that the transient response of a closed loop linear system can be represented by a sum of terms such as $Ae^{\sigma t}\sin(\omega_0 t + \beta)$ where A and β are constants determined by the initial conditions and the form of the input signal and σ is real and negative for stable systems. The value of σ determines the decay time of oscillations of frequency $\omega_0/2\pi$, the damping factor α being defined as the dimensionless ratio σ/ω_0. The main mode of the transient response is that term in which σ in s^{-1} has the smallest absolute value. The quantities ω_0 and σ are expressed in terms of the slopes of the gain–frequency and phase–frequency plots, the phase margin and the unity gain frequency ω_c. Their values are measured from oscillograms similar to those of Fig. 6.14, i.e. $\omega_0/2\pi$ is measured directly from the photograph by comparing it with the calibration trace, and the smallest value of σ is found by plotting the logarithm of the successive amplitudes against time: the amplitudes can also be determined from the calibration trace. This method is particularly useful in experimental situations where the measurement of the open loop frequency response function is too tedious or too complicated.

Another labour-saving technique for obtaining direct quantitative data about the closed loop response is a **Nichol's chart**, which is an x–y plot with phase shift as the x-axis and log of gain as the y-axis. Superimposed on this graph are two sets of loci: one set represents various values of the closed loop gain, and the other, values of the closed loop phase shift. When the open loop characteristics of a system are plotted on such a chart, with frequency as a parameter, intersections with the constant gain and phase loci indicate the closed loop behaviour. Details of Nichol's chart analysis may be found in other texts.[2]

6.6 Stabilisation techniques

There are several techniques for stabilising feedback systems, all based on the fact that the magnitude and phase characteristics of unstable systems can be reshaped by using compensators so as to satisfy the stability conditions. The compensators need not necessarily be electrical networks. Mechanical, electrical or electromechanical devices with suitable transfer functions are often used. The three general methods of compensation are:

1) lead compensation;
2) lag compensation;
3) pole-zero or lag-lead compensation.

Lead compensation

If the open loop transfer function of a feedback system is given by

$$\bar{y}_L(s) = \frac{100}{s(0.2s+1)} \tag{6.77}$$

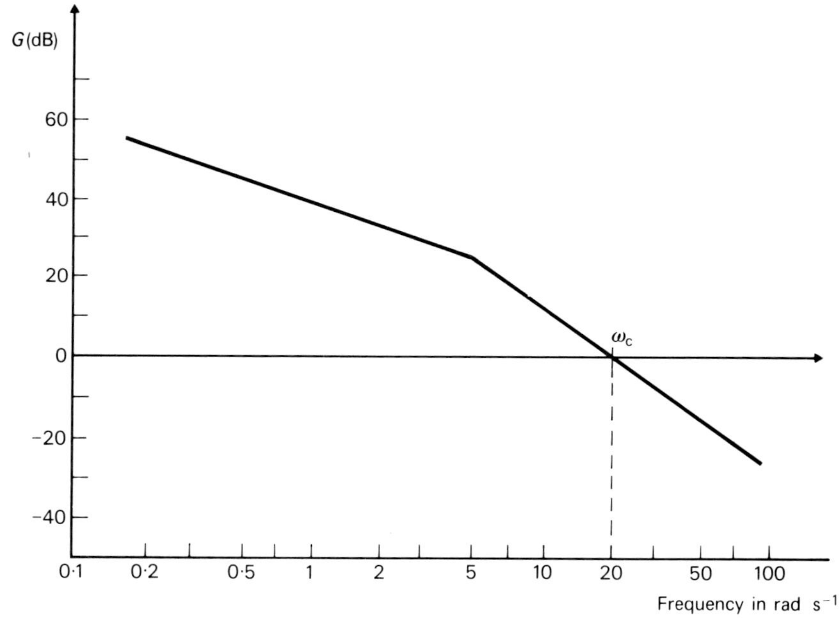

Fig. 6.25 Bode plot of an uncompensated system with $\omega_c \approx 20\,\mathrm{rad\,s^{-1}}$

then the Bode plot of Fig. 6.25 shows that at the crossover frequency $\omega_c \approx 20\,\mathrm{rad\,s^{-1}}$, the slope is about 20 dB per octave and the system is unstable. For stabilising the system, a lead compensation with transfer function

$$\bar{y}_e(s) = \frac{\tau_z s + 1}{a(\tau_p s + 1)} \tag{6.78}$$

is added, where $\tau_z = 1/\omega_z$ is the time constant of a zero; $\tau_p = 1/\omega_p$ is the time constant of a pole; and a is the attenuation of the compensation. To make up for this attenuation, the forward gain must be increased by a factor numerically equal to a. In practice, it is regarded as a good compromise to set $\tau_z = 10\tau_p$ and $a = \tau_z/\tau_p = 10$. In this case, since the crossover frequency $\omega_c = 20\,\mathrm{rad\,s^{-1}}$, the zero frequency ω_z is set to $10\,\mathrm{rad\,s^{-1}}$, so that with $\omega_z = 10\,\mathrm{rad\,s^{-1}}$ and $\omega_p = 100\,\mathrm{rad\,s^{-1}}$,

$$\bar{y}_e(s) = \frac{(0.1s + 1)}{10(0.01s + 1)} \tag{6.79}$$

the transfer function of the compensated system becomes

$$\begin{aligned} \bar{y}_{Lc}(s) = \bar{y}_L(s)\bar{y}_e(s) &= \left(\frac{100 \times 10}{s(0.2s + 1)}\right)\left(\frac{0.1s + 1}{10(0.01s + 1)}\right) \\ &= \frac{100(0.1s + 1)}{s(0.2s + 1)(0.01s + 1)} \end{aligned} \tag{6.80}$$

Figure 6.26 shows the Bode plot of the compensated system which has $\omega_c = 50\,\mathrm{rad\,s^{-1}}$ and a slope of about 6 dB per octave, which represents a stable system. Lead compensation increases the bandwidth of the system and the crossover frequency ω_c moves to a higher value.

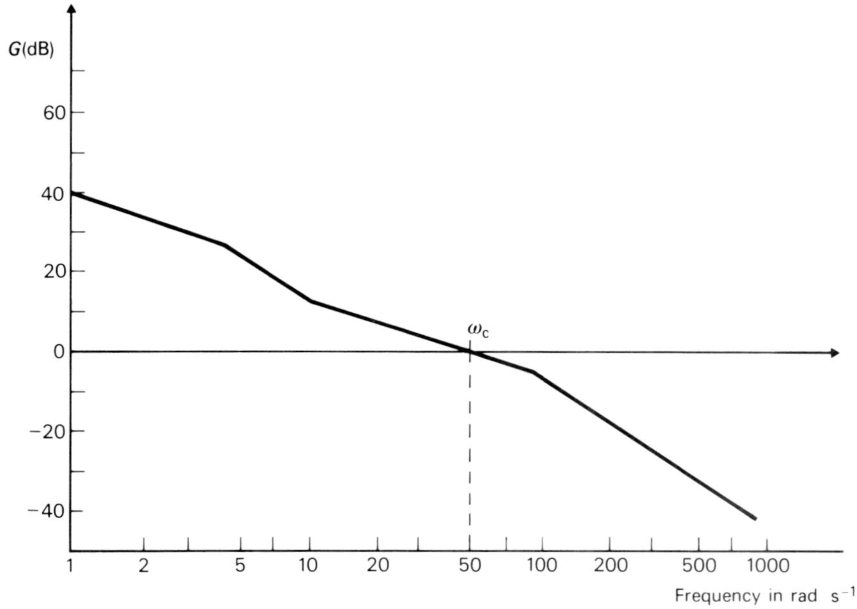

Fig. 6.26 Bode plot of the system in Fig. 6.25 with lead compensation and $\omega_c \approx 50\,\text{rad s}^{-1}$

Lag compensation

Lag compensation decreases the bandwidth of the system and eliminates the effect of high-frequency poles.

If the open loop transfer function of a feedback system is given by

$$\bar{y}_L(s) = \frac{100}{s(0.02s+1)(0.01s+1)} \tag{6.81}$$

then the Bode plot of Fig. 6.27 shows that at the crossover frequency $\omega_c \approx 60\,\text{rad s}^{-1}$, the slope is about 20 dB per octave and the system is unstable. For stabilising the system, a lag compensator with a transfer function

$$\bar{y}_a(s) = \frac{\tau_z s + 1}{\tau_p s + 1} \tag{6.82}$$

is added, where $\tau_p > \tau_z$ or $\omega_p < \omega_z$ so that the pole occurs at a lower frequency than the zero. In practice, setting $\tau_p = 10\tau_z$ with the pole frequency $\omega_p = 0.2\,\text{rad s}^{-1}$ means that the zero frequency $\omega_z = 2\,\text{rad s}^{-1}$ and the transfer function is given by

$$\bar{y}_a(s) = \frac{0.5s + 1}{5s + 1} \tag{6.83}$$

The transfer function of the compensation system becomes

$$\bar{y}_{Lc}(s) = \bar{y}_L(s)\bar{y}_a(s) = \left(\frac{(0.5s+1)}{(5s+1)}\right)\left(\frac{100}{s(0.02s+1)(0.01s+1)}\right) \tag{6.84}$$

Figure 6.28 shows the Bode plot of the compensated system which has $\omega_c = 10\,\text{rad s}^{-1}$ and a slope of about 6 dB per octave, which represents a stable system.

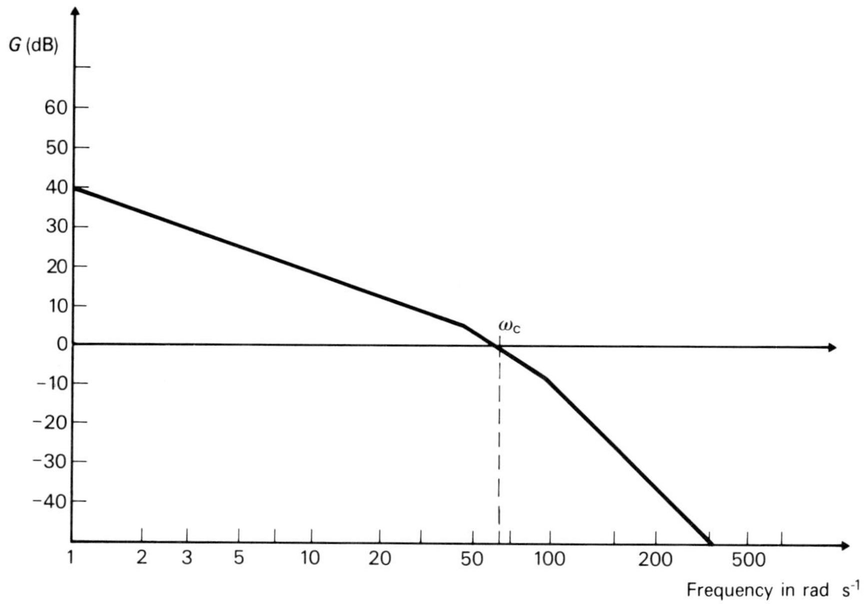

Fig. 6.27 Bode plot of an uncompensated system with $\omega_c \approx 60\,\text{rad s}^{-1}$

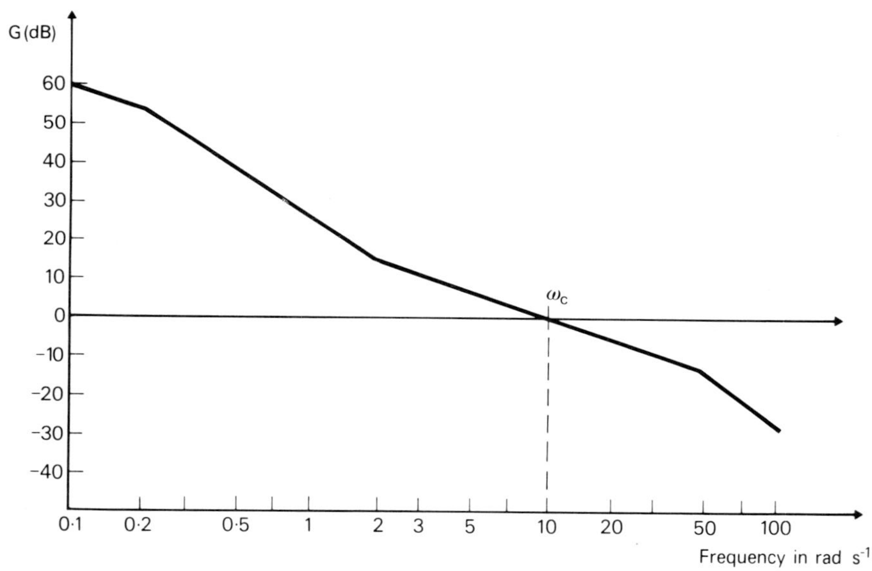

Fig. 6.28 Bode plot of the sytem of Fig. 6.27 with lag compensation and $\omega_c = 10\,\text{rad s}^{-1}$

Pole-zero or lag-lead compensation

In some cases a lead or a lag alone is not sufficient, or a lag-lead combination is more convenient to use. The lag compensator reduces the bandwidth and the lead compensator adjusts the phase margin.

If the open loop transfer function of a feedback system is given by

$$\bar{y}_L(s) = \frac{1000}{s(0.1s+1)(0.01s+1)} \tag{6.85}$$

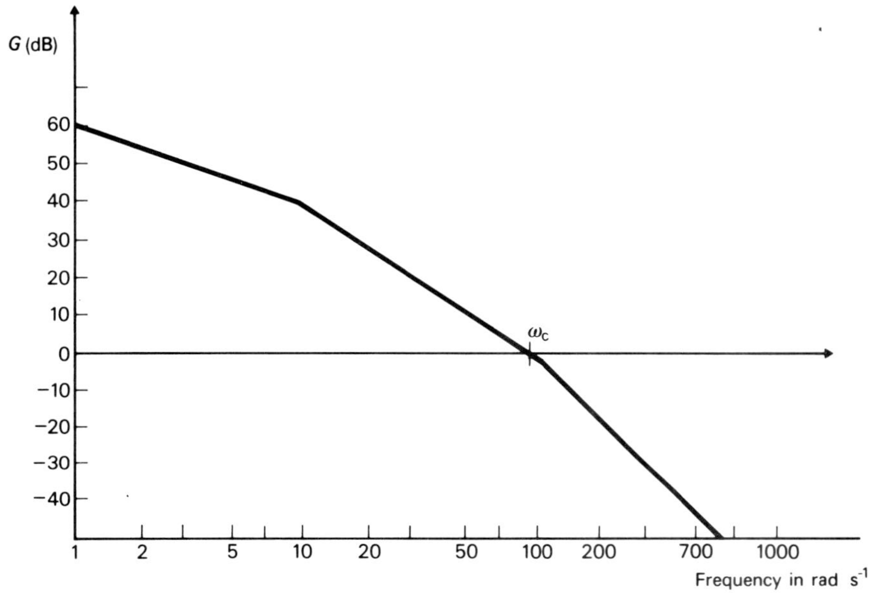

Fig. 6.29 Bode plot of an uncompensated system with $\omega_c = 80 \text{ rad s}^{-1}$

then the Bode plot of Fig. 6.29 shows that at the crossover frequency $\omega_c = 80 \text{ rad s}^{-1}$, the slope is about 18 dB per octave and the system is unstable. If a lag compensation is chosen with a transfer function

$$\bar{y}_a(s) = \frac{\tau_z s + 1}{\tau_p s + 1} \tag{6.86}$$

then to obtain a reduction in gain at high frequencies the ratio τ_p/τ_z is set to 33 dB, or 50, and ω_z is set to 10 rad s^{-1}. Hence

$$\bar{y}_a(s) = \frac{0.1s+1}{(50 \times 0.1s)+1} = \frac{0.1s+1}{5s+1} \tag{6.87}$$

and
$$\begin{aligned}\bar{y}_{Lc_1} &= \bar{y}_L(s)\bar{y}_a(s)\\ &= \left(\frac{0.1s+1}{5s+1}\right)\left(\frac{1000}{s(0.1s+1)(0.01s+1)}\right)\\ &= \frac{1000}{s(5s+1)(0.01s+1)}\end{aligned} \tag{6.88}$$

Figure 6.30 shows the Bode plot with $\omega_{c_1} = 15 \text{ rad s}^{-1}$. If a lead compensator is introduced with a transfer function given by

$$\bar{y}_e(s) = \frac{(0.2s+1)}{10(0.02s+1)} \tag{6.89}$$

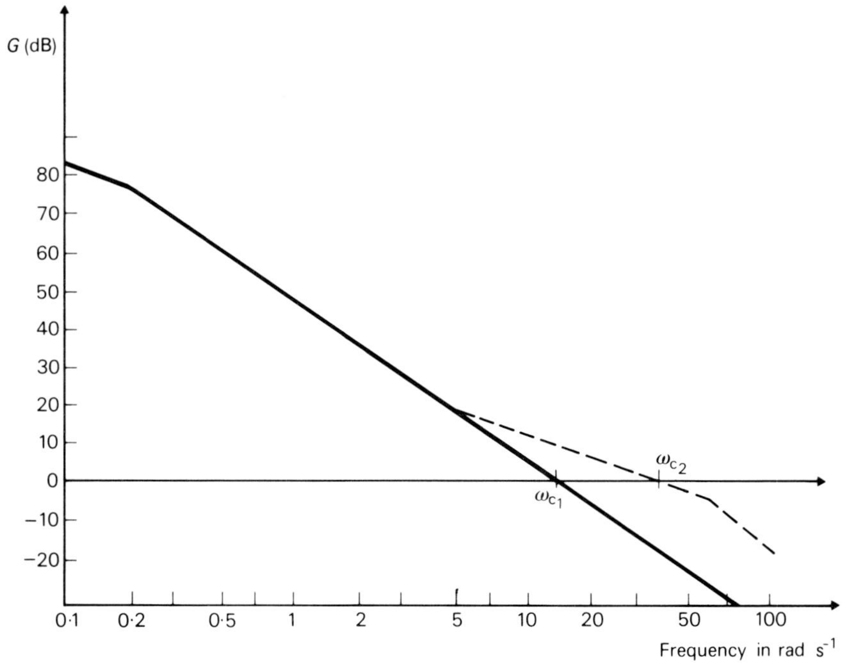

Fig. 6.30 Bode plot of the system of Fig. 6.29 with a lag compensation giving $\omega_{c_2} = 12\,\text{rad s}^{-1}$ and a lead compensation giving $\omega_{c_2} = 30\,\text{rad s}^{-1}$

then

$$\bar{y}_{Lc_2}(s) = \left(\frac{0.2s+1}{10(0.02s+1)}\right)\left(\frac{1000}{s(5s+1)(0.01s+1)}\right)$$

giving

$$\bar{y}_{Lc_2}(s) = \frac{1000(0.2s+1)}{s(5s+1)(0.02s+1)(0.01s+1)} \tag{6.90}$$

The Bode plot of Fig. 6.30 then gives a crossover frequency $\omega_{c_2} = 30\,\text{rad s}^{-1}$ and a slope of about 6 dB per octave, which represents a stable system.

6.7 Practical applications

The feedback theory of linear systems outlined so far is adequate for most experimental situations, some of which are illustrated by the following typical examples.

Control mechanisms

- Open loop or non-feedback
- Closed loop or feedback
  - Discontinuous or regulators
  - Continuous or proportional

Control mechanisms are either of non-feedback, i.e. open loop, type, or of feedback, i.e. closed loop, type. The latter can be further divided into continuous or proportional controls and discontinuous controls or regulators.

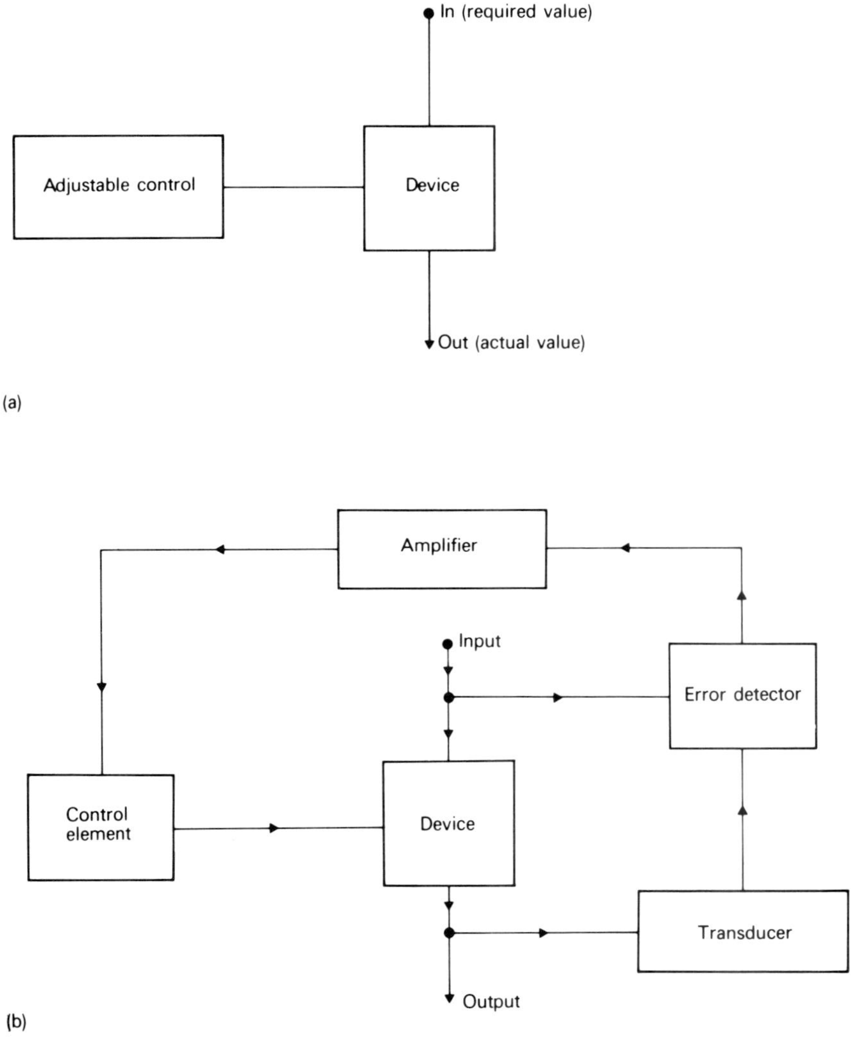

Fig. 6.31 (a) Open loop system; (b) closed loop system

Open loop systems, Fig. 6.31(a), give reliable results only when the calibration conditions of the adjustable control are maintained. **Closed loop** systems, Fig. 6.31(b), are used mainly in the following applications.

(i) An 'error' measuring element whose function is to produce a signal whose magnitude and sign are proportional to the difference between the actual value, i.e. the device output, usually obtained from a transducer, and the required value, i.e. the device input.
(ii) Signal and power amplifiers with provision for stabilising the overall system by techniques described in Section 6.6.

(iii) Control elements or effectors such as motors, relays, thyristors or power transistors, i.e. circuits that carry out the correction so as to reduce the error to zero.

'Error activated' feedback systems

Systems such as those described in (i) on p. 181 are known as **servomechanisms**. The discontinuous controls, including the 'on-off' relay types such as those used in thermostatic control of boilers, and sampled controllers such as those used in radar tracking schemes, are essentially nonlinear regimes although they can often be crudely approximated to linear feedback systems.

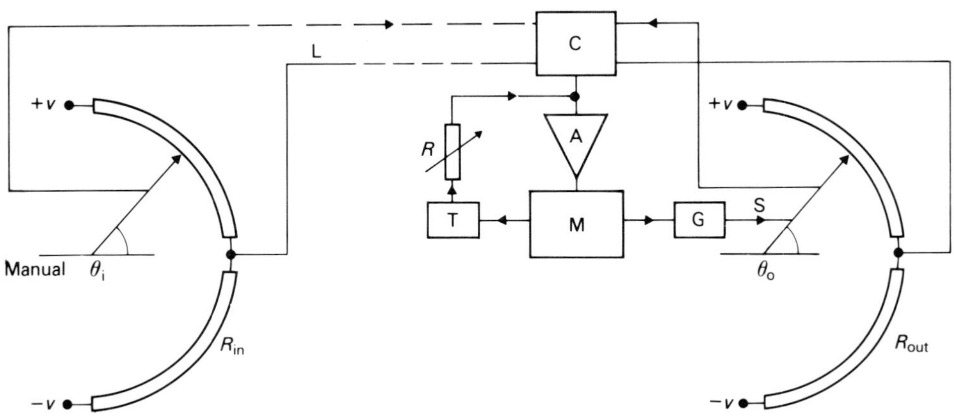

Fig. 6.32 A closed loop feedback control mechanism

A simple example of a closed loop feedback control mechanism is the remote position control shown in Fig. 6.32. The angular position of the output shaft S rotated by the geared servo-motor M-G is θ_o. The servo-motor is driven from an amplifier with gain A and has a tachogenerator T attached to it. The output of the tachogenerator is proportional to the velocity $d\theta_o/dt$ of the shaft, and is used in a subsidiary feedback loop to stabilise the overall system. The angular position of the input shaft, which is accessible for manual control, is θ_i. The centre-tapped variable resistors R_{in} and R_{out} act as transducers producing voltages proportional to the magnitudes and signs of θ_i and θ_o respectively. These voltages are connected via transmission lines L to a potentiometric arrangement C which computes the error signal that actuates the motor. The differential equation of motion is given by

$$\frac{1}{\omega_n^2}\frac{d^2\theta_o}{dt^2}+\tau\frac{d\theta_o}{dt}+\theta_o=\theta_i \tag{6.91}$$

where ω_n^2 is a constant of the system depending on the amplifier gain A, the gear ratio G, the current-acceleration constant of the motor, the position error ($\pm(\theta_o \approx \theta_i)$ – error signal), etc. The coefficient τ of the velocity damping term is adjusted by the resistor R. The transient response is governed by the roots of the auxiliary equation

$$\frac{s^2}{\omega_n^2}+\tau s+1=0 \tag{6.92}$$

where $s \equiv d/dt$ are given by

$$s = -\frac{\tau\omega_n^2}{2} \pm \omega_n\sqrt{\frac{\omega_n^2\tau^2}{4} - 1} \tag{6.93}$$

A critically damped response is obtained when $\tau\omega_n = 2$.

Another common example of a closed loop control mechanism is a speed control system known as the 'Velodyne'.[12] Briefly, a voltage proportional to the actual speed of the shaft is derived by using a transducer. A voltage from a manual control similar to that of Fig. 6.32 gives a voltage proportional to the desired value of the shaft speed. An error signal is derived by comparing these voltages, and is then amplified and fed to the motor so that its speed changes in a sense that tends to reduce the error signal to zero.

In general, feedback control mechanisms must have a compromise between the following three basic requirements:

1) accuracy of tracking – this implies a high loop gain so that small error signals can be detected;
2) a rapid speed of response;
3) a high stability over the entire working range.

The general techniques for stabilisation, outlined in Section 6.6, can often be used; the compensation may not necessarily take the form of an electrical network because damping due to frictional, eddy current or 'wane in a viscous liquid' is sometimes more convenient.

Power supplies for electromagnets

In instruments such as electron spin resonance spectrometers, the current through an electromagnet must be held constant to one part in 10^6 over values of currents ranging from 0–30 A, for periods of about one hour. The current through the electromagnet is sensed with a standard resistance R_s, Fig. 6.33, which is small compared to the ohmic resistance of the coils of the electromagnet. The voltage across the standard resistance is compared with a constant reference voltage via a manual control whose setting determines the value of the current through the electromagnet. The error signal from the comparator is amplified with a chopper stabilised amplifier and the output is used to actuate the control elements, which usually consist of power transistors. Overall stability is achieved by including lead-lag networks in the voltage amplifier stage.

Temperature controllers

A typical temperature controller using a closed feedback loop, which may be discontinuous (relay type) or proportional, is shown in Fig. 6.34. The transducer for obtaining a voltage proportional to the temperature is usually a thermocouple or a platinum resistance thermometer for temperatures around 1000 °C, a semiconductor device like a thermistor or a p-n junction for temperatures around 100 °C, and a carbon resistor for temperatures well below 0 °C. The output of the transducer is compared with a reference voltage whose value is set by a manual control and which corresponds to the required value of the temperature in the enclosure. The output of the comparator is amplified by an amplifier with suitable gain and frequency response and is used to actuate the power control elements.

Electric heaters have long thermal time constants, and two types of AC power control elements, i.e. saturated reactors and thyristors or TRIACs, are most suitable for proportional

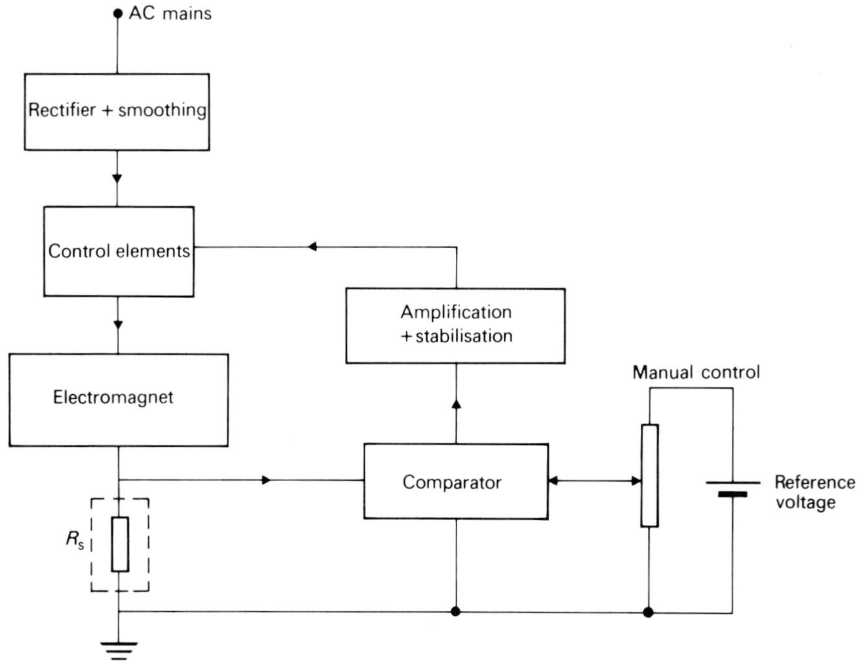

Fig. 6.33 Block diagram of a typical electromagnet power supply

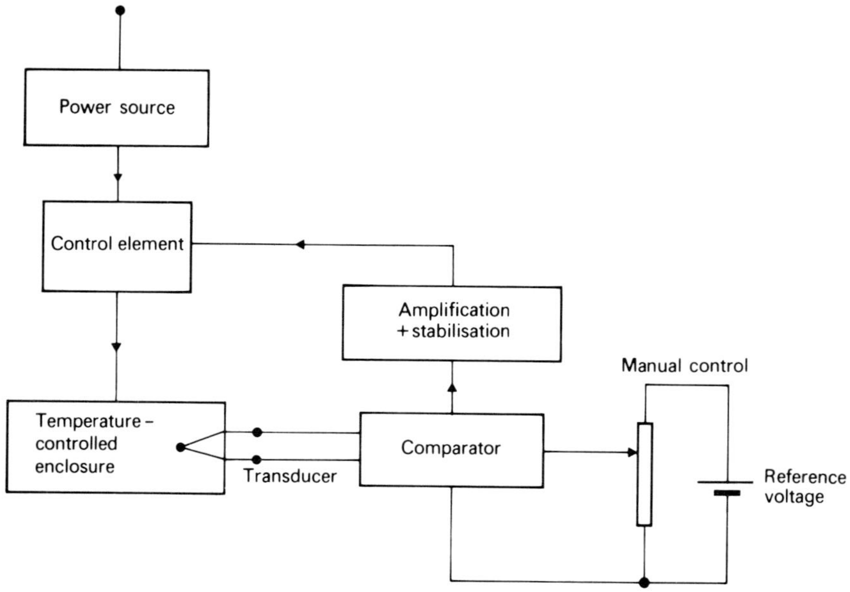

Fig. 6.34 Block diagram of a typical temperature controller

control. Saturated reactors are inductances whose values depend on the state of saturation of their ferromagnetic cores. A small direct current through a separate winding on the core, derived from the error signal amplifier, can therefore control the value of the inductive reactance and hence the power through the heating element which is usually connected in series with it. Thyristor or TRIAC controls are of two types:

a) phase controls;
b) burst firing or zero voltage switching controls.

Phase control circuits

These circuits are currently available as linear integrated circuit chips such as L 120 (SGS/ATES Ltd). The general principle is shown in Fig. 6.35(a), where a half-wave version is used for simplicity.

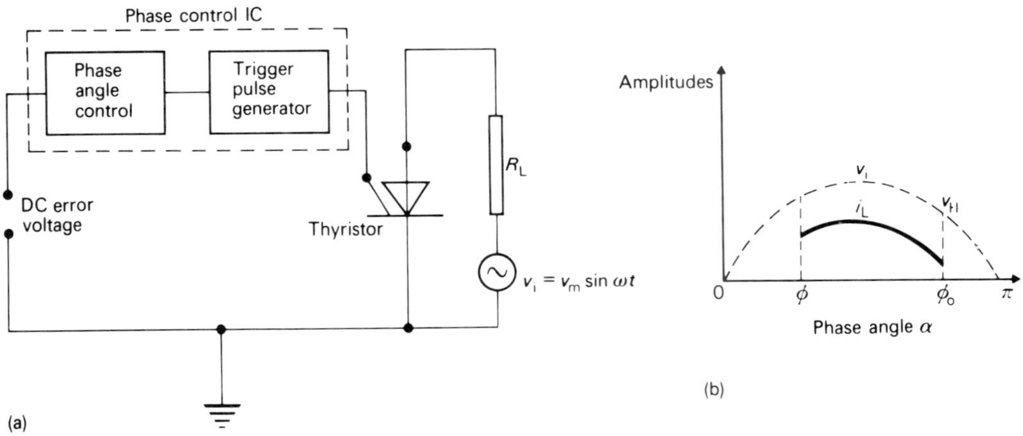

Fig. 6.35 (a) Principle of 'phase control'; (b) voltage and current waveforms;

The thyristor blocks the current through the load R_L until it is triggered at an angle ϕ during the positive half of a cycle, Fig. 6.35(b). The trigger pulse generator produces pulses that are synchronous with the AC supply frequency and the phase angle ϕ is controlled by the amplified DC error voltage. Thus conduction starts at ϕ and stops at $(\pi - \phi_o)$. If the AC supply voltage is represented by $v_i = v_m \sin \omega t$, then the average current through the load R_L will be given by

$$I_{dc} = \frac{1}{2\pi} \int_{\phi}^{\pi - \phi_o} i_L \, d\alpha$$

where $\alpha = \omega t$, and

$$i_L = \frac{(v_m \sin \omega t) - v_H}{R_L}$$

where $v_H = v_m \sin \phi_o$; see Fig. 6.35(b). This gives

$$I_{dc} = \frac{v_m}{2\pi R_L} \int_{\phi}^{(\pi - \phi_o)} \left(\sin \alpha - \frac{v_H}{v_m} \right) d\alpha \tag{6.94}$$

Since $v_H = v_m \sin \phi_o$ and v_H/v_m is small, ϕ_o may be taken as zero and equation 6.94 will reduce to

$$I_{dc} = \frac{v_m}{2\pi R_L}(1 - \cos \phi) \tag{6.95}$$

In practice, the LM 120 produces dual pulses suitable for triggering a TRIAC so that controlled conduction can take place over both the positive and negative halves of a cyle.

Burst control systems

This method avoids the generation of RF interference due to step changes in current through the load, because suppression of such interference is difficult and expensive especially when the power requirements of the load exceed 500 W. The principle of a burst control circuit, also available as a linear integrated circuit chip such as L 121 (SGS/ATES Ltd), is shown in Fig. 6.36.

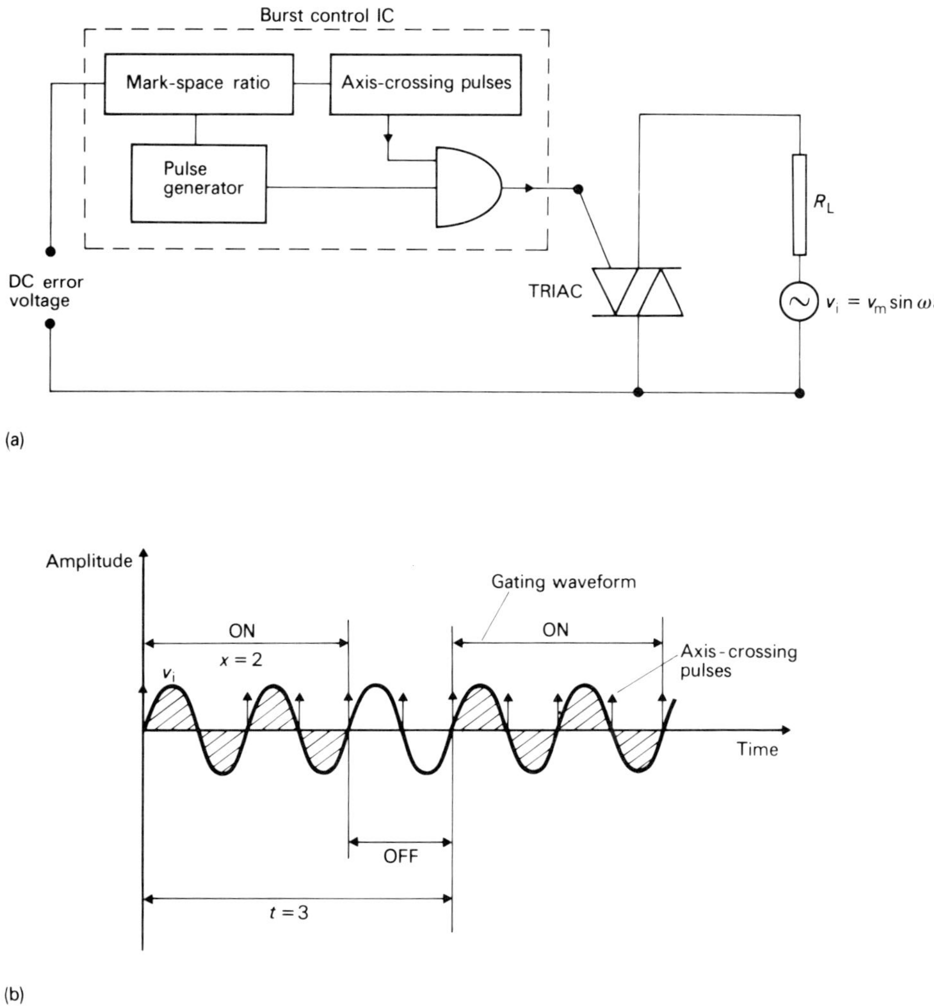

Fig. 6.36 (a) Principle of 'burst control; (b) input and control wave forms

If a sinusoidal waveform power source (e.g. AC mains) is assumed, then pulses corresponding to points in the waveform when the voltage is nearly zero are obtained as shown in Fig. 6.36(b). These pulses are gated by a pulse generator whose repetition rate is preselected and whose mark-to-space ratio is controlled by a DC voltage which in this case is the amplified error signal. The output of the AND gate, Fig. 6.36(a), comprises bursts of pulses synchronous with the AC supply waveform at zero crossing points; this output fires the TRIAC which otherwise blocks the current through the load. The on-off time of the pulse generator thus controls the number of integral cycles during which the TRIAC is conducting and the number of integral cycles during which the TRIAC is non-conducting. If P_0 is the maximum uncontrolled power and P is the power through the load, then

$$P = P_0 x/t \tag{6.96}$$

where x is the number of full conducting cycles, and t is the number of full cycles contained in one period of the pulse generator. Thus, in Fig. 6.36(b), $x = 2$ and $t = 3$. The power through the load is controlled by gating an integral number of cycles, so that switching always occurs when the AC supply voltage is near zero and does not therefore give rise to transients or RF interference.

The burst control system is not suitable for controlling power in loads with a small time constant such as lamps or motors, because it would cause 'flicker'. However, the system is ideal for furnaces or electric heaters, which have large thermal time constants and high power consumption.

Open loop controls with microprocessors and stepping motors

A stepping motor is essentially a digital device that can produce precise bidirectional angular displacements under the control of a pulsed input. It is therefore ideally suited to use in conjunction with microprocessors[13] and, in general, as an effector or control element for open loop controls.

The most common type of stepping motor consists of a permanent magnet rotor with several pairs of poles and two stator coils. The stator fields are set up by a direct current flowing through the stator coils, and step rotation is obtained by alternately reversing the field direction of the two stators. Each reversal of the direction of the field of one of the stators causes the spindle to rotate through an angle defined by the number of pairs of poles and the number of field reversals per cycle. For example, a stepping motor with twelve pairs of poles and four field reversals per cycle will rotate $360/(12 \times 4) = 7.5°$ per input pulse.

To simplify the switching sequence, each stator consists of a centre-tapped coil with one half of the coil always in use. Between each switching period, there is direct current passing through the stator coils which supplies the torque to hold the rotor in position. The four sections of the stators are activated in a 1, 2, 3, 4 sequence for a forward drive and a 4, 3, 2, 1 sequence for a reverse drive. In the simplest case, this may be achieved by using a four-position reversible ring counter, driving four power transistors[14]. The input pulses to the ring counter are obtained from a gated clock pulse oscillator under manual or automatic programme control. Integrated circuits such as SAA 1027 (Mullard Ltd) for driving stepping motors can be used to deliver about 300 mA at 18 V per phase, without the use of external power transistors.

The microprocessor–stepping motor combination is particularly useful in experimental situations where accurate position control and provision for data acquisition are required.

References

(1) Bode, H. W., Feedback – the history of an idea. Proceedings of the symposium on *Active Networks and Feedback Systems*, **X**, Polytechnic Press of the Polytechnic Institute of Brooklyn (1960)
(2) Distefano, J. J., et al., *Feedback and Control Systems*, Schaum's Outline Series. McGraw-Hill (1967)
(3) Chestnut, H., and Mayer, R. W., *Servomechanisms and Regulating Systems Design*, **I** and **II**. Wiley, New York (1963)
(4) Green, S. L., *The Theory and Use of the Complex Variable*. Pitman (1948)
(5) Bode, H. W., Relation between attenuation and phase in feedback amplifier design. *Bell System Technical J.*, **19**, pp. 421–54 (July 1940)
(6) Bode, H. W., *Network Analyses and Feedback Amplifier Design*. Van Nostrand (1947)
(7) Tanner, J. A., Minimum phase networks. *Electronic Engineering*, p. 148 (November 1951)
(8) Black, H. S., Stabilised feedback amplifiers. *Bell System Technical J.*, **13**, p. 1, (January 1934)
(9) Van der Ziel, A., *Noise*. Prentice-Hall (1956)
(10) Nyquist, H., Regeneration theory. *Bell System Technical J.*, **13**, p. 126 (July 1932)
(11) West, J. C., and Potts, J., A simple connection between closed loop transient response and open loop frequency response. *Proc. IEE*, **100** (II) *Power Engineering*, 75, pp. 201–12 (June 1953)
(12) Williams, F. C., and Uttley, A. M., The velodyne. *J. Inst. Elec. Engrs*, **93** (IIIa) (1946)
(13) Wilson, D., Heron, K., de Sa, A., and O'Reilly, W., An automatic rotating-head torque magnetometer. *J. Phys. E. Sci. Instrum.*, **10**, pp. 1214–16 (1977)
(14) McCartan, D. G., and de Sa, A., A stepping motor drive for small telescopes. *J. Brit. Astr. Ass.*, **85**, 4, pp. 324–8 (1975)

Further reading

Budek, Jurek, Burst firing techniques using traces. *Applications Report B86*, Texas Instruments Ltd
Mullard Ltd, *Techniques of thyristor power control*, **1** (1970)
Whiteley, A. L., Theory of servo systems with particular reference to stabilisation. *J. IEE*, **93** (II), p. 353 (1946)

7

Transducers

7.1 Introduction

A transducer is generally used as a means of converting one form of signal into another. Conversion of non-electrical signals into electrical ones is sometimes effected for ease of processing. The term 'transducer' is often used to describe the entire input device, as in the case of a strain gauge–diaphragm combination used for measuring pressure. Instrument transducers used for measuring physical quantities by electrical means are referred to as **sensors** or **detectors**. Transducers may be genuine energy converters or they may require an auxiliary source of energy to effect a signal conversion. Digital transducers, or transducers that give a direct digital output, are preferred in experimental situations involving digital signal processing because they avoid the use of A/D converters.

The range of possible transducers for each physical phenomenon is very extensive. For this reason, the physical principles, merits and disadvantages of a few selected transducers, commonly used in electronic instrumentation, will be discussed with a view to helping the user to optimise his choice of transducer for a given application.

7.2 Transducer terminology

The following terms are used to specify the performance of a transducer.

Accuracy is the term used to relate the output of an instrument to the true value of its input, with a specified standard deviation.

Repeatability is the closeness of agreement of a group of output values for a constant input, under given environmental conditions.

Resolution is the smallest increment in the input that can be detected with certainty by the transducer.

Sensitivity is the ratio of the change in the magnitude of the output to the corresponding change in the magnitude of the input.

Linearity is a measure of the constancy of the sensitivity of the transducer over the entire useful range of input values.

Dead zone is the largest input change to which the transducer fails to respond.

Hysteresis is the algebraic difference between the average output errors corresponding to input values, when the latter are approached from the maximum and minimum possible input settings.

Drift is the unidirectional variation in the transducer output which is not caused by any changes in its input.

Zero stability is the ability of the transducer to restore its output to zero when its input is returned to zero.

Monotonicity implies that the output of a transducer, in response to a continuously increasing input signal, should not at any point decrease or skip a value.

Currently, more and more transducers based on Hall generators, photodetectors, temperature sensors, piezo-crystals, etc. tend to have amplifiers, detectors and other signal processing circuits built in integrated form as parts of their mountings. This eliminates the use of cables for carrying low level signals from the transducers to the processing circuitry, thereby minimising noise and pick-up problems.

The primary requirement of any transducer is its accuracy and efficiency of signal conversion. This implies that its signal-to-noise ratio must be high, its static calibration must not change and it must be fairly immune to variations in the ambient temperature and often in environmental conditions. A linear relationship between input and output and a fast speed of response are desirable. These, however, can often be enhanced by making the transducer part of a closed feedback loop.

Bearing in mind the above basic requirements, each transducer application must be considered on its merits, because complex and expensive signal processing can often be avoided by choosing a suitable transducer for a given task.

7.3 Types of transducers

In general, transducers may be divided into two main groups: passive transducers and active transducers.

Passive transducers

These transducers do not require an external power supply to effect the conversion of one form of signal into another. They can be subdivided into the following commonly-used types.

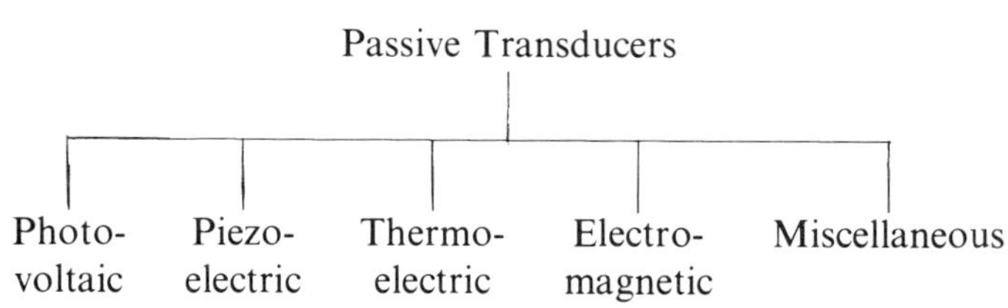

Photovoltaic cells
Photovoltaic cells (see Chapter 1) are used in applications such as portable exposure meters, where simplicity is more important than accuracy and speed of response. Their output currents are large enough to drive moving coil galvanometers and their sensitivity is at a maximum when the resistance of the galvanometer is equal to the internal resistance of the cell. However, the internal resistance decreases with illumination and despite the resultant mismatch, adequate calibration allows the cells to be used over a wide range of illuminations.

Photovoltaic devices consisting of a p-n junction with a large surface area are used as solar cells to convert solar radiation directly into electrical energy with an overall efficiency of 10% or more.

Piezoelectric transducers
Modern piezoelectric transducers are usually made of ceramics such as lead zinconate

titanate (PZT) with a well-defined piezoelectrically sensitive axis. The piezoelectric '*d* constant' may be defined as

$$d_{ij} = \frac{\text{charge produced per unit area}}{\text{force applied per unit area}}$$

where $i = 1, \ldots, 3$ and $j = 1, \ldots, 6$. The value of i defines the axis across which the charge appears and j the axis along which the force is applied; 4, 5 and 6 represent the shear around the 1, 2 and 3 axis respectively.

The electric charge produced is thus linearly related to the applied stress so that piezoelectric transducers can be used to measure in terms of an electrical output such mechanical effects as force, pressure and acceleration. The properties of piezoelectric elements are temperature-dependent and the choice of piezoelectric material for a transducer is fundamental to its performance.

There are many mechanical designs available for various applications, but the two most commonly-used forms are the compression mode and the thickness shear mode. The **compression mode** is mainly used for applications such as gas igniters, in which a mechanical stress, applied to the crystal, generates a high voltage (10–20 V) that activates a spark gap. The flexure elements have a wide range of application: transducers for converting a mechanical stress into a voltage are used in microphones, gramophone pick-up cartridges, accelerometers, stress and strain gauges. Sound generators, optical scanners, choppers and deflection devices are examples where the inverse piezoelectric effect is used for converting electrical impulses into mechanical movement. In applications such as ultrasonic transducers, low-frequency filters and liquid level sensors, both the electromechanical and mechanoelectrical transducer actions are exploited.

Passive thermoelectric transducers

Thermocouples are a good example of passive thermoelectric transducers. They are widely used in thermometry, especially in applications requiring a fast response time. The relation between emf e and temperature t for any practical thermocouple with the reference junction at 0 °C may be represented as

$$e = a + bt + ct^2 + dt^3 \qquad (7.1)$$

where the coefficient d and those of higher order terms are usually so small that terms beyond ct^2 are seldom needed.

Thermopiles consist of a number of thermocouples connected in series and arranged so that radiation falls on one set of junctions which is blackened, while the other set is protected from radiation. Thermopiles are generally used for detecting infrared radiation, the thermojunctions being made of Bi-Sn and Bi and Sb alloys.

Thermocouples are also used in wattmeters and in microwave power measurement, in which the thermocouple emf is proportional to the average heating effect. Sometimes a thermocouple and a heater wire, sealed in an evacuated glass envelope, are used to measure the rms value of a signal, irrespective of its waveform. The thermocouple output is proportional to the square of the heater current produced by the signal.

Electromagnetic passive transducers

The basic principle of these transducers is that the emf induced in a coil is proportional to the rate of change of flux through it. For example, in tachometers used for measuring the angular velocity of a shaft, an armature moving in the field of a permanent magnet is mechanically coupled to it and designed so as to minimise mechanical loading. A direct current proportional to the angular velocity is often obtained by using a slip-ring commutator. Moving coil electromagnetic pick-ups used in sound reproduction and magnetic tape replay heads may also be classed as passive transducers. The principle is also widely used in magnetometry. In spinner magnetometers, the intensity and direction of the magnetisation of a rock sample are determined by spinning the sample at a constant rate inside a pick-up coil and measuring the amplitude and phase of the induced emf. In vibration magnetometers, the magnetic moment of a sample is determined by vibrating it at a constant rate inside an array of pick-up coils and measuring the magnitude of the induced emf.

Variable reluctance tachometers are used for shaft speed measurements. A toothed rotor made of ferromagnetic material moves in the field of a horse-shoe permanent magnet carrying a pick-up coil. The magnetic field surrounding the coil is distorted by the movement of the toothed rotor and a voltage pulse is induced in the pick-up coil. The rms value of the output voltage for a given tooth size is proportional to the rotor speed, provided that the clearance between the pick-up coil and the rotor is kept constant. The frequency of the output pulses is dependent on the number of teeth and the speed of the rotor.

Miscellaneous passive transducers

There are some passive transducers which convert signals of interest into mechanical deflections which can be observed directly, without the aid of additional power sources. Such transducers can also work in an active mode when they are used to convert mechanical signals into electrical ones. For example, bimetal strips, consisting of two metals of different coefficients of linear expansion bonded together, are used as passive temperature transducers. The two metals are constrained to the same length at their common surface so that any temperature change causes the strip to bend in a circular arc. The resultant displacement is used to actuate a pointer on a dial. They are also used as temperature sensing elements in thermostatic controls.

There are several types of transducers which convert a fluid pressure signal into a mechanical deflection which may be observed directly or converted into an electrical signal. Apart from the U-type mercury manometers, Fig. 7.1(a), there are three basic methods for converting pressure to deflection.

Bourdon tubes, mainly the C-type, spiral type, twisted and helical types; such tubes translate pressure into a displacement whose direction is indicated by the arrows in Fig. 7.1(b). The spiral type, because of its increased arc length, is most sensitive while the twisted type is more rugged and is used for relatively greater pressures.

Diaphragms, Fig. 7.1(c), are often corrugated in order to increase their deflection range.

Bellows, Fig. 7.1(d), are made of a single piece of metal and used in cases where large deflections are required.

On account of their large mechanical moving parts, the types of fluid pressure transducers described above have response times that are slow compared to other types based on piezoelectric and strain gauge techniques.

Two commonly-used fluid flow rate sensing elements are Venturi tubes and Pitot tubes. Both types are based on Bernoulli's theorem on ideal fluids under streamlined flow conditions. In the **Venturi tube**, Fig. 7.1(e), the volumetric rate of flow is given by

$$V = K_1 K_2 K_3 A_y \left(\frac{2\Delta p}{\rho_x}\right)^{1/2} \tag{7.2}$$

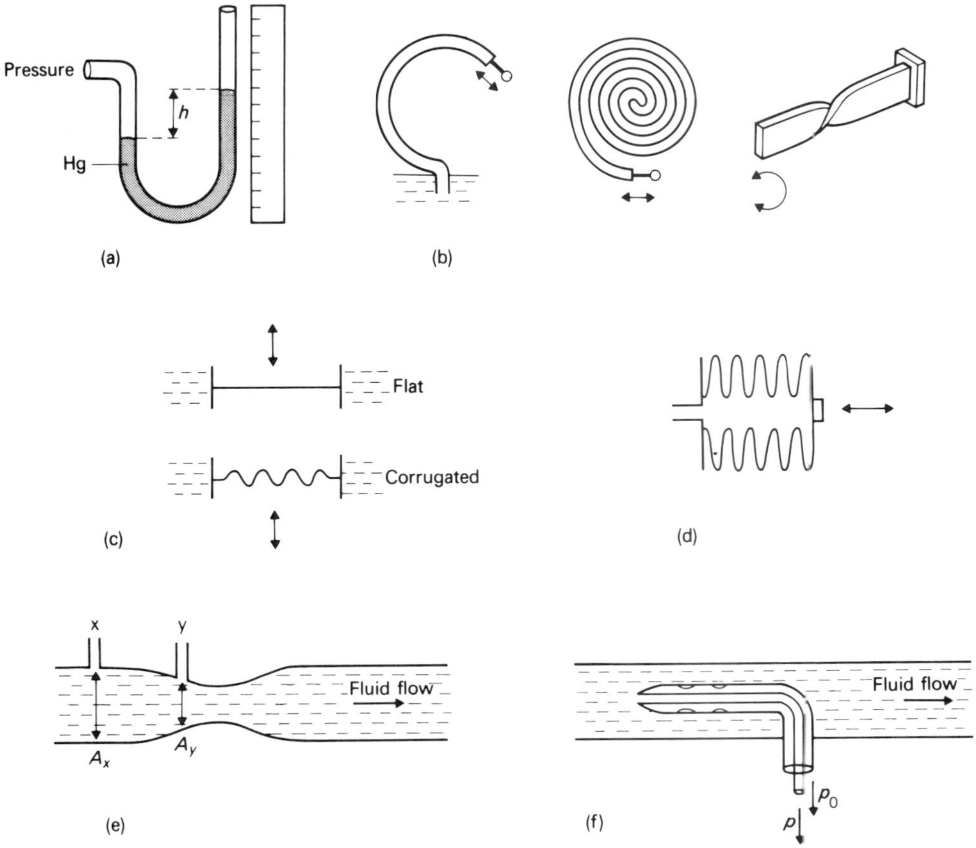

Fig. 7.1 (a) U-type mercury manometer; (b) Bourdon tubes; (c) diaphragm; (d) bellows; (e) Venturi tube; (f) Pitot tube

in which K_1 is the coefficient of discharge, a function of the Reynolds number $(vd\rho/\eta)$; v is the velocity of flow; d is the diameter of the constriction; ρ is the density of the fluid; and η is the dynamic viscosity of the fluid. The quantity K_2 is given by $[1-(A_x/A_y)^2]^{-1/2}$, where A_x is the cross-sectional area at point x and A_y is the cross-sectional area at point y. The quantity K_3 is the expansion ratio, which is unity for liquids and less than unity for compressible fluids; ρ_x is the density of the fluid at point x; and Δp is the pressure difference between points x and y.

The **Pitot tube**, Fig. 7.1(f), consists of two concentric tubes, the inner tube having its open end facing the flow. The outer tube has a closed end, but a number of holes in its wall. The pressure in the outer tube is the static pressure, but the pressure in the inner tube in the sum of the static pressure and the pressure due to the impact of the fluid stream on the stationary fluid in the tube. The velocity of the fluid, v, at this point is given by

$$v = \left(\frac{2(p-p_0)}{\rho}\right)^{1/2} \tag{7.3}$$

where p is the pressure at the entrance to the inner tube; p_0 is the static pressure; and ρ is the density of the fluid.

The velocity of the fluid at any point may be determined by measuring the pressure difference $(p - p_0)$ generated by the Pitot tube. Pitot tubes are mainly used for flow rate measurements in gases.

Active transducers

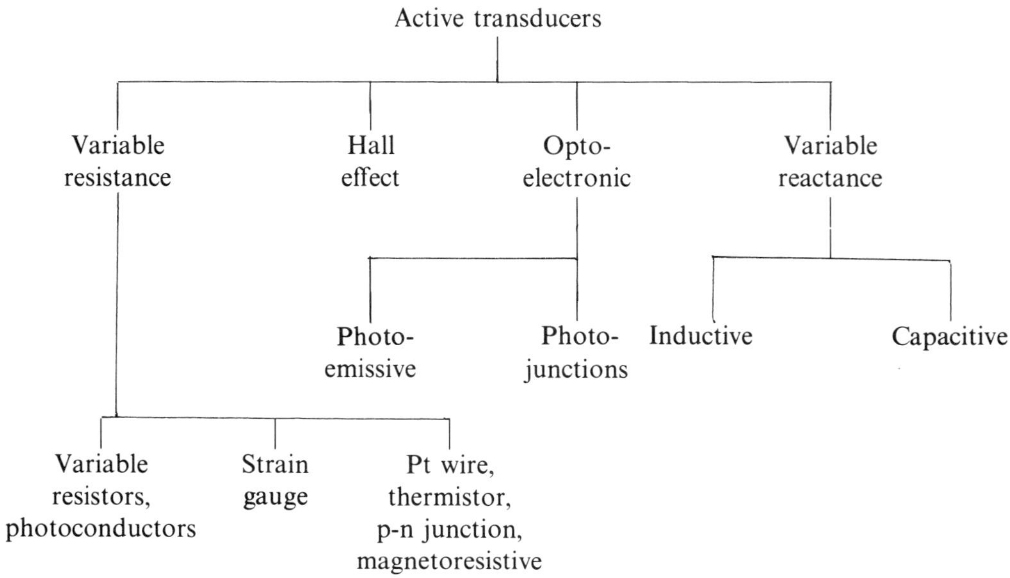

Active transducers require an external energising source like a DC power supply or an audio oscillator on which they rely for their performance. They can be divided into the following commonly-used types.

Variable resistance transducers

Variable resistances, with their rectilinear or rotary wipers mechanically coupled to the moving object, are the most common examples of linear and angular position transducers. The energising source may be AC or DC. In Fig. 7.2 (a) and (b), if $V_{out} = 0$ between points A and B represents the initial position of the object coupled to the wiper, then the subsequent open circuit values of V_{out} are proportional to its displacement. If C is the centre tap of the resistance and $V_{out} = 0$ between B and C represents the initial position, then the sign of V_{out} will indicate the direction of the displacement.

Variable resistors are constructed by two main methods, either by winding fine high-resistance wire on an insulating former or by depositing a thin film of metal oxide or carbon on it. In both cases, because of the complex manufacturing processes involved, the resistance per unit length does not remain constant and gives rise to linearity errors. The thin-film track resistors give a far higher resolution than the wire wound ones in which the limit is dictated by the number of turns per unit length.

A wide range of position transducers using **photoresistors** (see Chapter 1) are currently available. The movement of a beam of light reflected from a moving object is detected by an arrangement of photoresistors such as 'photopotentiometer' or 'photobridge' (Photain Controls Ltd), the mechanisms of which are obvious from Fig. 7.2 (c) and (d) and therefore do not merit further description. Figure 7.2(e), however, is a transducer consisting of four discrete photoconducting elements and is convenient for sensing the movement of a light spot

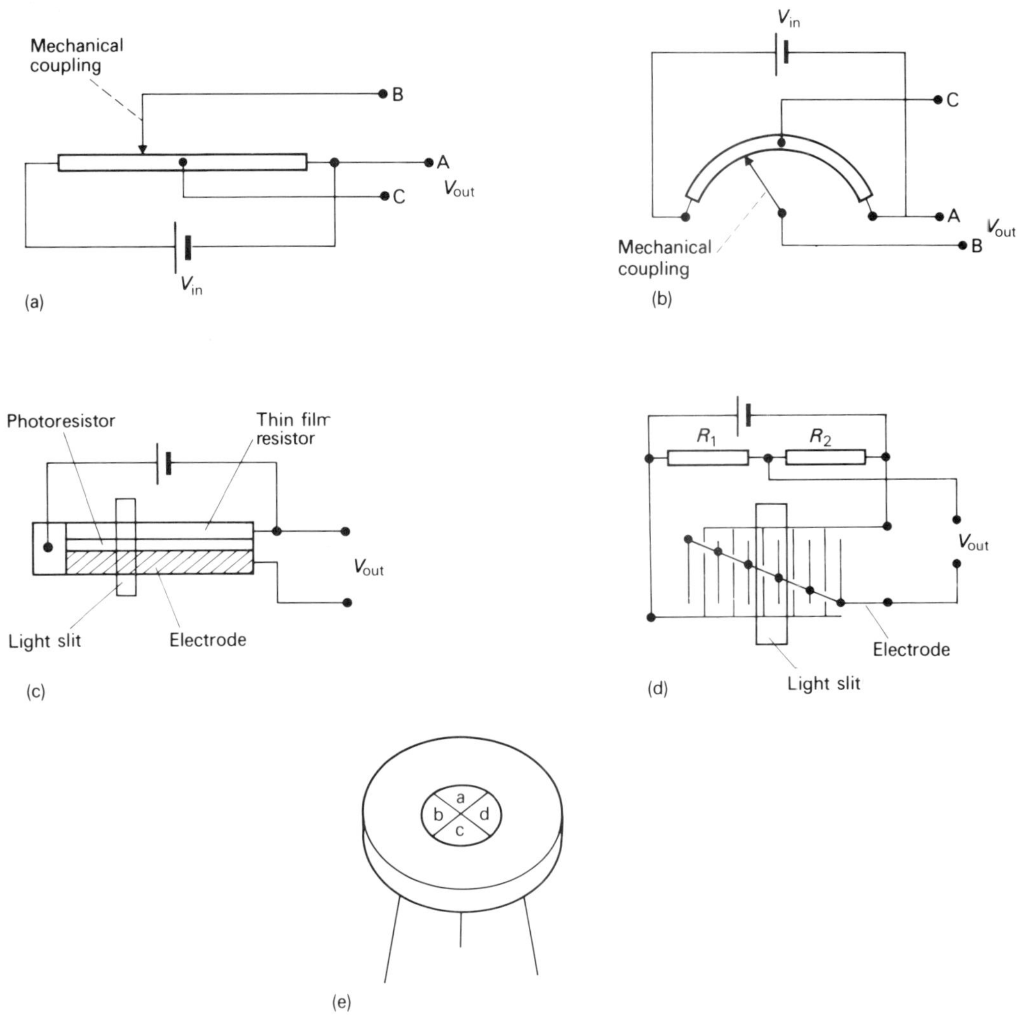

Fig. 7.2 (a) Linear resistive transducer; (b) angular resistive transducer; (c) photopotentiometer; (d) photobridge; (e) a photodetector with four photoconducting elements a, b, c, d: one for each quadrant

about two axes. As there are no mechanical contacts, these devices eliminate wiper noise, have a high resolution and an extended service life. They have a high output but their response time is inherently slow. They are commonly used as null detectors for servomechanisms in instruments such as light spot followers.

Strain gauges

Strain gauges operate on the principle that small dimensional changes in a material alter its electrical resistance. The change is usually measured by some form of Wheatstone bridge energised by an AC or DC supply. There are three main types of strain gauge: etched foil, wire and semiconductor. **Foil** gauges are manufactured by depositing a thin constantan film on an insulating resin substrate, the different patterns being produced by photographic and etching techniques. **Wire** gauges are made from constantan alloy resistance wire, wound so as to form a grid and fastened to a cresol resin based carrier. **Semiconductor** strain gauges consist of a p-type silicon wafer, typically less than 0.05 mm thick and 0.3 mm wide, mounted on a cresol resin substrate and provided with suitable electrodes. Etched foil gauges are used for static and dynamic strain measurements, requiring an accuracy of a few per cent, whereas wire gauges are used in applications requiring a higher degree of accuracy and linearity. Both types can be used up to 200 °C. The semiconductor type work over a lower temperature range and are less robust, but offer a far greater output which can often be used directly without further amplification.

Transducers for temperature measurement
Variable resistance transducers are widely used for the measurement and control of temperature. Several types are suitable for such applications.

Platinum resistance thermometers are an example of transducers which rely on the variation of resistance of a metal with temperature and consist simply of a piece of Pt wire wound on a ceramic former. The changes in resistance with temperature are measured by some form of Wheatstone bridge. This is a very stable and reproducible type of temperature transducer, and can be used over a wide range of temperatures and environments because Pt is chemically inert and resistant to contamination. Its temperature coefficient of resistance is small compared to metals such as Ni, and hence its signal output is small.

The bulk semiconductors used in various types of **thermistor** (see Chapter 1) have a far higher temperature coefficient of resistance. However, for normal NTC (negative temperature coefficient) thermistors, the dependence of resistance on temperature follows an exponential law. Themistors are commonly used in the temperature range −100 °C to +300 °C, often in conjunction with linearising electronic circuitry for absolute temperature measurement or in feedback loops for temperature control.

Forward biased diodes made of Si, Ge and GaAs have been used for quite some time in thermometry in the range 4–300 K[1,2]. Integrated circuits such as the LM 3911 (National Semiconductors Ltd) use the differences in the emitter-base voltage of transistors operating at different current densities as the basic temperature-sensitive element. Fabricated on single monolithic chips they include temperature sensors, reference voltages and operational amplifiers. They give a linear output of about $10\,\mathrm{mV\,^\circ C^{-1}}$ over a temperature range of −25 °C to +85 °C.

The magnetic diodes currently available, such as AHY10 (Telefunken Ltd), are a good example of modern **magnetoresistive transducers**. Basically they consist of a block of intrinsic material i with p and n zones alloyed at either end, Fig. 7.3. The carrier lifetime in the intrinsic material is longer than that in the destroyed surface zone, r. Hence, during current flow, if carriers are deflected into this region by an external magnetic field, higher recombination takes place and the electrical resistance of the diode will increase.

The variation in resistance with the applied magnetic field is non-linear and temperature-dependent. Temperature compensation may be achieved as in the AHY10A by connecting two diodes placed close together in series, Fig. 7.3, and subjecting one of them to the magnetic field.

Hall effect transducers

As shown in Chapter 1, the Hall voltage is proportional to the product of the excitation current and the magnetic field normal to the crystal wafer. An electromagnet operating in the linear region of its *B–H* characteristic is often used to generate the magnetic field, so that the

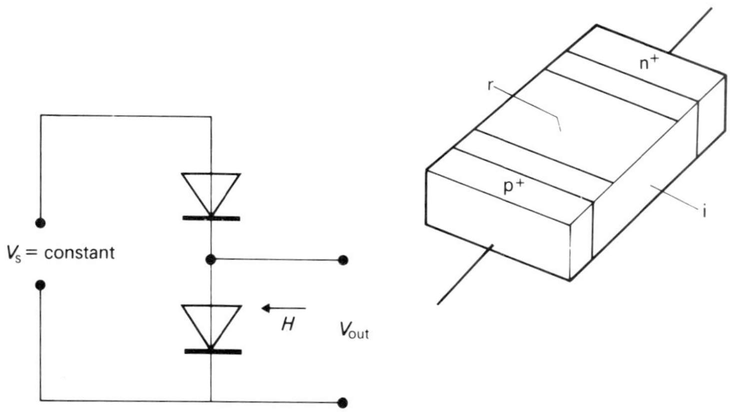

Fig. 7.3 Temperature-compensated magnetic diode

Hall voltage is proportional to the product of the excitation current and the magnet current. The following examples illustrate the use of the Hall effect in transducer applications.

Miniature **proximity switches** for 'bounce free' switching, incorporating a Hall crystal and switching transistors, are commercially available in IC form. A small permanent magnet activates the device from a distance of about 4 mm.

In **gaussmeters**, a Hall crystal probe is excited with a constant current and placed perpendicular to the magnetic field to be measured. The Hall voltage generated is proportional to the field. Careful calibration is necessary because a common occurrence is an offset voltage due to misalignment of electrodes and non-linearity errors at high fields.

Hall crystals are often used in **analogue computation**. For example, by varying both the crystal current (x variable) and the magnet current (y variable) over a linear range, a simple analogue multiplier can be made, because the Hall voltage will be proportional to xy.

In **AC wattmeter** applications, the crystal current is made proportional to the line voltage and the magnet current proportional to the line current. As the crystal is a pure resistance of a few ohms, the crystal current I_H will be in phase with the line voltage, i.e.

$$I_H = K_1 V_0 \sin \omega t \tag{7.4}$$

$$B = K_2 I_0 \sin(\omega t + \phi) \tag{7.5}$$

so that

$$V_H = K V_0 I_0 \sin \omega t \sin(\omega t + \phi) \tag{7.6}$$

where $K = K_1 K_2$.

The DC component of this expression is given by

$$\frac{1}{2\pi}\int_0^{2\pi} V_H \, d\theta = \frac{K}{2} V_0 I_0 \cos \phi$$

in watts, where $\theta = \omega t$.

Optoelectronic transducers

In addition to light flux measurements, Section 7.4, optoelectronic transducers are mainly used for non-contact position sensing, measurement of displacement and optical coupling.

They consist of light emitters, i.e. photosensors with a compatible spectral response in the visible or infrared region. The emitters are usually incandescent lamps and visible or infrared light emitting diodes (LEDs). The sensors are either photoemissive (vacuum or gas filled) cells and a wide variety of photodiodes and phototransistors. The light beam is interrupted by, or reflected from, the moving object and the photosensors used to determine its position or displacement. The following examples illustrate some of the uses of these transducers in modern instrumentation.

Photoemissive cells

Two simple arrangements of photocells are used to convert small angular displacements linearly into voltage analogues. A parallel beam of light, reflected from a small mirror attached to the moving object, is allowed to fall on a 'beam split' prism, and two photocells such as CV90 (Mullard Ltd), Fig. 7.4 (a), or a 'split cathode' photocell such as RCA920, Fig. 7.4 (b). The size of the beam is small compared to the size of the photocathodes and the beam is initially positioned so as to illuminate both the photocathodes equally, thereby giving a zero differential output. Any displacement of the light beam illuminates one photocathode more than the other and produces a differential voltage output whose magnitude is proportional to the displacement and whose sign indicates its direction. In view of the small area of the photocathodes, these transducers are only useful for direct measurement of small displacements. They are used more often for the generation of an error signal for closed feedback loops (see Chapter 6) in which their fast response time is a distinct advantage[3].

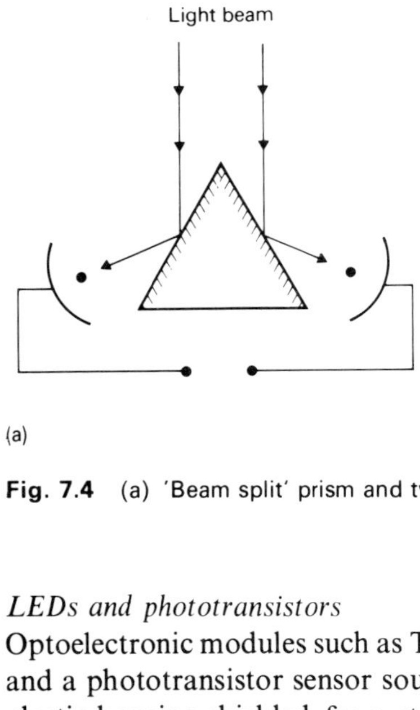

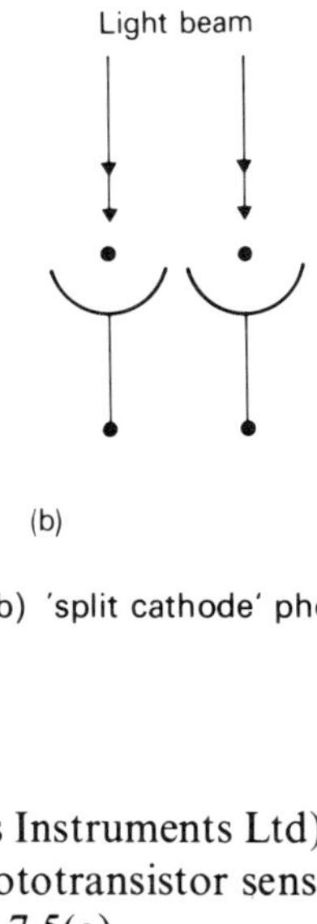

Fig. 7.4 (a) 'Beam split' prism and two photocells; (b) 'split cathode' photocell

LEDs and phototransistors

Optoelectronic modules such as TIL 138 (Texas Instruments Ltd) consist of an LED source and a phototransistor sensor source with a phototransistor sensor mounted in an opaque plastic housing shielded from stray light, Fig. 7.5(a).

These devices are designed for transmissive sensing applications such as shaft encoders, sector sensors, level indicators etc., in which the light beam from the source to the sensor is mechanically interrupted by the moving object. They are very convenient to use in conjunction with mechanical light choppers for generating a reference for signal recovery (see Chapter 10).

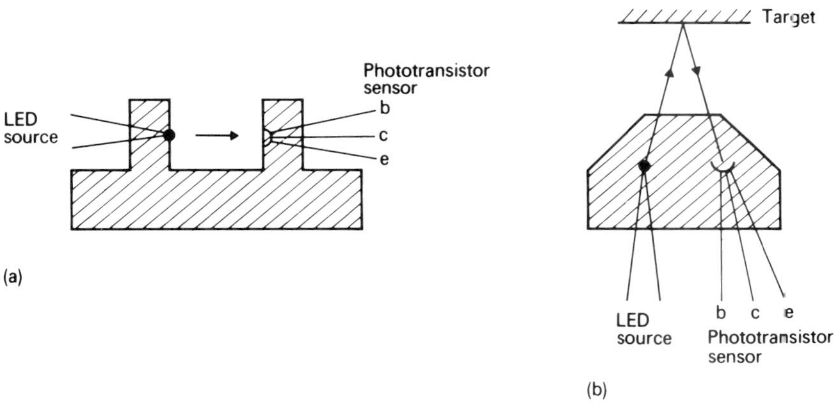

Fig. 7.5 (a) Optoelectronic module for transmissive sensing; (b) optoelectronic module for reflective sensing

The TIL 139 also consists of an LED source and phototransistor sensor, but their arrangement in the housing, Fig. 7.5(b), is meant for reflective sensing applications such as batch counters and situations where a reflecting surface can be easily attached to the moving target.

Arrays of selected source–sensor pairs mounted in a row, such as SDA 13 (Texas Instruments Ltd), are used for punched tape and card 'read heads' in which the optoelectronic modules have several advantages over the filament-lamp–photocell assemblies. They have a longer useful life, are more vibration resistant, require lower power resulting in lower heat dissipation in the head and do not have any of the ageing problems such as filament sag.

Optically-coupled isolators

Another common application of source–sensor transducers is in optically-coupled isolators, which offer an alternative solution to electrical isolation which is conventionally achieved by isolation transformers and relays. Basically an opto-isolator consists of a GaAs LED at the input stage and an npn phototransistor at the output stage, Fig. 7.6. The medium between the diode and sensor is usually an infrared transmitting glass as in TIXL 102/103 (Texas Instruments Ltd). The source and sensor are spectrally matched for optimum transfer characteristics, the electrical isolation will stand about 2 kV and the response time is a few microseconds.

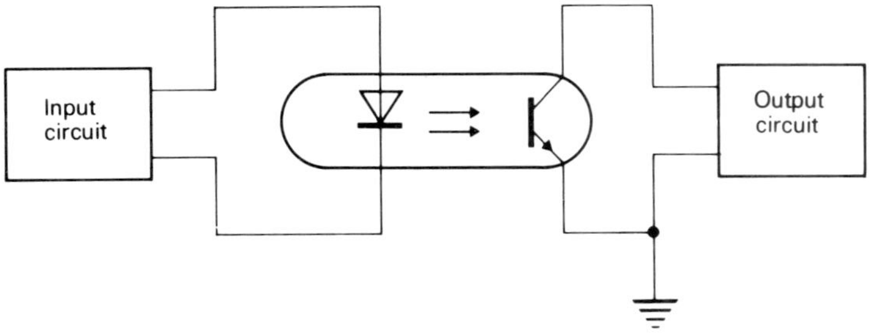

Fig. 7.6 Optically-coupled isolator

These devices are most useful in situations where high voltage and noise isolation are required as well as small size. Because the two systems are coupled together via a light beam, the necessity for a common ground is eliminated and apart from use as interface elements between digital systems, devices are often used as low-level signal choppers to isolate the switching circuitry from the output, thereby reducing output 'spikes'.

Variable reactance transducers

Variable reactance transducers can be divided into two distinct groups:

a) variable inductance type;
b) variable capacitance type.

Variable inductance transducers

This type is mainly used as displacement transducers in two basic forms.

In the first, a coil is wound on a core c of high permeability material, Fig. 7.7 (a). A plate b, made up of the same material, is attached to the moving object and separated from the core by a gap *d*. The inductance is inversely proportional to the reluctance of the magnetic circuit, which varies with *d*. If the inductance is energised by an alternating voltage with constant amplitude and frequency, then any change in the position of the object will produce a change in *d* and a corresponding change in the value of the inductance and the current in the circuit. Over a finite range, the variations in the current will be linearly related to the displacement. This principle can be implemented in a large number of ways, differing mainly in the physical construction of the magnetic circuit and methods of detecting changes in the inductance of the transducer.

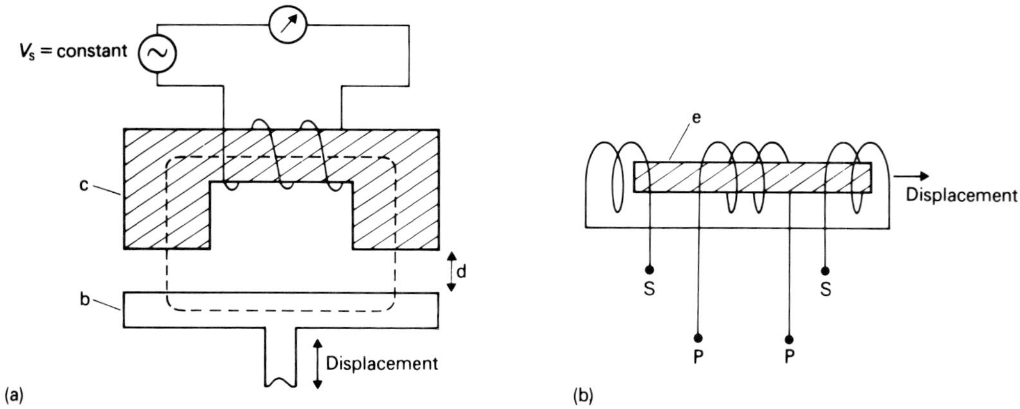

Fig. 7.7 (a) A variable inductance transducer; (b) a differential transformer

The second type of variable inductance transformers is the **differential transformer**, Fig. 7.7(b), which consists of a primary winding PP and two secondary windings connected in series opposition (SS). The primary is coupled to the two secondaries via a movable high permeability core or 'slug' e, and is driven from an alternating voltage source of constant amplitude and frequency. When the position of the slug is such that the induced emf's in the two secondaries are equal, the output at SS is zero. Any displacement of the slug from this initial position produces an output at SS since the coupling of one secondary will increase and that of the other will decrease. Over a finite range of displacements, the amplitude of the output is linearly related to the displacement. Its direction is indicated by the phase of the output relative to that of the excitation voltage. Transducers of this type are ideally suited to use in conjunction with a phase sensitive detector (see Chapter 10) using the excitation voltage for reference. The DC output of the phase sensitive detector will then indicate, in magnitude and sign, the magnitude and direction of the displacement.

Commercially available displacement transducers based on this principle have integral solid-state drive oscillators and demodulators. The units are powered by a DC supply and give a DC output voltage proportional to the displacement. Those used for rectilinear displacement are known as **linear variable differential transducers** (LVDT) and have a wide variety of mechanical mountings for the slug and a stroke of up to several inches. Their sensitivity, millivolts per mm displacement, and linearity vary with the range of displacements for which the transducers are designed. Transducers for angular displacement, known as **rotary variable differential transformers** (RVDT) are also available.

Variable capacitance transducers

Capacitive transducers rely on the fact that the capacitance C_T of a parallel plate condenser is given by

$$C_T = \varepsilon_0 \varepsilon_r A/d \tag{7.7}$$

where ε_0 is the absolute permittivity, ε_r is the relative permittivity of the dielectric between the plates, A their area and d their separation. Transducers that depend on the area overlap, Fig. 7.8(a), or position of the dielectric, Fig. 7.8 (c), give a change in capacitance that varies linearly with the displacement; these devices are suitable for the measurement of large displacements. On the other hand, those devices that depend on a change in distance d give a change in capacitance that is inversely proportional to the displacement and are used for the measurement of small displacements (Fig. 7.8 (b)).

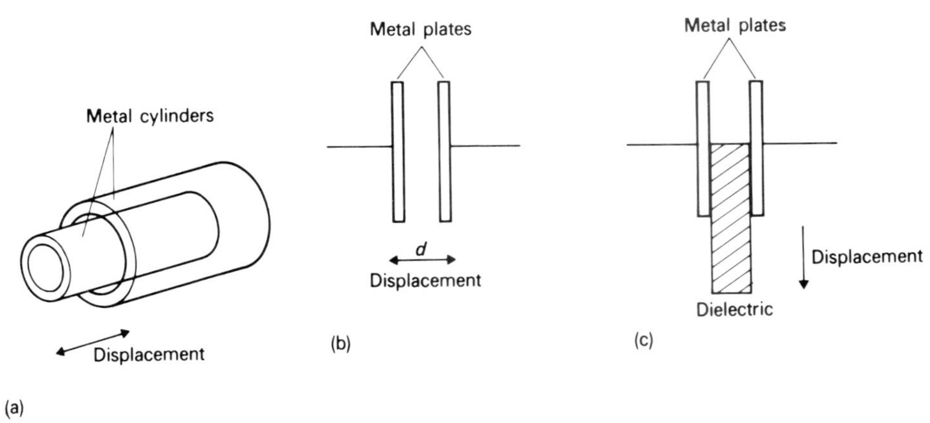

Fig. 7.8 (a) Capacitive transducer depending on the area overlap; (b) capacitive transducer depending on the change of distance d; (c) capacitive transducer depending on the position of the dielectric

There are two important techniques for the measurement of the changes in capacitance in these transducers. In Fig. 7.9 (a), the transducer C_T is connected in the feedback loop of an

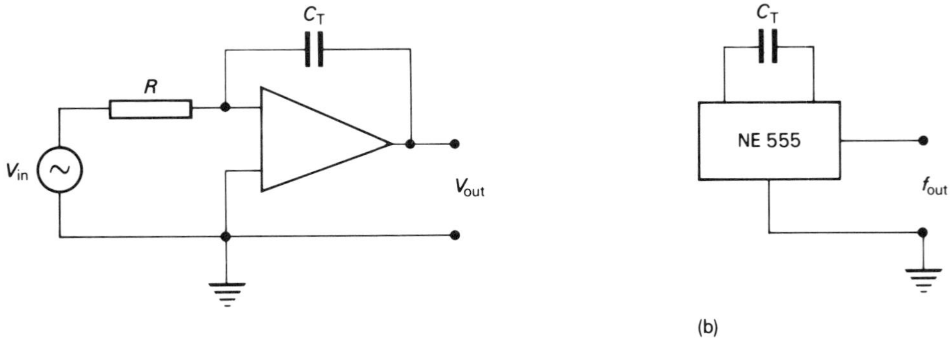

Fig. 7.9 (a) A capacitive transducer C_T in a feedback loop of an operational amplifier; (b) a capacitive transducer C_T controlling the frequency of an RC oscillator

operational amplifier. The output voltage is then given by

$$V_{out} = \frac{Z_{C_T}}{R} V_{in}$$

If R and the amplitude and frequency of V_{in} are kept constant, then $V_{out} \propto d$, the distance between the capacitor plates. In the second method, Fig. 7.9(b), the capacitor controls the frequency of an RC oscillator such as NE555 (Signetics Ltd). For small variations of C_T, the output frequency is given by

$$f_{out} \propto \frac{1}{C_T} \propto d$$

The output may be processed in digital form or converted to an analogue voltage by a frequency demodulator such as a phase-locked loop.

7.4 Conversion of optical signals into electrical ones

Optical signals in the infrared, visible and ultraviolet region are converted into electrical signals for processing or storage by a wide range of photoelectric devices. For low-level signals, the signal-to-noise ratio of the device is of primary importance, and therefore a device having an optimum response over the range of signal wavelengths must be selected. The spectral characteristics of most sources and detectors do not match and the effectiveness of a source A on a detector B is represented in Fig. 7.10:

$$\text{effectiveness} = \frac{\text{Area under curve C}}{\text{Area under curve A}}$$

$$= \frac{\int^{\lambda} f(A) f(B)\, d\lambda}{\int^{\lambda} f(A)\, d\lambda} \tag{7.8}$$

Photodiodes comprising reverse biased p-n junctions give rise to very small signal currents in the submicroampere to tens of microamperes range. They have low temperature coefficients and response times in the submicrosecond range. Their spectral response and

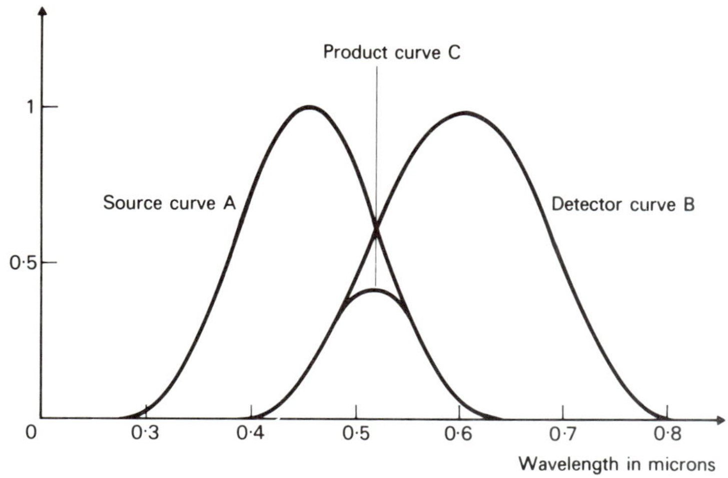

Fig. 7.10 Effectiveness of a source A on a detector B

speed are tailored by the geometry and doping of the junction. Photodiodes are usually made on a single chip of silicon, along with their amplifier sections, utilising the same doping processes for both. The simplest photodiode–amplifier combination is the phototransistor, in which base current is generated when light falls on the reverse biased base-collector p-n junction and is amplified by the current gain of the transistor.

The following are among the photodiode arrays currently available.

An array of 1024 photodiodes in a line about 2.7 cm long on a single IC may be used for edge and width sensing. The array is sequentially addressed and a voltage level, depending on the light intensity falling on each diode, is available in serial form.

A 64 × 64 photodiode array arranged in a square is available for area measurement.

Charge-coupled device arrays (see Chapter 2) are available for digitising and storing optical images.

Lastly, photomultipliers with bialkali and trialkali cathodes are invariably used for low level light signals[4] on account of their low noise and high gain (see Chapter 9).

7.5 Modern digital transducers

Apart from analogue transducers with an A/D converter combined in a single unit, there are several transducers that give an output suitable for direct digital processing or display. Some of the transducers commonly encountered in modern electronic instrumentation belong to this category and may be briefly described as follows.

Encoders

Digital encoders are mainly used for obtaining in digital form an absolute value of the position of rotating shafts. A disc or a drum with a digitally coded pattern is attached to the shaft and its position is sensed with brushes or optoelectronic transducers. For example, the encoder disc of Fig. 7.11 consists of segments, some of which are opaque and others that are transparent. They form a four-bit natural binary code which can be converted to electrical

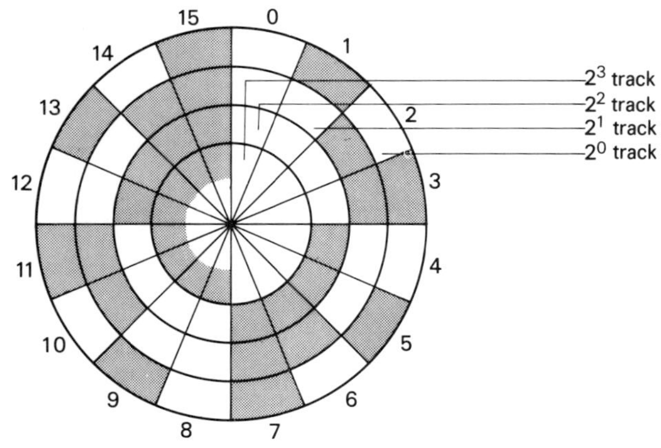

Fig. 7.11 A four-bit natural binary encoder

form by using four source–sensor optoelectronic modules. In practice, optical encoders use elaborate codes such as the ten-digit Gray cyclic code, giving 1024 counts per revolution with an accuracy of ± 1 count.

Moiré fringe transducers

Moiré fringes are interference fringes produced when a beam of parallel light is projected through two transparent diffraction gratings having the same pitch. The gratings are mounted face-to-face with their rulings slightly inclined to one another. The fringes are formed at right angles to the rulings on the gratings and the ratio of the distance between the fringes to the pitch of the gratings depends on the angle between the two gratings. If one grating is moved relative to the other, in a direction perpendicular to their rulings, then the fringes move in the field of view. The displacement of a moving object connected to one of the gratings can be measured by counting the number of fringes crossing a fixed point in the field of view. For example, a photodetector may be used to convert the fluctuations in the light intensity, caused by the moving bright and dark fringes, into pulses that can be amplified, shaped and counted by a digital counter.

Moiré fringe transucers with gratings having 200 to 2000 lines per cm are mainly used in situations where large displacements (≈ 1 m) must be measured accurately ($\approx \pm 0.001\,\%$). As the moving grating need only be short, the maximum displacement that can be measured accurately is determined by the length of the fixed grating that can be supported rigidly.[5]

Doppler shift transducers

When a narrow radio-frequency, microwave, laser light or ultrasonic beam at a frequency f and wavelength λ is aimed at an object moving towards the source at a velocity v, then the apparent frequency f' of the signal at the object is higher than f, and

$$f' - f = (v/c)f$$

where $c = f\lambda$ is the velocity of the transmitted beam. If the beam at frequency f' is reflected back to the source, then the difference between the frequency f'' of the received signal and that of the transmitted signal is

$$f'' - f = 2(v/c)f \tag{7.9}$$

provided $v/c \ll 1$. For example, of a microwave beam at 10 GHz is directed towards a car approaching at $15\,\mathrm{m\,s^{-1}}$ then the Doppler shift will be observed between the incident and reflected beams, i.e.

$$f'' - f = \frac{2 \times 10 \times 10^9 \times 15}{3 \times 10^8} = 1000\ \mathrm{Hz} \tag{7.10}$$

In practice, a fraction of the incident beam and the reflected beam are mixed and the resultant difference or **beat frequency** signal, after adequate amplification and filtering, is displayed on a digital frequency meter, preferably calibrated in metres per second.

Piezoelectric transducers usually in the range of 1 to 5 MHz are used to measure the velocity of moving fluids capable of scattering an ultrasonic beam. Part of the signal from the transmitter is mixed with the Doppler shifted scattered signal so that the resultant beat frequency signal has its frequency proportional to the velocity of the fluid. For example, the Doppler shift in ultrasonic beams scattered by blood flowing in arteries is utilised to locate blocked arteries or to estimate the velocity of blood flow in them.

Laser flow meters rely on Doppler shift in the light scattered by a flowing fluid. A laser beam is split in such a way that most of it shines on the fluid and a small fraction of it on a light detector. The scattered light also falls on the detector and combines with the direct beam to produce an output, at a beat frequency proportional to the velocity of flow of the fluid. The technique is used in situations where rapid fluctuations in the flow must be monitored without inserting a transducer into the flow.

References

(1) Cohen, B. G., et al., GaAs-p-n junction diodes for wide range thermometry. *Rev. of Sci. Instrum.*, **34**, 10, pp. 1091–3 (October 1963)

(2) Selar, N., and Pollock, D. B., On diode thermometers. *Solid State Electronics*, **15**, pp. 473–80 (1972)

(3) De Sa, A., and Widdowson, J., An astatic magnetometer with negative feedback. *J. Sci. Instrum.*, **7**, pp. 266–8 (April 1974)

(4) De Sa, A., and McCartan, D. G., A digitally controlled scanning device for a high resolution spectrograph. *J. Sci. Instrum.*, **5**, pp. 1183–5 (December 1972)

(5) Guild, J., *Diffraction Gratings as Measuring Scales*. OUP (1960)

Further reading

Jones, B. E., *Instrumentation Measurement and Feedback*. McGraw-Hill (1977)

Thomson, J., Instrument transducers. *J. Sci. Instrum.*, **34**, pp. 217–21 (June 1957)

8

Frequency analysis

8.1 Introduction

Electronic signals commonly encountered in experimental situations may be broadly divided into periodic, aperiodic and random types. Simple waveforms can be described in terms of parameters such as amplitude, frequency, phase, rise times, decay times, etc. Complex periodic waveforms are often described in terms of the amplitudes and phases of their Fourier components, whereas aperiodic signals are best analysed in terms of their power spectra. Random signals, however, cannot be completely specified by a finite number of parameters and can only be given statistical descriptions. Their behaviour is often predicted in terms of their power spectra, their probability density functions and their mean square values.

In the following sections, the mathematical principles and experimental techniques for describing and distinguishing such signals using frequency-domain and time-domain concepts are outlined. A brief review of the properties and applications of Fourier transforms is included at the beginning along with other useful mathematical concepts.

8.2 Some properties of Fourier transforms and their applications

The Fourier transform pair is used to transform a function of time $f(t)$ into a function of angular frequency $F(\mathrm{j}\omega)$ and vice versa, and is given by

$$f(t) = \frac{1}{2\pi}\int_{-\infty}^{\infty} F(\omega)\ {}^{\mathrm{j}\omega t}\,\mathrm{d}\omega \tag{8.1}$$

$$F(\mathrm{j}\omega) = \int_{-\infty}^{\infty} f(t)\,\mathrm{e}^{-\mathrm{j}\omega t}\,\mathrm{d}t \tag{8.2}$$

Provided that the Fourier transform of a function of time exists, and noting that $F(\omega)$ is real when $f(t)$ is an even function of time, i.e. $f(-t) = f(t)$, and $F(\omega)$ is imaginary when $f(t)$ is an odd function of time, i.e. $f(-t) = -f(t)$, the following properties are useful in signal analysis.

a) **Linearity property** If $F_1(\omega) = F[f_1(t)]$ and $F_2(\omega) = F[f_2(t)]$, and a_1 and a_2 are constants, then

$$F[a_1 f_1(t) + a_2 f_2(t)] = a_1 F_1(\omega) + a_2 F_2(\omega) \tag{8.3}$$

b) **Scaling property** If $F(\omega) = F[f(t)]$, and a is a real constant, then

$$F[f(at)] = \frac{1}{|a|} F\left(\frac{\omega}{a}\right) \tag{8.4}$$

c) **Time-shifting property** If $F(\omega) = F[f(t)]$ then

$$F[f(t-t_0)] = F(\omega)e^{-j\omega t_0} \tag{8.5}$$

d) **Frequency-shifting property** If ω_0 is a real constant and $F(\omega) = F[f(t)]$ then

$$F[f(t)e^{j\omega_0 t}] = F(\omega - \omega_0)$$

This is known as the **modulation theorem.**

e) If $F(\omega) = F[f(t)]$ and $f(t) \to 0$ as $t \to \infty$, then

$$F\left[\frac{d(f)}{dt}\right] = j\omega F(\omega) \tag{8.6}$$

i.e. differentiation in the time domain corresponds to multiplication of the Fourier transform by $j\omega$. Similarly,

$$F\left[\int f(t)\,dt\right] = \frac{F(\omega)}{j\omega} \tag{8.7}$$

i.e. integration in the time domain corresponds to division of the Fourier transform by $j\omega$.

The **duality theorem** is of particular interest in signal analysis. This theorem is a direct consequence of the scaling property of Fourier transforms. Thus, if equation 8.4 is expressed in terms of dimensionless quantities such as t/T = constant and T = constant, then it can be rewritten as

$$F[f(t/T)] = TF(\omega T)$$

where $a = 1/T$, therefore

$$F(\omega T) = \frac{1}{T}\int_{-\infty}^{\infty} f\left(\frac{t}{T}\right)e^{-j\omega t}\,dt \tag{8.8}$$

from equation 8.2. The complex conjugate of $F(\omega T) = F^*(\omega T)$ is obtained by changing the sign of j, i.e.

$$F^*(\omega T) = \frac{1}{T}\int_{-\infty}^{\infty} f\left(\frac{t}{T}\right)e^{j\omega t}\,dt \tag{8.9}$$

Interchanging the dimensionless quantities t/T and ωT gives

$$F^*(t/T) = T\int_{-\infty}^{\infty} f(\omega T)e^{j\omega t}\,d\omega \tag{8.10}$$

Comparing equation 8.10 with 8.1 gives

$$F^*(t/T) = F^{-1}[2\pi T f(\omega T)] \tag{8.11}$$

Replacing T by $T/2\pi$ gives

$$F^*\left[\left(\frac{2\pi t}{T}\right)\right] = F^{-1}\left[Tf\left(\frac{\omega T}{2\pi}\right)\right] \tag{8.12}$$

Equation 8.12 is the duality theorem, in which the roles of t/T and ωT are interchanged in the Fourier transform and its inverse. Its physical meaning can be illustrated as follows.

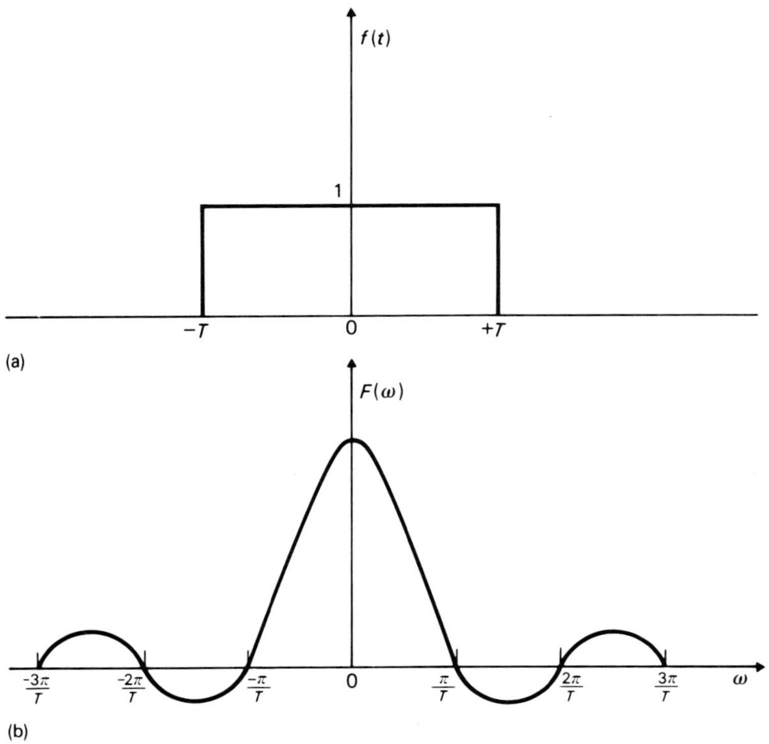

Fig. 8.1 (a) A rectangular pulse; (b) frequency spectrum of (a)

Consider a rectangular pulse defined as

$$f(t) = 1 \quad \text{for } -T \leqslant 0 \leqslant T$$
$$f(t) = 0 \quad \text{outside these limits, Fig. 8.1(a).}$$

Then

$$F[f(t)] = F(\omega) = \int_{-T}^{T} \mathrm{e}^{-\mathrm{j}\omega t}\mathrm{d}t = \frac{1}{\mathrm{j}\omega}\left[\mathrm{e}^{-\mathrm{j}\omega t}\right]_{-T}^{T}$$

Therefore

$$F(\omega) = 2T\left(\frac{\sin \omega T}{\omega T}\right) \tag{8.13}$$

The spectrum is shown in Fig. 8.1(b). The definition $(\sin \omega T)/\omega T$ is called the **sampling function** because of its use in sampled data systems.

If a spectrum is uniform but confined to a frequency range of $-\Omega \leqslant \omega \leqslant \Omega$, Fig. 8.2(b), then $f(t)$ may be replaced by $f(t/T)$ so that the Fourier transform of equation 8.13 becomes

$$F(\omega T) = 2\left(\frac{\sin \omega T}{\omega T}\right) \tag{8.14}$$

The time function that represents the spectrum of Fig. 8.2(a) can be written from the duality

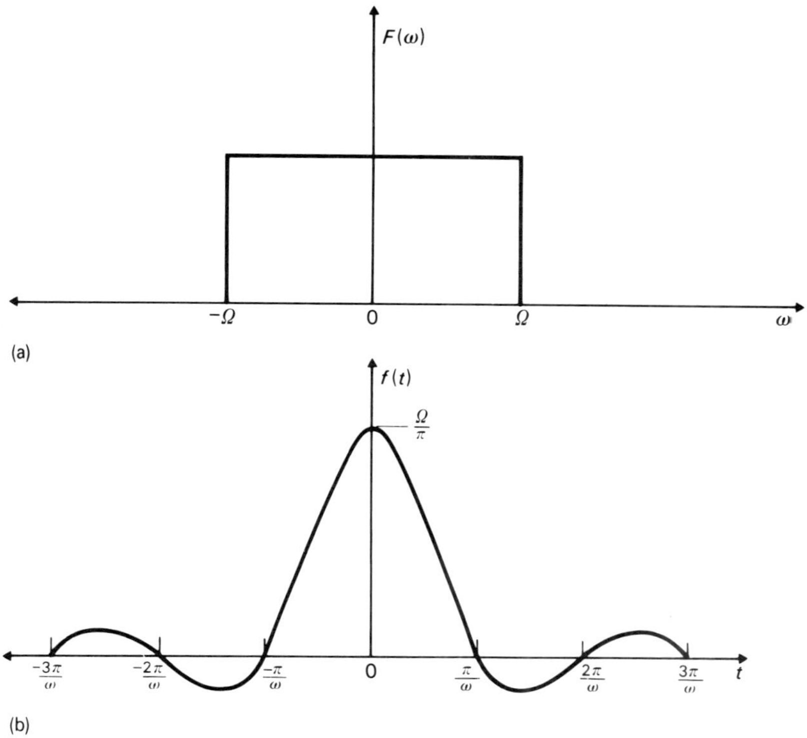

Fig. 8.2 (a) A uniform spectrum confined to the frequency range $\pm\Omega$; (b) a time function $f(t)$ that represents the spectrum of (a)

theorem of equation 8.12 as

$$f(t) = \frac{2}{T}\left(\frac{\sin(2\pi t/T)}{2\pi t/T}\right) \tag{8.15}$$

which is illustrated by Fig. 8.2(b) in which $2\pi/T = \Omega$, i.e.

$$f(t) = \frac{\Omega}{\pi}\left(\frac{\sin\Omega t}{\Omega t}\right) \tag{8.16}$$

Another example is the spectrum of a pair of δ-functions of magnitude $\frac{1}{2}$ represented by

$$f(t) = \tfrac{1}{2}\delta(t+t_1) + \tfrac{1}{2}\delta(t-t_1) \tag{8.17}$$

The Fourier transform of $f(t)$ is given by

$$F(\omega) = \int_{-\infty}^{\infty} \tfrac{1}{2}[\delta(t+t_1)+\delta(t-t_1)]\,e^{-j\omega t}\,dt$$

Therefore

$$F(\omega) = \tfrac{1}{2}(e^{j\omega t_1} + e^{-j\omega t_1}) = \cos\omega t_1 \tag{8.18}$$

Conversely, the spectrum of $\cos\omega_1 t$ consists of a pair of lines at frequencies $\pm\omega_1$.

8.3 Spectral representation of signals

The term 'spectral' is used to describe the properties of any physical system in the frequency domain. As applied to electronic signals, it is used to describe their information content in terms of their amplitude spectra, phase spectra and energy spectra as follows.

Spectral representation of periodic signals

A periodic function defined as any function for which $f(t) = f(t+T)$, the smallest value of the constant T being its period, can be represented by a Fourier series

$$f(t) = C_0 + \sum_{n=1}^{\infty} [C_n \cos(n\omega_0 t - \theta_n)] \tag{8.19}$$

Graphs of the values of $|C_n|$ and θ_n plotted against the frequency $n\omega_0$ are the amplitude and phase spectra of the periodic function. They are often plotted for integral multiples of the fundamental frequency ω_0 and are known as **discrete** spectra. However, it must be noted that a Fourier series representation of a given function $f(t)$ is possible only if the following conditions are satisfied:

i) the function $f(t)$ has a finite number of discontinuities in one period;
ii) the function $f(t)$ has a finite number of maxima and minima in one period;
iii) the function $f(t)$ is absolutely integrable over one period, i.e. $\int_{-T/2}^{T/2} |f(t)|\,dt$ must be finite.

The orthogonality relations between the terms of a Fourier series are as follows.

$$\int_{-T/2}^{T/2} \cos m\omega_0 t\,dt = 0 \qquad \text{for } m \neq 0 \tag{8.20}$$

$$\int_{-T/2}^{T/2} \sin m\omega_0 t\,dt = 0 \qquad \text{for all } m \tag{8.21}$$

$$\int_{-T/2}^{T/2} \cos m\omega_0 t \cos n\omega_0 t\,dt = \begin{cases} 0 & \text{for } m \neq n \\ T/2 & \text{for } m = n \neq 0 \end{cases} \tag{8.22}$$

$$\int_{-T/2}^{T/2} \sin m\omega_0 t \sin n\omega_0 t\,dt = \begin{cases} 0 & \text{for } m \neq n \\ T/2 & \text{for } m = n \neq 0 \end{cases} \tag{8.23}$$

$$\int_{-T/2}^{T/2} \sin m\omega_0 t \cos n\omega_0 t\,dt = 0 \qquad \text{for all } m \text{ and } n \tag{8.24}$$

It is important to remember these relations. For example, the power content of a periodic waveform $f(t)$ in the period T is defined as the mean squared value

$$\overline{P^2} = \frac{1}{T}\int_{-T/2}^{T/2} [f(t)]^2\,dt \tag{8.25}$$

This means that if $f(t)$ represents the voltage across or a current through a 1 Ω resistance then $\overline{P^2}$ is the average power delivered to it. If a Fourier series representation of $f(t)$ is possible, i.e.

$$f(t) = \sum_{n=-N}^{N} C_n e^{jn\omega_0 t} = \sum_{n=0}^{\infty} C_n e^{jn(2\pi/T)t} \tag{8.26}$$

then it can be shown that the mean square value can be approximated to

$$\overline{P^2} = \sum_{n=-N}^{N} C_n C_n^* \tag{8.27}$$

where C_n^* is the complex conjugate of C_n. When $n \to \infty$ then, from equation 8.25,

$$\overline{P^2} = \frac{1}{T}\int_{-T/2}^{T/2} [f(t)]^2\,dt = \sum_{-\infty}^{\infty} C_n C_n^* \tag{8.28}$$

If the trigonometric form of the Fourier series of $f(t)$ given by equation 8.19 is used, then

$$\overline{P^2} = C_n^2 + \sum_{n=1}^{\infty} \overline{\frac{1}{2} C_n^2} \tag{8.29}$$

The mean squared value of a periodic function is the sum of the mean squared values of the amplitudes of its harmonics and is independent of their phases.

Spectral representation of aperiodic signals

An aperiodic signal may be considered as a limiting case of a periodic one when the period $T \to \infty$. Thus

$$F(\omega) = \int_{-\infty}^{\infty} f(t)e^{-j\omega t}dt$$

is analogous to

$$C_n = \frac{1}{T}\int_{-T/2}^{T/2} f(t)e^{-jn\omega t}dt$$

and $$f(t) = \frac{1}{2\pi}\int_{-\infty}^{\infty} F(\omega)e^{j\omega t}d\omega$$

is analogous to

$$f(t) = \sum_{-\infty}^{\infty} C_n^{jn\omega_0 t}$$

However, the units of $F(\omega)$ are the units of C_n multiplied by units of time, i.e. if the units of $f(t)$ and C_n are in volts, then $F(\omega)$ will be in volts/radians per second. The quantity $F(\omega)$ is therefore called the **harmonic (amplitude) density function of** $f(t)$ at an angular frequency ω. The terms **spectral density** and **frequency distribution function** are the other synonyms of $F(\omega)$.

Just as the Fourier series of an **even** periodic function contains only **cosine** terms and that of an **odd** periodic function contains only **sine** terms,

for $f(-t) = f(t)$ (even function)

$$F(\omega) = 2\int_{0}^{\infty} f(t)\cos \omega t\,dt \tag{8.30}$$

for $f(-t) = -f(t)$ (odd function)

$$F(\omega) = -2j\int_{0}^{\infty} f(t)\sin \omega t\,dt \tag{8.31}$$

From equation (8.27), for a **periodic** function, the energy content per period referred to a 1 Ω resistor is given by

$$\int_{-T/2}^{T/2} [f(t)]^2 \mathrm{d}t = T\overline{P^2} = T \sum_{-\infty}^{\infty} C_n C_n^* \tag{8.32}$$

For an **aperiodic** function, the total signal energy referred to a 1 Ω resistance, also known as its **quadratic content** if $T = 1/\Delta f$ where Δf is the spacing between harmonics, is given by

$$C_n C_n^* = \frac{C_n}{\Delta f} \frac{C_n^*}{\Delta f} (\Delta f)^2 \tag{8.33}$$

and equation 8.32 can be rewritten as

$$\int_{-T/2}^{T/2} [f(t)]^2 \mathrm{d}t = T\Delta f \sum_{-\infty}^{\infty} \frac{C_n}{\Delta f} \frac{C_n^*}{\Delta f} \Delta f \tag{8.34}$$

As $T \to \infty$, $\Delta f \to \mathrm{d}f = \mathrm{d}\omega/2\pi$, and since $T\Delta f = 1$, $C_n/\Delta f$ becomes $F(\omega)$. Therefore

$$\int_{-\infty}^{\infty} [f(t)]^2 \mathrm{d}t = \frac{1}{2\pi} \int_{-\infty}^{\infty} F(\omega) F^*(\omega) \mathrm{d}\omega \tag{8.35}$$

Equation 8.35 is called **Parseval's theorem**, and since $F(\omega)F^*(\omega) = |F(\omega)|^2$ it can be written as

$$\int_{-\infty}^{\infty} [f(t)]^2 \mathrm{d}t = \frac{1}{2\pi} \int_{-\infty}^{\infty} |F(\omega)|^2 \mathrm{d}\omega = P_q \tag{8.36}$$

which is a relationship between the spectral density and the quadratic content P_q.

If $f(t)$ is a real function and $F[f(t)] = F(\omega)$, then $|F(\omega)|^2$ is an even function of ω and equation 8.36 can be written as

$$P_q = \frac{1}{\pi} \int_0^{\infty} |F(\omega)|^2 \mathrm{d}\omega \tag{8.37}$$

The quantity $|F(\omega)|^2 = S(\omega)$ is called the **energy density function** since $S(\omega)\mathrm{d}\omega$ will represent the energy in an infinitesimal band of frequencies $\mathrm{d}\omega$, and P_q can be finally written as

$$P_q = \frac{1}{\pi} \int_0^{\infty} S(\omega) \mathrm{d}\omega \tag{8.38}$$

Similarly, the energy in a band of frequencies ω_1 to ω_2 is given by

$$\int_{\omega_1}^{\omega_2} S(\omega) \mathrm{d}\omega \tag{8.39}$$

8.4 Some statistical techniques in signal analysis

The average behaviour of signals that cannot be completely specified by a finite number of parameters can often be predicted by using some of the following statistical techniques.

Averaging

If a continuous function of time $x(t)$ is recorded simultaneously at observation stations A_1,

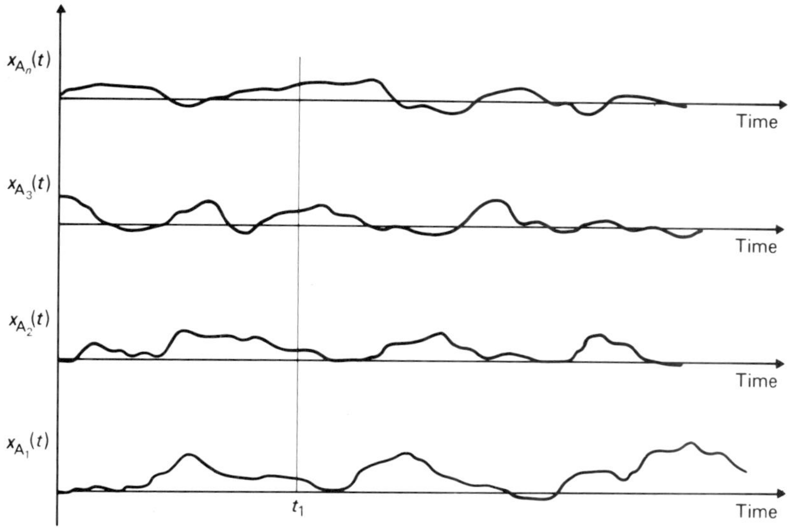

Fig. 8.3 A continuous function of time $x(t)$ recorded simultaneously at observation stations A_1, A_2, A_3, A_n

A_2, A_3, A_N (Fig. 8.3), then the average value of $x(t)$ calculated from such an ensemble at time t_1 will be

$$\mu_x(t_1) = \frac{1}{N}\sum_{k=1}^{N} x_{A_k}(t_1) \tag{8.40}$$

and will depend upon the instant of measurement t_1. If the value of $\mu_x(t_1)$ is time-invariant, i.e. $\mu_x(t_1) = \mu_x(t_2)$, then the ensemble is said to be **stationary**. Non-stationary ensembles of recordings of repetitive signals are often used in signal averaging (see Chapter 10) by dividing the recordings into equal lengths and synchronising them to an 'artificial time zero'.

When a single record of $x(t)$ over a long time period is available, then its average value, known as the **time average**, can be calculated as

$$\mu_x(k) = \lim_{T\to\infty} \frac{1}{T}\int_0^T x_k(t)\mathrm{d}t \tag{8.41}$$

and gives specific information about that recording.

Consider a stationary system in which the ensemble average at any time t_j is equal to the time average of any recording x_k, i.e.

$$\frac{1}{N}\sum_{m=1}^{N} x_m(t_j) = \lim_{T\to\infty}\frac{1}{T}\int_0^T x_k(t)\mathrm{d}t$$

or
$$\mu_x(t_j) = \mu_x(k) \tag{8.42}$$

for all j and k. This is called an **ergodic system**. For example, thermal noise recordings from several identical resistors kept at the same constant temperature would comprise an ergodic ensemble.

Probability density functions (PDF)

The energy density spectrum of a signal cannot specify that signal uniquely because it contains no phase information. It merely shows how the energy is distributed in the frequency domain. For example, two periodic signals will have the same power spectrum if they both contain the same frequency components at the same amplitudes. If, however, the phase of just one component of one signal is shifted with respect to the phase of the corresponding component of the other, the two signals can have drastically different waveforms. The PDF is a statistic of a signal that gives waveform information and is independent of its power spectrum. For example, the PDF of the signal in Fig. 8.4(a) may be computed by the technique illustrated in the block diagram of Fig. 8.4(b).

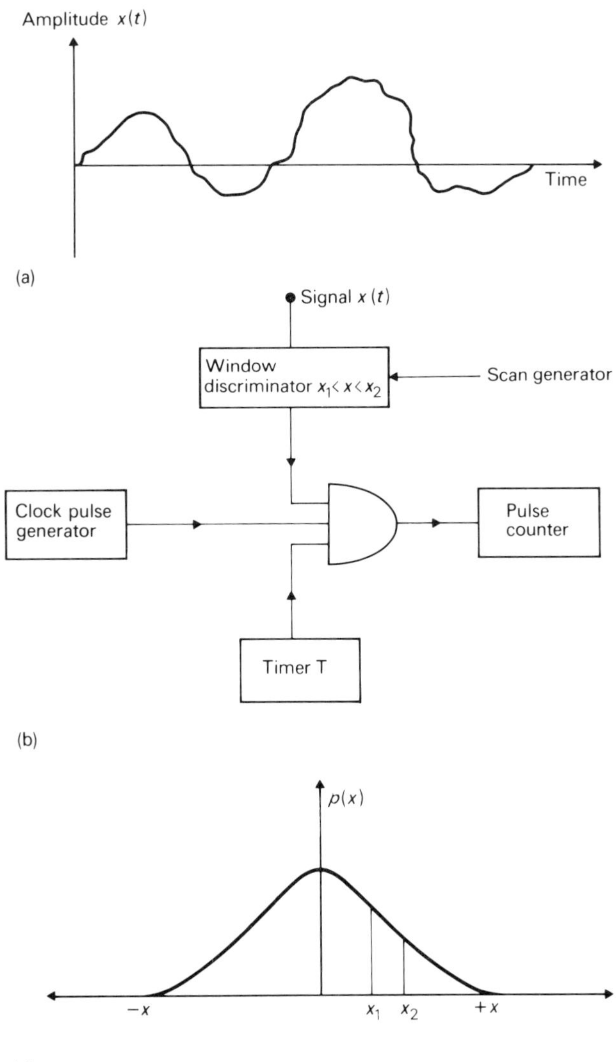

Fig. 8.4 (a) A signal $x(t)$ whose PDF is to be computed; (b) a technique for computing the PDF of signals such as $x(t)$ in (a); (c) PDF of the signal $x(t)$

A high-frequency clock pulse generator is connected via a three-input AND gate to a pulse counter. One input is connected to a timer and the other to a 'window discriminator' (see p. 73). Pulses are counted over an interval T, defined by the timer, only when the signal amplitude $x(t)$ lies in the range $x_1(t) < x(t) < x_2(t)$, as detected by the window discriminator. A scan generator is used to vary both the reference levels of the discriminator simultaneously so that the quantity $[x_2(t) - x_1(t)]$ is constant over the entire amplitude range of $x(t)$. The count accumulated in T sec at any given values of $x_1(t)$ and $x_2(t)$ is proportional to the PDF, $p(x)$, and can be plotted as a function of the amplitude of $x(t)$ as in Fig. 8.4(c). This technique is slow and tedious, and an accurate measurement of the PDF requires a long averaging time T at each reading. More elaborate instruments having 100 or more parallel channels can give in a very short time an on-line display of the complete PDF curve.

The PDF curve can be used to calculate the mean value μ_x of the amplitude of a random signal from the expression

$$\mu_x = \int_{-\infty}^{\infty} xp(x)\mathrm{d}x \tag{8.43}$$

i.e. each amplitude of $x(t)$ is weighted by a factor of $p(x)\mathrm{d}x$ which is the probability that a particular value of $x(t)$ will occur within the infinitesimal interval $\mathrm{d}x$; the integral is computed for all possible values of x. Similarly, the mean square value of a signal can be computed as

$$\overline{\psi^2}(x) = \int_{-\infty}^{\infty} x^2 p(x)\mathrm{d}x \tag{8.44}$$

In practical situations where a signal is quantised into discrete amplitude levels, the cumulative PDF is often used, defined as the integral of the density function or

$$P(x) = \int_{u=-\infty}^{u=x} p(u)\mathrm{d}u \tag{8.45}$$

Estimation of mean values

In a wide variety of experimental situations, such as the evaluation of cross-correlation functions or the mean value of fluctuating signals, averaging is limited to a finite interval of time instead of the theoretical infinite time required to produce precise results. Hence the measured value will differ from the true value and some statistical estimate of the errors involved must be made.

If the probability distribution function of the variable in question is known, then the averaging time necessary to reduce the errors to an acceptable level can be predicted. For example, in the case of band-limited Gaussian noise, the standard deviation σ is given approximately by

$$\sigma^2 = \frac{\sigma_0{}^2}{2BT} \tag{8.46}$$

where σ_0 is the rms value of the fluctuating component of the noise, T is the interval of time over which accurate integration is done by an instrument such as an integrating digital voltmeter instead of an RC filter and B is the noise bandwidth where $BT > 1$. A 95% confidence limit occurs when the actual error is less than 2σ. Conversely, if the accuracy of a result is specified, then the integrating time T required to attain this accuracy with 95%

certainty can be calculated using equation 8.46 and the inequality

$$\text{actual error} < 2\sigma \leqslant = 2\sqrt{\sigma_0^2/2BT} \tag{8.47}$$

since σ_0 and B are known.

8.5 z-transforms and sampled signals

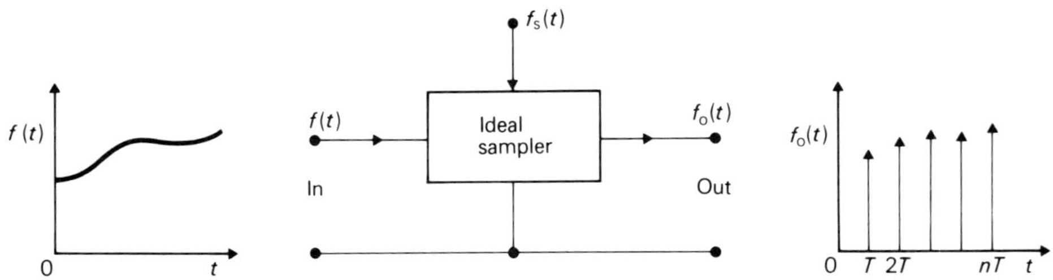

Fig. 8.5 An ideal sampler

The z-transformation technique is used to convert sampled signals in the time domain to an algebraic representation in the z-domain, just as the Laplace transformation technique is used to convert continuous signals from the time domain to the s-domain. Consider a signal $f(t)$, to be sampled by an 'ideal sampling device', Fig. 8.5, so that its output is a train of impulses instead of a train of pulses of finite width. The impulses will have an area $f(nT)$ equal to the magnitude of $f(t)$ at the sampling instants $t = 0, T, 2T, \ldots, nT$. Ideal sampling of a function $f(t)$ with a continuous train of unit impulses denoted by

$$f_s(t) = \sum_{n=-\infty}^{\infty} \delta(t - nT) \tag{8.48}$$

will give an output

$$f_o(t) = f_s(t) f(t) = \sum_{n=-\infty}^{\infty} f(t)\delta(t - nT) \tag{8.49}$$

Now, $\quad L[\delta(t - nT)] = e^{-snT}$ (8.50)

and if $f(t) = e^{s_k t}$ for $t > 0$ and $f(t) = 0$ for $t < 0$, then the output $f_o(t)$ of the samples will be given by

$$f_o(t) = \delta(t) + e^{s_k T}\delta(t - T) + e^{2s_k T}\delta(t - 2T) + \ldots \tag{8.51}$$

and its Laplace transform will be given by

$$L[f_o(t)] = \overline{f}_o(s) = 1 + e^{s_k T}e^{-sT} + e^{2s_k T}e^{-2sT} \tag{8.52}$$

Substituting

$$z = e^{sT} \tag{8.53}$$

produces the z-transform of the function $f_o(t)$ denoted by $f(z)$, i.e.

$$f(z) = 1 + e^{s_k T}z^{-1} + e^{2s_k T}z^{-2} \tag{8.54}$$

Therefore

$$f(z) = \frac{1}{1 - e^{s_k T} z^{-1}} = \frac{z}{z - e^{s_k T}} \tag{8.55}$$

Thus $f(z)$ has a pole $z = e^{s_k T}$ in the z-plane, whose position changes with the sampling interval T. Also, e^{-sT} corresponds to a delay of one sampling interval T in the s-domain while z^{-1} corresponds to a similar delay in the z-domain. Replacing $s = \sigma + j\omega$ in equation 8.53 gives

$$z = e^{(\sigma + j\omega)T} \tag{8.56}$$

Therefore

$$|z| = e^{\sigma T}$$

and its phase angle is equal to ωT so that a point s_k in the s-plane transforms to a point z_k in the z-plane, Fig. 8.6. Since $\sigma = 0$ along the imaginary axis, $|z| = 1$ so that the path along the imaginary axis in the s-plane transforms into a unit circle in the z-plane. Lastly, in order to revert to the time domain, the inverse z-transform has to be found by using techniques such as partial fraction expansion or contour integration.

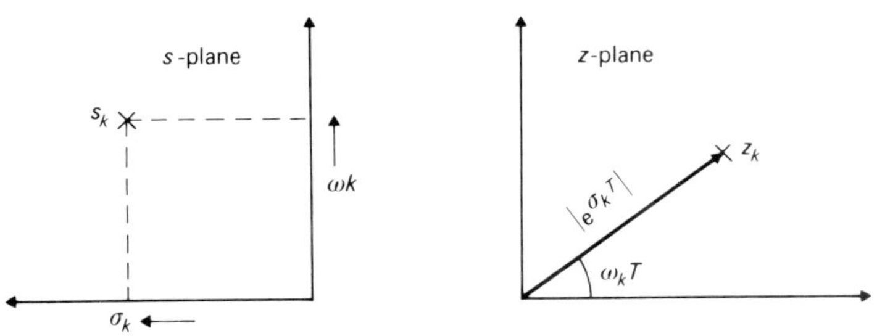

Fig. 8.6 The s-plane and the z-plane

The z-transform of the impulse response of a system is its z-transfer function. If the input to a system is represented by

$$U(z) = u_0 + u_1 z^{-1} + u_2 z^{-2} \tag{8.57}$$

and if the transform of its impulse response is

$$H(z) = h_0 + h_1 z^{-1} + h_2 z^{-2} \tag{8.58}$$

then the output will be given by

$$\begin{aligned} X(z) = {} & u_0(h_0 + h_1 z^{-1} + h_2 z^{-2} + \ldots) \\ & + u_1(h_0 + h_1 z^{-1} + h_2 z^{-2} + \ldots) \\ & + u_2(h_0 + h_1 z^{-1} + h_2 z^{-2} + \ldots) \end{aligned} \tag{8.59}$$

Therefore

$$X(z) = U(z)H(z)$$

or the z-transfer function is given by

$$H(z) = X(z)/U(z) \tag{8.60}$$

The z-transform technique operates on the assumption that the only signals to a system are impulses, and their application to the analysis of digital filters is outlined in Section 8.9, p. 230.

8.6 The time domain and frequency domain

Every function of time has its counterpart in the frequency domain and vice versa. It was shown in Chapter 6 that the concept of the unit impulse response of a linear system leads to the definition of its transfer function $\overline{y}(s)$ which is the generalised relationship between its response and stimulus.

In the time-domain approach, the input signal is considered to be composed of a succession of impulse functions, each of which generates a weighted version of the impulse response at the output of the linear system. The output signal $R(t)$ is found by the superposition of all the responses and is represented by the convolution integral of equation 6.29, i.e.

$$R(t) = \int_0^t s(t-\tau)y(\tau)\mathrm{d}\tau$$

The time-domain approach is ideally suited to situations where a linear system is disturbed by an isolated pulse or step function with well-defined initial conditions, as in the case of step response investigation of the stability of feedback systems (Chapter 6). It is also useful in sampled data systems where the input is a series of pulses, each separated in time from another.

In the frequency-domain approach, the frequency spectrum of the input time function is found by taking its Fourier transform, and the output spectrum is obtained by multiplying the input function by the frequency response function of the linear system. The output time function is finally found by taking the inverse Fourier transform. This procedure can be lengthy and difficult, but there is no similar method for signals in the time domain. For example, the response of a linear system to an aperiodic input signal having a continuous spectrum can be obtained as follows.

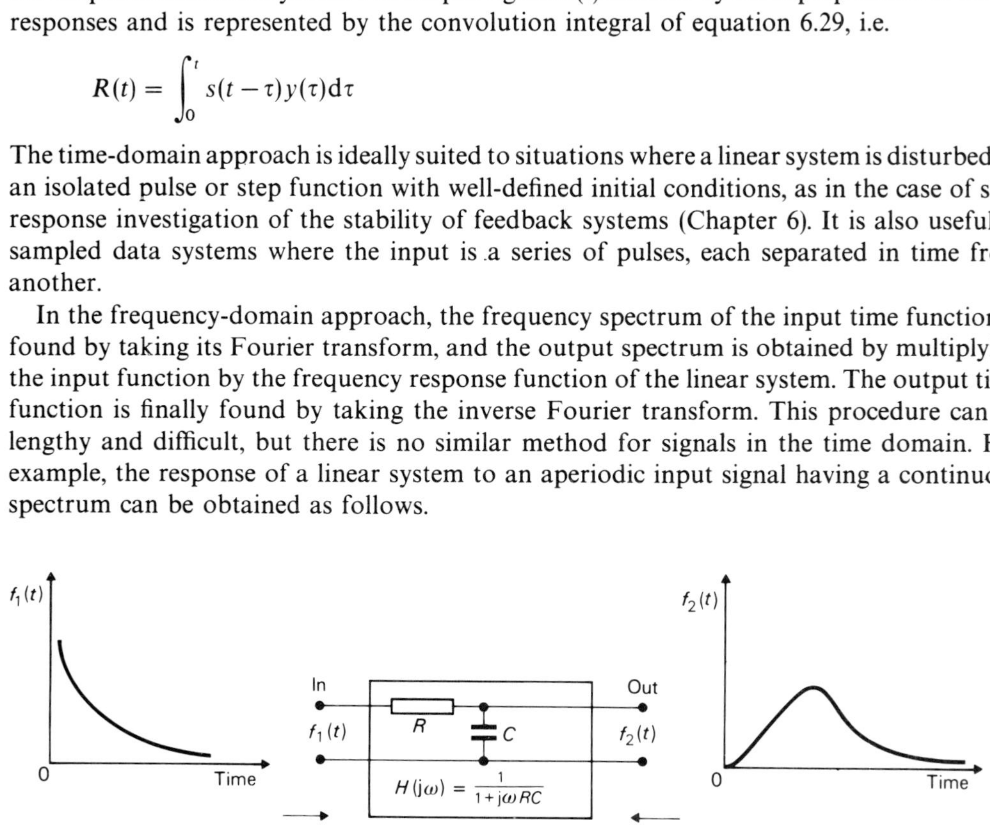

Fig. 8.7 Response of a linear system to an aperiodic input signal

Let the input be given by

$$\begin{aligned} f_1(t) &= \mathrm{e}^{-\alpha t} && \text{for } 0 < t < \infty \\ &= 0 && \text{for } -\infty < t < 0 \end{aligned} \tag{8.61}$$

Its frequency spectrum is given by

$$F_1(j\omega) = F[f_1(t)] = \frac{1}{\alpha + \mathrm{j}\omega} \tag{8.62}$$

Let the frequency response function of the system, Fig. 8.7, be

$$H(\mathrm{j}\omega) = \frac{1}{1+\mathrm{j}\omega t_1} \tag{8.63}$$

where $t_1 = RC$. The frequency spectrum of the output is given by

$$F_2(\mathrm{j}\omega) = F_1(\mathrm{j}\omega)\,H(\mathrm{j}\omega) \tag{8.64}$$

$$= \left(\frac{1}{\alpha+\mathrm{j}\omega}\right)\left(\frac{1}{1+\mathrm{j}\omega t_1}\right)$$

The output signal $f_2(t)$ is $F^{-1}[F_2(\mathrm{j}\omega)]$, the inverse Fourier transform of $F_2(\mathrm{j}\omega)$, and will be given by

$$f_2(t) = \frac{1}{(1-\alpha t_1)}\left(\mathrm{e}^{-\alpha t} - \mathrm{e}^{-t/t_1}\right) \tag{8.65}$$

It must be noted that $H(\mathrm{j}\omega)$ could be the Fourier transform of a signal $h(t)$ and would therefore describe a signal instead of a system, so that signals and systems are analogous in the frequency domain.

The above technique in the frequency domain can also be used to specify the properties of a linear system for performing a particular signal processing task. For example, if the signal $f_1(t)$ is required to be converted to a different waveform $f_3(t)$ using linear processing, then the frequency response function of the linear processor must be

$$H(\mathrm{j}\omega) = F_3(\mathrm{j}\omega)/F_1(\mathrm{j}\omega) \tag{8.66}$$

where $F_3(\mathrm{j}\omega) = F[f_3(t)]$. The quantity $F_3(\mathrm{j}\omega)$ must not be required to contain any frequency components that are absent in $F_1(\mathrm{j}\omega)$ if $H(\mathrm{j}\omega)$ is to be a linear system.

As pointed out in Section 6.2, p. 149, the Laplace transform gives a more general frequency-domain description of a signal than the Fourier transform. It allows one to deduce a frequency-domain description of certain signals for which the Fourier integral fails to converge, and it also allows one to represent the signal by a set of poles and zeros in the s-plane. The Fourier transform technique outlined above is referred to as a **steady state** sinusoidal analysis, because sinusoidal functions are assumed to exist at all times.

Lastly, the **convolution theorem** states that 'the multiplication of two frequency functions, such as $F_1(\mathrm{j}\omega)\,H(\mathrm{j}\omega)$, is equivalent to the convolution of their corresponding time functions. This can be easily verified as follows.

If the convolution integral of equation 6.29, which gives the response to a stimulus in the time domain, is written as

$$f_2(t) = \int_{-\infty}^{\infty} f_1(t-\tau)h(\tau)\,\mathrm{d}\tau \tag{8.67}$$

(see Fig. 8.7) where $f_2(t)$ is the result of convoluting $f_1(t)$, the input, and $h(\tau)$ where

$$h(\tau) = F^{-1}[H(\mathrm{j}\omega)]$$
$$f_1(t) = F^{-1}[F_1(\mathrm{j}\omega)]$$
$$f_2(t) = F^{-1}[F_2(\mathrm{j}\omega)]$$

then in the frequency domain, the output function $F_2(\mathrm{j}\omega)$ is found by substituting for $f_2(t)$

from equation 8.67, giving

$$F_2(\mathrm{j}\omega) = \int_{-\infty}^{\infty} f_2(t)\,\mathrm{e}^{-\mathrm{j}\omega t}\,\mathrm{d}t$$

$$= \int_{-\infty}^{\infty} \left(\int_{-\infty}^{\infty} f_1(t-\tau)\,h(\tau)\,\mathrm{d}\tau \right) \mathrm{e}^{-\mathrm{j}\omega t}\,\mathrm{d}t$$

Therefore

$$F_2(\mathrm{j}\omega) = \int_{-\infty}^{\infty} (F_1(\mathrm{j}\omega)\,\mathrm{e}^{-\mathrm{j}\omega\tau})\,h(\tau)\,\mathrm{d}\tau \tag{8.68}$$

by the time displacement property of Fourier transforms, and finally

$$F_2(\mathrm{j}\omega) = F_1(\mathrm{j}\omega) \int_{-\infty}^{\infty} h(\tau)\,\mathrm{e}^{-\mathrm{j}\omega\tau}\,\mathrm{d}\tau = F_1(\mathrm{j}\omega)\,H(\mathrm{j}\omega) \tag{8.69}$$

8.7 Convolution, correlation and power spectra of waveforms

The process of convolution or folding can be illustrated graphically by considering a triangular waveform $f_1(t)$, Fig. 8.8(a). If this waveform is folded about the $t = 0$ axis, the result

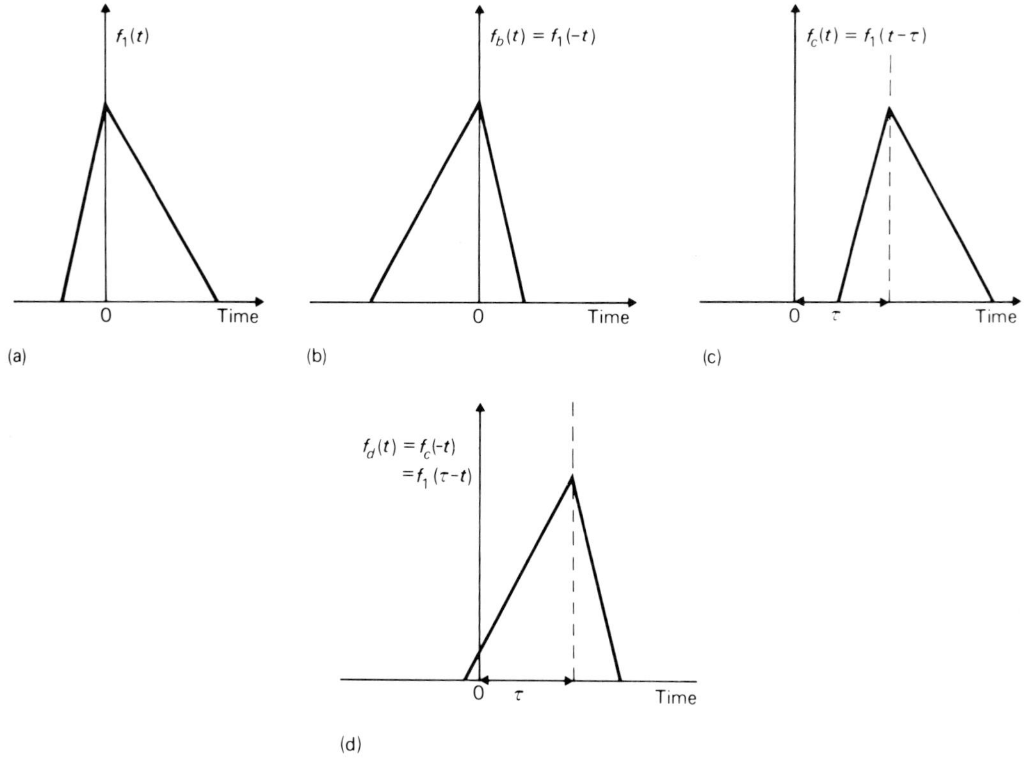

Fig. 8.8 (a) A triangular waveform $f_1(t)$; (b) $f_1(t)$ folded over $t = 0$; (c) $f_1(t)$ delayed by an interval τ; (d) $f_c(t)$ convoluted about an axis $t = \tau$

would be $f_b(t) = f(-t)$, Fig. 8.8(b). If $f_1(t)$ is delayed by a time interval τ, then the result is $f_c(t) = f_1(t-\tau)$, Fig. 8.8(c). Lastly, if $f_c(t)$ is convoluted about an axis $t = \tau$, then the result is $f_d(t) = f_c(-t) = f_1(\tau - t)$, Fig. 8.8(d). The convolution integral of equation 8.67 thus involves the multiplication of $h(\tau)$ with a time-reversed version of $f_1(t)$, followed by an integration over all time to give a value of $f_2(t)$ relevant to a particular instant t. To evaluate the integral graphically for any value of t, one must plot a graph of $h(\tau) f_1(t-\tau)$ versus τ and calculate the area under the graph.

Cross-correlation between two waveforms is a process very similar to convolution and is used to indicate the degree of similarity between two waveforms (see Chapter 10). In general, the cross-correlation function of $f_1(t)$ and $h(t)$ can be written as

$$R_{f_1h}(\tau') = \int_{-\infty}^{\infty} f_1(t-\tau')\,h(t)\,\mathrm{d}t \tag{8.70}$$

The process is illustrated in Fig. 8.9(b) where $h(t)$ and a time-delayed version of $f_1(t)$, i.e. $f_1(t-\tau')$, are multiplied and integrated to give the value $R_{f_1h}(\tau')$ relevant to the particular value of the time delay τ'. A plot of $R_{f_1h}(\tau')$ versus τ' is called a **correlogram**, Fig. 8.9(a). If at any value of τ' the waveforms have a similarity, then the value of the cross-correlation function is positive. One waveform is searching to find itself in the other, and when it does so the cross-correlation function shows a positive peak. If it finds an inverted version, a negative peak is produced.

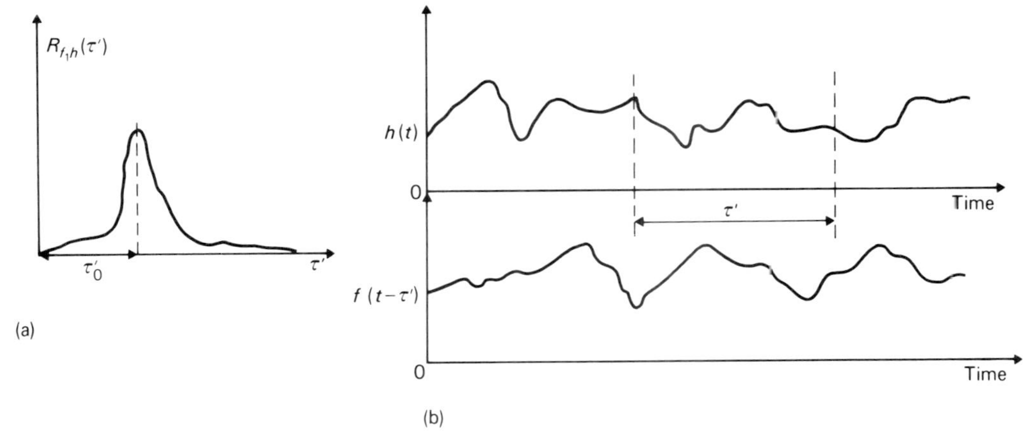

Fig. 8.9 (a) A typical correlogram; (b) a function of time $h(t)$ and a time-delayed function of time $f_1(t-\tau')$

The only substantial difference between the convolution integral of equation 8.67 and the correlation integral of equation 8.70 is that the convolution integral requires the reversal of one of the time functions. Also, from equations 8.67 and 8.69, the convolution function given by

$$f_2(t) = \int_{-\infty}^{\infty} f_1(t-\tau)\,h(\tau)\,\mathrm{d}\tau$$

has its Fourier transform given by

$$F[f_2(t)] = F_2(\mathrm{j}\omega) = F_1(\mathrm{j}\omega)\,H(\mathrm{j}\omega) \tag{8.71}$$

since

$$F[f(-t)] = F(-j\omega)$$

Thus

$$F(-j\omega)H(j\omega) = F\left(\int_{-\infty}^{\infty} f_1(\tau - t)\,h(\tau)\,d\tau\right) \tag{8.72}$$

$$F^{-1}[F(-j\omega)H(j\omega)] = \frac{1}{2\pi}\int_{\omega=-\infty}^{\omega=\infty} F(-j\omega)H(j\omega)\,e^{j\omega t}\,d\omega$$

$$= \frac{1}{2\pi}\int_{\omega=-\infty}^{\omega=\infty} F\left[\int_{-\infty}^{\infty} f_1(\tau - t)\,h(\tau)\,d\tau\right] e^{j\omega t}\,d\omega \tag{8.73}$$

by substitution from equation 8.72. Interchanging the variables t and τ gives

$$F^{-1}[F(-j\omega)H(j\omega)] = \frac{1}{2\pi}\int_{\omega=-\infty}^{\omega=\infty} F\left[\int_{-\infty}^{\infty} f_1(t - \tau)\,h(t)\,dt\right] e^{j\omega\tau}\,d\omega$$

$$= F^{-1}[F\{R_{f_1h}(\tau)\}]$$

from equation 8.70. Therefore

$$F[R_{f_1h}(\tau)] = F(-j\omega)H(j\omega) \tag{8.74}$$

Comparing equations 8.74 and 8.71, if the functions $f_1(t)$ and $h(t)$ are identical then the cross-correlation function $R_{f_1h}(\tau)$ is known as the **autocorrelation function** $R_{ff}(\tau)$ given by

$$R_{ff}(\tau) = \int_{-\infty}^{\infty} f(t - \tau)f(t)\,dt \tag{8.75}$$

and the convolution function becomes the **autoconvolution function** given by

$$f_{o2}(t) = \int_{-\infty}^{\infty} f_1(t - \tau)f_1(\tau)\,d\tau \tag{8.76}$$

From equation 8.74,

$$F[R_{ff}(\tau)] = F(-j\omega)F(j\omega) = |F(j\omega)|^2 \tag{8.77}$$

In other words, the Fourier transform of the autocorrelation function is the energy density function $|F(j\omega)|^2$, the amplitude spectrum of a non-periodic function. From equation 8.71,

$$F[f_{o2}(t)] = F_1(j\omega)F_1(j\omega) = [F_1(j\omega)]^2 \tag{8.78}$$

Thus the Fourier transform of the autoconvolution function gives the complex spectrum, unlike the amplitude spectrum of equation 8.77.

From equation 8.73, if $F(j\omega) = H(j\omega)$ then

$$F^{-1}[F(-j\omega)F(j\omega)] = \frac{1}{2\pi}\int_{\omega=-\infty}^{\omega=\infty} F(-j\omega)F(j\omega)\,e^{j\omega t}\,d\omega$$

$$= \frac{1}{2\pi}\int_{\omega=-\infty}^{\omega=\infty} F\left(\int_{-\infty}^{\infty} f_1(\tau - t)f_1(\tau)\,d\tau\right) e^{j\omega t}\,d\omega$$

Therefore

$$\frac{1}{2\pi}\int_{\omega=-\infty}^{\omega=\infty}|F(\mathrm{j}\omega)|^2\,\mathrm{e}^{\mathrm{j}\omega t}\,\mathrm{d}\omega = \frac{1}{2\pi}\int_{\omega=-\infty}^{\omega=\infty} F\left[\int_{-\infty}^{\infty} f_1(\tau - t)f_1(\tau)\,\mathrm{d}\tau\right]\mathrm{e}^{\mathrm{j}\omega t}\,\mathrm{d}\omega$$

At $t = 0$,

$$\frac{1}{2\pi}\int_{\omega=-\infty}^{\omega=\infty}|F(\mathrm{j}\omega)|^2\,\mathrm{d}\omega = \int_{-\infty}^{\infty}\left[f_1(\tau)\right]^2\mathrm{d}\tau \tag{8.79}$$

Equation 8.79 is Parseval's theorem, deduced as a special case of convolution.

Lastly, the equation that gives the Fourier transform of the product of two time functions $[f_1(t)f_2(t)]$ is often called the convolution formula in the frequency domain, and is useful in spectral analysis. It can be shown that

$$F[f_1(t)f_2(t)] = \frac{1}{2\pi}\int_{-\infty}^{\infty} F_1(\mathrm{j}x)F_2(\mathrm{j}\omega - \mathrm{j}x)\,\mathrm{d}x \tag{8.80}$$

where $F[f_1(t)] = F_1(\mathrm{j}\omega)$, $F[f_2(t)] = F_2(\mathrm{j}\omega)$ and x is a dummy variable. The proof of equation 8.80 is as follows. From the expression

$$f_1(t) = \frac{1}{2\pi}\int_{-\infty}^{\infty} F_1(\mathrm{j}\omega)\,\mathrm{e}^{\mathrm{j}\omega t}\,\mathrm{d}\omega$$

or

$$f_1(t) = \frac{1}{2\pi}\int_{-\infty}^{\infty} F_1(\mathrm{j}x)\,\mathrm{e}^{\mathrm{j}xt}\,\mathrm{d}x$$

the Fourier transform of the product is given by

$$F[f_1(t)f_2(t)] = \int_{t=-\infty}^{t=\infty}\left(\frac{1}{2\pi}\int_{x=-\infty}^{x=\infty} F_1(\mathrm{j}x)\,\mathrm{e}^{\mathrm{j}xt}\,\mathrm{d}x\right)f_2(t)\,\mathrm{e}^{-\mathrm{j}\omega t}\,\mathrm{d}t$$

Interchanging the order of integration gives

$$F[f_1(t)f_2(t)] = \frac{1}{2\pi}\int_{x=-\infty}^{x=\infty}\int_{t=-\infty}^{t=\infty} f_2(t)\,\mathrm{e}^{-\mathrm{j}(\omega - x)t}F_1(\mathrm{j}x)\,\mathrm{d}t\,\mathrm{d}x$$

but

$$\int_{t=-\infty}^{t=\infty} f_2(t)\mathrm{e}^{-\mathrm{j}(\omega - x)t}\,\mathrm{d}t = F_2(\mathrm{j}\omega - \mathrm{j}x) \tag{8.81}$$

Therefore

$$F[f_1(t)f_2(t)] = \frac{1}{2\pi}\int_{x=-\infty}^{x=\infty} F_1(\mathrm{j}x)F_2(\mathrm{j}\omega - \mathrm{j}x)\,\mathrm{d}x \tag{8.82}$$

8.8 Sampling and reconstitution of electronic signals

It is often necessary to represent a continuous signal $f_1(t)$, Fig. 8.10(a), by samples of its amplitude at known intervals of time. These samples may be represented by pulses of varying amplitude, as shown in Fig. 8.10(b), commonly used in communications systems. They may

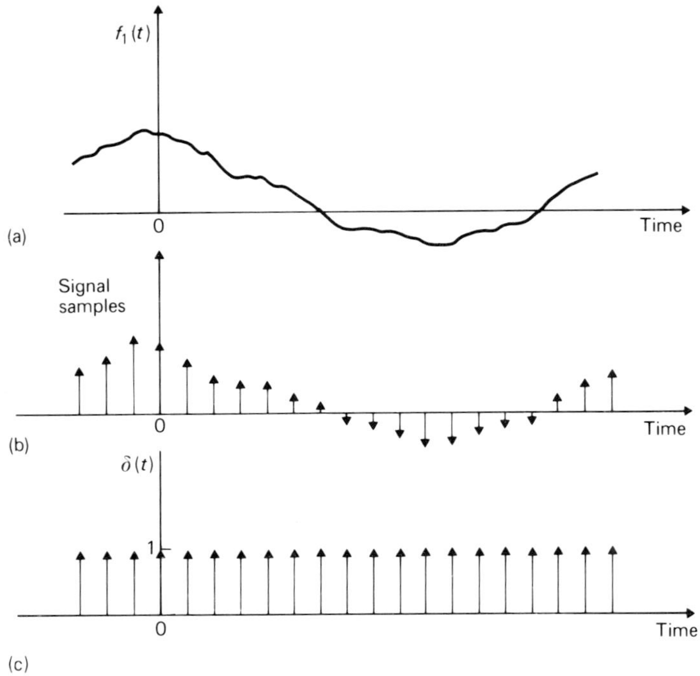

Fig. 8.10 (a) A continuous signal $f_1(t)$; (b) signal samples represented as pulses of varying amplitude; (c) a train of δ–functions

also be converted into some other form, for example in the A/D converters used in digital processing of signals (see Chapter 5). The basic problem of such a representation is to determine the minimum number of samples required to define the signal uniquely, as too low a sampling rate will lead to a loss of information about its waveform and too high a sampling rate will produce an unnecessarily large number of samples.

Initially each sample may be considered as a weighted δ-function proportional to the value of the amplitude of the waveform at the relevant instant, Fig. 8.10(b). The sampled signal is thus a result of multiplying the continuous signal $f_1(t)$ with a train of δ-functions, Fig. 8.10(c). Its spectrum is found by convoluting their respective spectra as follows.

Firstly, the spectrum of a unit δ-function symmetrical about $t = 0$ is

$$F[\delta(t)] = F(\omega) = 1 \tag{8.83}$$

i.e. its spectrum has a constant unit amplitude and extends over all values of ω, Fig. 8.11.

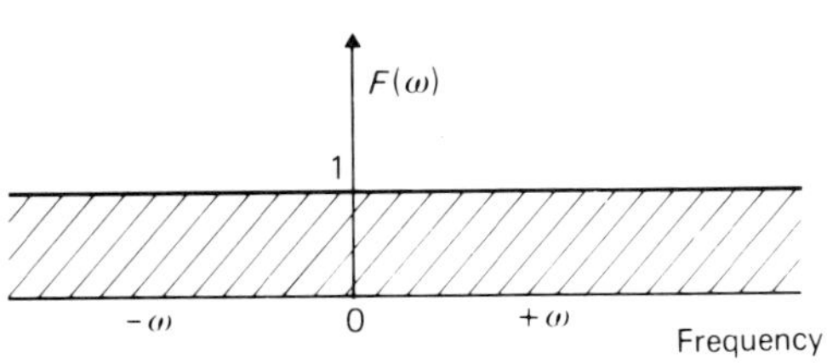

Fig. 8.11 Spectrum of a unit δ–function symmetrical about $t = 0$

If we now consider an infinite train of unit δ-functions which is strictly periodic with a period T, the sampling interval, then its spectrum can be represented as

$$F(\omega) = \int_{-\infty}^{\infty} \delta(t) \cos n\omega t \, dt \tag{8.84}$$

since $\delta(t)$ is an even function. Therefore,

$$F(\omega) = \cos n\omega T \tag{8.85}$$

where $n = 1, 2, 3, \ldots$ and $\omega = 2\pi/T$.

In other words, the spectrum of a periodic δ-pulse train will consist of an infinite set of cosine harmonics, all of the same unit amplitude, separated by $\omega = 2\pi/T$, Fig. 8.12. Comparing Figs 8.10(c) and 8.12, it is obvious that the time-domain waveform is identical to its frequency spectrum.

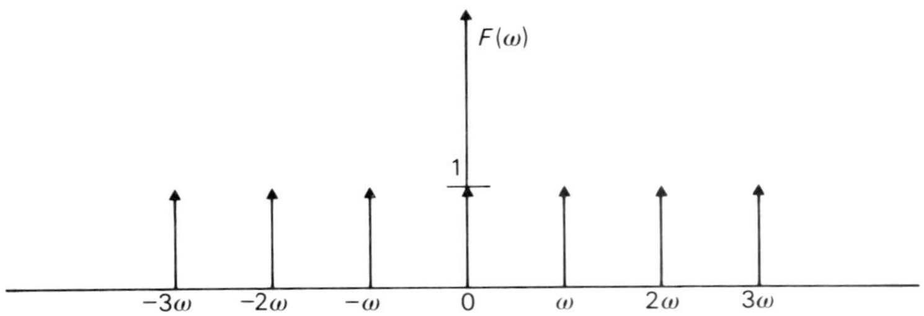

Fig. 8.12 Spectrum of a periodic δ-pulse train

Secondly, if we consider $f_1(t)$ to be a real function, although in principle it can be a complex one, then the spectrum will be

$$F[f_1(t)] = F_1(\omega) \tag{8.86}$$

The result of convoluting $F_1(\omega)$ and $F(\omega)$ will be

$$F_2(\omega) = \int_{-\infty}^{\infty} F_1(\omega - \omega') F(\omega') \, d\omega' \tag{8.87}$$

where ω' is a dummy variable.

Now $F(\omega)$ can be represented by a set of equally-spaced spectral lines or δ-functions in the frequency domain, Fig. 8.12, i.e.

$$F(\omega) = \ldots \delta\left(\omega + \frac{4\pi}{T}\right) + \delta\left(\omega + \frac{2\pi}{T}\right) + \delta(\omega) + \delta\left(\omega - \frac{2\pi}{T}\right) + \delta\left(\omega - \frac{4\pi}{T}\right) \ldots \tag{8.88}$$

Therefore

$$F_2(\omega) = \int_{-\infty}^{\infty} F_1(\omega - \omega')\,\delta\!\left(\omega' + \frac{4\pi}{T}\right) d\omega' + \int_{-\infty}^{\infty} F_1(\omega - \omega')\,\delta\!\left(\omega' + \frac{2\pi}{T}\right) d\omega' + \int_{-\infty}^{\infty} F_1(\omega - \omega')\,\delta\omega'\, d\omega' + \ldots \tag{8.89}$$

or $$F_2(\omega) = F_1\!\left(\omega + \frac{4\pi}{T}\right) + F_1\!\left(\omega + \frac{2\pi}{T}\right) + F_1(\omega) + F_1\!\left(\omega - \frac{2\pi}{T}\right) + \ldots \tag{8.90}$$

It will be noted that all the values on the right-hand side of equation 8.90 are replicas of the signal spectrum $F_1(\omega)$ bodily shifted by $2\pi/T$ radians per second along the frequency axis.

If $F_1(\omega)$ has significant frequency components in the range $\pm\omega_n$, as sampling causes the spectrum to repeat every $2\pi/T$ radians per second, ω_n must be less than π/T in order to avoid an overlap between adjacent repetitions of $F_1(\omega)$. In the absence of such an overlap the original waveform can, in principle, be reconstructed from its sampled version by merely passing it through an ideal linear filter that transmits equally, without phase shift, all components in the frequency range $-\omega_n$ to $+\omega_n$ and rejects all others.

The minimum sampling rate for adequate representation of a continuous signal is formally specified by the sampling theorem, which states that a continuous signal that contains no significant frequency components above f Hz may in principle be recovered from its sampled version if the sampling interval is less than $(1/2f)$ seconds. In practice, the reconstitution of the waveform of such a signal from its δ-function samples is not an exact replica of the original, mainly because no physically realisable filter has an infinitely sharp cut-off frequency range, neither does it transmit all the frequency components equally without introducing a phase shift. The similarity between a signal and its reconstituted version may be enhanced by using a sampling rate somewhat higher than that specified by the sampling theorem. The spectrum of the sampled signal then contains a band of frequencies with a negligible information content, so that the reconstituting filter does not need to have an infinitely sharp cut-off frequency range.

Periodic signals are often sampled with a rectangular pulse so that the waveform is observed over a limited interval of time, i.e. the duration of the pulse sometimes called the **observation window**. If $f(t)$ is an even periodic function of time with period T seconds, then its spectrum $F(\omega)$ will consist of lines $2\pi/T$ apart, the height of each line representing the magnitude of a cosine harmonic. The spectrum of the rectangular pulse is of the form $(\sin x)/x$, and the process of sampling a waveform with a 'time window' of known characteristics is mathematically equivalent to multiplying the signal function $f(t)$ by the function of the rectangular pulse. However, the multiplication of two time functions is equivalent to the convolution of their spectra, so that when the line spectrum of the signal is convoluted with a spectrum of form $(\sin x)/x$, the result $F_0(\omega)$ is a repeated weighted version of the $(\sin x)/x$ pattern about each of the spectral lines, Fig. 8.13. Narrow sampling windows cause a spreading of the spectral energy, which is undesirable if it is required to distinguish between the relative magnitudes of closely-spaced frequency components of the signal. However, the spreading can be exploited to smooth out short-term fluctuations in the spectrum.

Finally, it is worthwhile noting that a signal that is time-limited, i.e. that only exists within

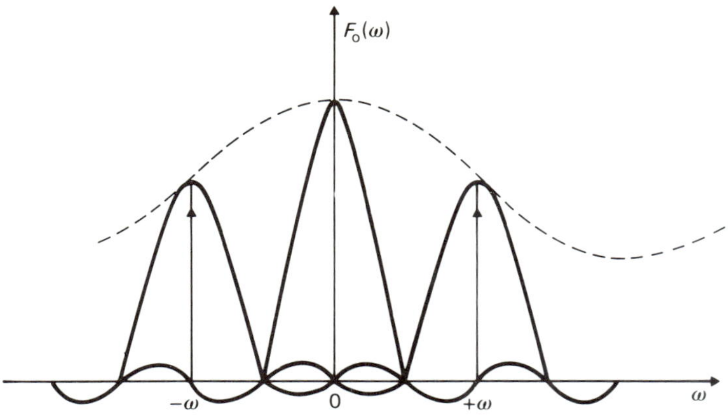

Fig. 8.13 Spectrum of a sampled waveform

the 'window', can be adequately represented by a set of spectral lines in the frequency domain. This is analogous to the fact that a frequency-limited signal can be represented by a set of samples in the time domain, the minimum number of samples necessary being stipulated by the sampling theorem.

8.9 Filters

In general, filters are transmission systems that alter the shape of the frequency spectra of the signals passing through them. The most commonly encountered requirements of such systems, in their idealised form, are as follows:

1) **low-pass** filters where no frequencies *above* a cut-off value ω_c should be transmitted;
2) **high-pass** filters where no frequencies *below* a cut-off value ω_c should be transmitted;
3) **band-pass** filters where only frequencies lying *between* two preset values ω_1 and ω_2 *should be* transmitted;
4) **rejection** filters where a single frequency or a band of frequencies *should not be* transmitted.

These functions can be realised by using analogue filters or digital filters.

Analogue filters

It is neither fruitful nor convenient to analyse specific circuits, because the range of both active and passive circuits is enormous and is adequately discussed in other texts.[1, 2]

However, it is worthwhile outlining the general principles of ideal linear filters and the main difficulties in their physical implementation.

Firstly, consider an ideal linear filter with a frequency response function (Fig. 8.14)

$$H(j\omega) = h(\omega)\,e^{j\theta(\omega)} \tag{8.91}$$

where $h(\omega)$ is the amplitude response function and $\theta(\omega)$ is the phase response function. This function should transmit a signal $v_1(t)$ without distortion, i.e. the output waveform $v_2(t)$ should be of the same shape as the input waveform $v_1(t)$. Except in the trivial case where

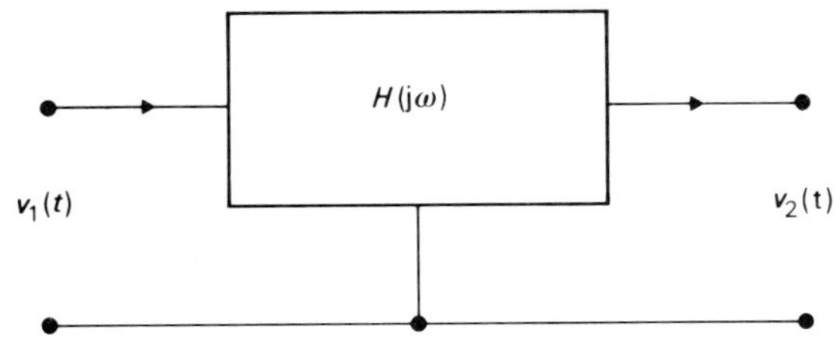

Fig. 8.14 An ideal filter

$H(j\omega) = 1$, the necessary and sufficient condition for distortion-free transmission can be deduced as follows.

If $F[v_1(t)] = V_1(j\omega)$ and $F[v_2(t)] = V_2(j\omega)$, then

$$V_2(j\omega) = H(j\omega)V_1(j\omega) = h(\omega)e^{j\theta(\omega)}V_1(j\omega) \tag{8.92}$$

Now, if

$$v_2(t) = h_0 v_1(t - T_d) \tag{8.93}$$

then the shapes of $v_2(t)$ and $v_1(t)$ will be the same provided that h_0 and T_d are constants, i.e. $v_1(t)$ will be scaled by a factor h_0 and delayed by an interval T_d. From equation 8.92,

$$v_2(t) = F^{-1}[V_2(j\omega)] = \frac{1}{2\pi}\int_{-\infty}^{\infty} h(\omega)e^{j\theta(\omega)}V_1(j\omega)e^{j\omega t}d\omega \tag{8.94}$$

Replacing t by $(t - T_d)$ gives

$$v_2(t - T_d) = \frac{1}{2\pi}\int_{-\infty}^{\infty} h(\omega)e^{j\theta(\omega)}V_1(j\omega)e^{j\omega(t-T_d)}d\omega$$

Therefore

$$v_2(t - T_d) = \left(\frac{1}{2\pi}\int_{-\infty}^{\infty} h(\omega)e^{j\theta(\omega)}V_1(j\omega)e^{j\omega t}d\omega\right)e^{-j\omega T_d} \tag{8.95}$$

Comparing equations 8.91, 8.94 and 8.95, distortion-free transmission requires

$$H(j\omega) = h_0 e^{-j\omega T_d} \tag{8.96}$$

i.e. the amplitude response function $h(\omega) = h_0$, which is independent of frequency, and the phase response function $\theta(\omega) = -\omega T_d$, which is proportional to ω.

In physically realisable networks, the above two conditions can seldom be satisfied simultaneously and the ideal frequency cut-off is never infinitely sharp. For example, consider an ideal band pass filter whose characteristics are defined by

$$H(\omega) = h_0 e^{-j\omega T_d} \quad \text{for } \omega_1 < |\omega| < \omega_2 \tag{8.97}$$

$$H(j\omega) = 0 \quad \text{for } \omega_1 > |\omega| > \omega_2 \tag{8.98}$$

This filter may be realised by using resistors, capacitors and active elements, and the ideal and practical characteristics are shown in Figs 8.15(a) and (b) respectively.

Better approximations to ideal differentiators and integrators that operate over a wide frequency range can be realised by the use of active filters, utilising the properties of modern, inexpensive, IC operational amplifiers.

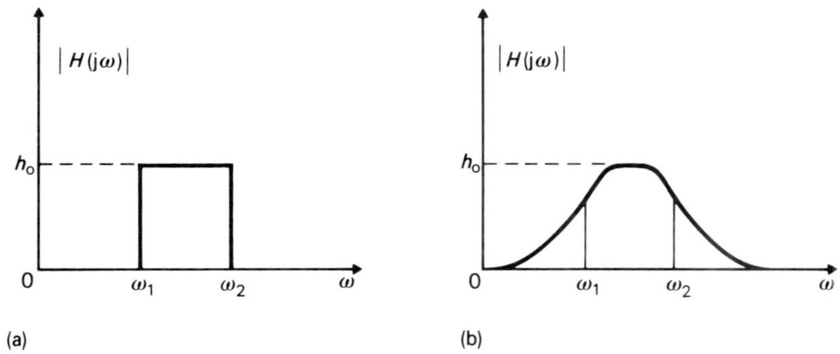

Fig. 8.15 (a) Ideal band-pass filter characteristics; (b) practical band-pass filter characteristics

Modern analogue filters are synthesised from certain known physically realisable voltage transfer functions, so that ideal filter characteristics can be closely approximated. Consider the case of a filter with a transfer function

$$H(s) = \frac{1}{(a_n s^n + a_{n-1}s^{n-1} + \ldots + a_1 s + a_0)} \tag{8.99}$$

where the factors of the denominator will represent the poles of $H(s)$. The square of the magnitude of the frequency response function $|H(j\omega)|^2 = H(j\omega)H(-j\omega)$ will obviously be a polynomial with even powers of ω. Hence, any desired value of $|H(j\omega)|^2$ can be approximated by choosing a suitable polynomial in ω^2 for the denominator in equation 8.99.

The two most commonly-used approximations are as follows.

The **Butterworth approximation** is defined by

$$|H_1(j\omega)|^2 = \frac{1}{1 + (\omega/\omega_0)^{2n}} \tag{8.100}$$

The **Chebychev approximation** is defined by

$$|H_2(j\omega)|^2 = \frac{1}{1 + \varepsilon^2 C_n^{\,2}(\omega/\omega_0)} \tag{8.101}$$

where C_n = nth order Chebychev polynomial and $\varepsilon^2/C_n^{\,2}(\omega/\omega_0)$ oscillates between 0 and ε^2 in the frequency range $-\omega_0 < \omega < \omega_0$ for any value of n which is equal to the number of poles that must be on the left-hand side of the s-plane for the filter to be stable (see Chapter 6). The location of the poles can be found by substituting $\omega = s/j$ in the denominator. For example, in equation 8.100,

$$|H_1(s)|^2 = \frac{1}{1 \pm (s/\omega_0)^{2n}}$$

depending on whether n, also called the **order** of the filter, is even or odd. Frequency response curves for a low-pass Butterworth filter are shown in Fig. 8.16(a), and those for a low-pass Chebychev filter are shown in Fig. 8.16(b).

Numerical examples of the design of both types of filters are given in other texts.[3]

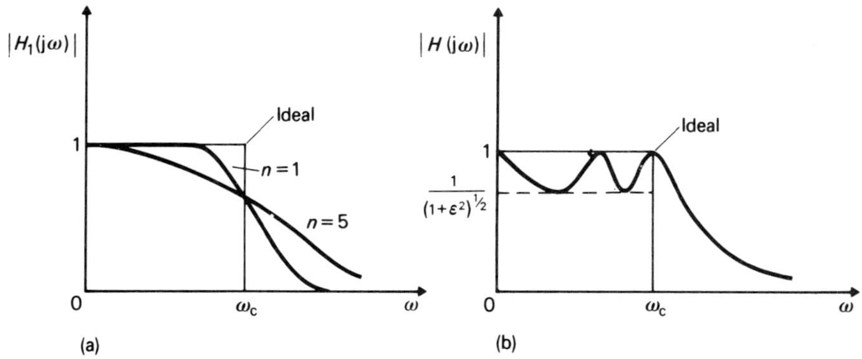

Fig. 8.16 (a) Frequency response curves of a low-pass Butterworth filter; (b) frequency response curves of a low-pass Chebychev filter

Digital filters

Digital filters are described in terms of input–output relationships instead of the voltage–current relationships used in the analysis of linear networks. The transfer function of a digital filter is defined as the ratio of the z-transform of the output signal to the z-transform of the input signal, or in terms of the z-transform of the impulse response as in equation 8.60. Let

u_k be the kth sample value of the input signal;
$u_{(k-r)}$ be the $(k-r)T$th sample value of the same signal;
v_k be the kth sample value of the output signal;
$v_{(k-r)}$ be the $(k-r)T$th sample value of the same signal.

If $n+1$ input samples are used along with m previous output samples, then the kth sample of the output sample will be

$$v_k = (a_0 u_k + a_1 u_{(k-1)} + \ldots + a_n u_{k-n}) - (b_1 v_{k-1} + \ldots + b_m v_{k-m}) \tag{8.102}$$

where the coefficients are real and constant. If $U(z)$ is the z-transform of the input signal and $V(z)$ is the z-transform of the output signal, the transfer function $G(z) = V(z)/U(z)$ describes a general linear digital filter, therefore

$$G(z) = \frac{V(z)}{U(z)} = \left(\frac{a_0 + a_1 z^{-1} + \ldots + a_n z^{-n}}{1 + b_1 z^{-1} + \ldots + b_m z^{-m}}\right) \tag{8.103}$$

This is the transfer function of a recursive digital filter in which the coefficients of the denominator are identically non-zero. Nonrecursive digital filters have a transfer function which is a polynomial in z^{-1}.

The polynomials of the numerator and denominator of equation 8.103 can be factorised as

$$G(z) = A\left(\frac{(z^{-1} - \alpha_1)(z^{-1} - \alpha_2)\ldots(z^{-1} - \alpha_n)}{(1 - \beta_1 z^{-1})(1 - \beta_2 z^{-1})\ldots(1 - \beta_n z^{-1})}\right) \tag{8.104}$$

where α_1 and β_1 may be real or complex conjugate pairs and A is a real constant. The quantities α_i, $i = 1, 2, 3, \ldots, n$ are the zeros and β_r, $r = 1, 2, 3, \ldots, m$ are the poles of the transfer function $G(z)$ located in the z^{-1}-plane. For stability, a digital filter must have the poles of its transfer function located outside the unit circle $|z^{-1}| = 1$. For example, a digital filter with a transfer function

$$G(z) = \frac{1+z^{-1}}{(1+0.5z^{-1})(1-0.4z^{-1})} \tag{8.105}$$

has one pole at $z^{-1} = -1/0.5 = -2$ and another at $z^{-1} = 1/0.4 = 2.5$. Both poles lie outside the unit circle on the z^{-1}-plane and therefore the filter is stable. However, a digital filter with a transfer function

$$G(z) = \frac{1+z^{-1}}{(1+5z^{-1})(1-4z^{-1})} \tag{8.106}$$

has one pole at $z^{-1} = -1/5 = -0.2$ and another pole at $z^{-1} = -1/4 = -0.25$. Both poles lie inside the unit circle on the z^{-1}-plane and therefore the filter is unstable.

The frequency response of the digital filter represented by equation 8.102 may be calculated as follows. Let $u_k = \mathrm{e}^{\mathrm{j}\omega kT}$, where $k = 0, 1, 2, 3, \ldots$. Since the system is linear, the output v_k is given by $H(\omega)\mathrm{e}^{\mathrm{j}\omega kT}$ where $H(\omega)$ is complex. Equation 8.102 will reduce to

$$\begin{aligned} H(\omega)\mathrm{e}^{\mathrm{j}\omega kT} = {} & (a_0\,\mathrm{e}^{\mathrm{j}\omega kT} + a_1\mathrm{e}^{\mathrm{j}\omega(k-1)T} + \ldots + a_n\mathrm{e}^{\mathrm{j}\omega(k-n)T} + \ldots) \\ & - (b_1 H(\omega)\mathrm{e}^{\mathrm{j}\omega kT} + \ldots + b_m H(\omega)\,\mathrm{e}^{\mathrm{j}\omega(k-m)T}) \end{aligned} \tag{8.107}$$

Therefore

$$H(\omega) = \frac{a_0 + a_1\mathrm{e}^{-\mathrm{j}\omega T} + \ldots + a_n\mathrm{e}^{-\mathrm{j}n\omega T}}{1 + b_1\mathrm{e}^{-\mathrm{j}\omega T} + \ldots + b_m\mathrm{e}^{-\mathrm{j}m\omega T}} \tag{8.108}$$

The modulus of $H(\omega)$ gives the amplitude characteristics of the digital filter and the argument of $H(\omega)$ gives its phase characteristics.

The basic building blocks in terms of which the transfer functions of digital filters may be implemented are:

i) multipliers;
ii) time delays;
iii) adders.

For example, the implementation of the transfer function of equation 8.103 is illustrated in Fig. 8.17, in which the time delays are represented by z^{-1} combined with the multipliers, and the summation Σ yields the output.

Lastly, rewriting equation 8.103 as

$$G(z) = \frac{\sum_r a_r z^{-r}}{\sum_r b_r z^{-r}} \tag{8.109}$$

and substituting $z^{-1} = \cos\omega T - \mathrm{j}\sin\omega T$ gives

$$G(\omega T) = \frac{\sum_r a_r \cos r\omega T - \mathrm{j}\sum_r a_r \sin r\omega T}{\sum_r b_r \cos r\omega T - \mathrm{j}\sum_r b_r \sin r\omega T}$$

Therefore

$$G(T) = \frac{\sum_r c_r \cos^r \omega T - \mathrm{j}\sin\omega T \sum_r d_r \cos^r \omega T}{\sum_r c_r \cos^r \omega T - \mathrm{j}\sin\omega T \sum_r d_r \cos^r (}$$

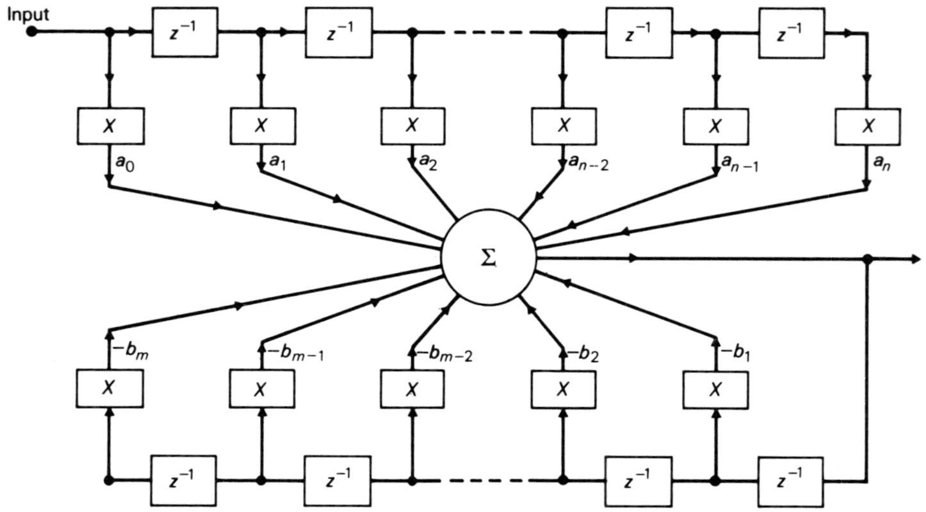

Fig. 8.17 Implementation of the transfer function $G(z) = \dfrac{(a_0 + a_1 z^{-1} + \ldots + a_n z^{-n})}{(1 + b_1 z^{-1} + \ldots + b_m z^{-m})}$

and $$|G(\omega T)|^2 = \frac{\left(\sum_r c_r \cos^r \omega T\right)^2 + (1 - \cos^2 \omega T)\left(\sum_r d_r \cos^r \omega T\right)^2}{\left(\sum_r e_r \cos^r \omega T\right)^2 + (1 - \cos^2 \omega T)\left(\sum_r f_r \cos^r \omega T\right)^2}$$

$$= = \frac{\sum_r g_r \cos^r \omega T}{\sum_r h_r \cos^r \omega T}$$

Substituting,

$$\cos \omega T = \frac{1 - \tan^2 (\omega T/2)}{1 + \tan^2 (\omega T/2)}$$

and $$|G(\omega T)|^2 = \frac{\sum_r \alpha_r \tan^{2r} (\omega T/2)}{\sum_r \beta_r \tan^{2r} (\omega T/2)} \tag{8.110}$$

Equation 8.110 may be expressed as

$$|G(\omega T)|^2 = \frac{1}{1 + \varepsilon^2 P^2(\omega)} \tag{8.111}$$

where ε is a real constant and $P^2(\omega)$ is a real rational function in $\tan^2 (\omega T/2)$ given by

$$P^2(\omega) = \sum_{r=0}^{n} a_r \tan^{2r}(\omega T/2) \tag{8.112}$$

where $\tan (\omega T/2)$ is the low-pass frequency variable and a_r are real constant coefficients.

In the simplest case,

$$P^2(\omega) = a \tan^{2n}(\omega T/2) \quad \text{and} \quad |G| = \frac{1}{[1 + a \tan^{2n} (\omega T/2)]^{1/2}}$$

which will fall by 3 dB at the cut-off frequency ω_c when the constant $a = \tan^{2n}(\omega_c T/2)^{-1}$, therefore

$$|G(\omega)|^2 = \frac{1}{\left(1 + \frac{\tan(\omega T/2)}{\tan(\omega_c T/2)}\right)^{2n}}$$

Thus

$$|G(0)| = 1 \qquad |G(\omega_c)| = 1/\sqrt{2} \qquad |G(\pi/T)| = 0$$

which resembles the analogue Butterworth-type low-pass filter.

In practice, digital filter transfer functions are realised by suitable programming of a digital computer, which is fed with a sampled version of the input signal. The use of the **Fast Fourier Transform** (FFT), as a computational tool for filter simulation and spectral analysis,[4] leads to substantial reductions in computing time, provided that some of the pitfalls in its use are avoided (see Chapter 10).

References

(1) Ghausi, M. S., *Principles and Design of Linear Active Circuits.* McGraw-Hill (1963)
(2) Ruston, H., and Bordogna, J., *Electric Networks.* McGraw-Hill (1966)
(3) Connor, F. R., *Networks.* Edward Arnold (1971)
(4) Cochran, William T., Cooley, J. W., *et al.*, What is the Fast Fourier Transform? *Proc. IEEE*, **55**, 10, pp. 1664–74 (October 1967)

Further reading

Bendat, J. S., and Piersol, A. G., *Random Data: Analysis and Measurement Procedures.* Wiley-Interscience (1971)
Bogner, R. E., and Constantinides, A. G., *Introduction to Digital Filtering.* Wiley (1975)
Javid, M., and Brunner, E., *Analysis, Transmission and Filtering of Signals.* McGraw-Hill (1963)
Lee, Y. W., *Statistical Theory of Communications.* Wiley (1960)

9

Noise

9.1 Introduction

The term **noise** is a generic word used to describe any unwanted signal. It is best defined in the context of a specific experimental situation. In general, the sources of noise can be classified in two distinct categories:

a) noise such as thermal noise or shot noise which are fundamental phenomena;
b) environmental noise, often man-made but sometimes from natural causes like thunderstorms.

The first of these categories sets a theoretical limit to measurement, whereas the second often determines the practical limit.

The following sections outline the physical properties of fundamental noise and some common causes of environmental noise. Lastly, several experimental techniques for minimising the noise level are described. These techniques, when used in conjunction with signal recovery instruments (Chapter 10), invariably enhance their performance.

9.2 Sources of fundamental noise

The two most important noise mechanisms used for the characterisation of fundamental noise are **thermal noise** and **shot noise**. The former is described in terms of a *continuous* random variable function, whereas the latter is described in terms of a *discrete* random variable function because the fluctuating quantities such as currents and voltages can assume a range of discrete values only.

Thermal noise

Thermal noise may be considered to be a direct consequence of the second law of thermodynamics. For example, the random movement of electrons in a resistance at a uniform temperature gives rise to noise. The open circuit rms value of this noise was determined experimentally by Johnson[1] and theoretically by Nyquist[2] as

$$(\overline{e_n^{\,2}})^{1/2} = (e_n)_{rms} = (4kTRB_n)^{1/2} \tag{9.1}$$

where k is the Boltzmann constant; R is the value of the resistance in ohms; T is the temperature of the resistance in degrees K; and B_n is the noise equivalent bandwidth of the device used to observe the noise.

The presence of an emf across a resistor means that power can be drawn by a load connected across it. Figure 9.1(a) shows the equivalent circuit with a 'noiseless' load R_L at 0 K

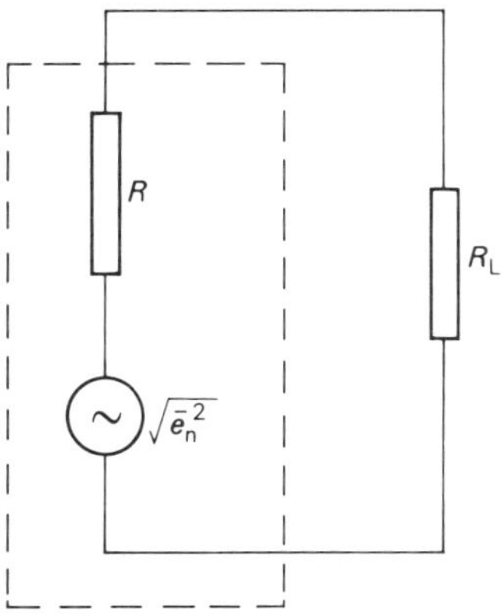

(a)

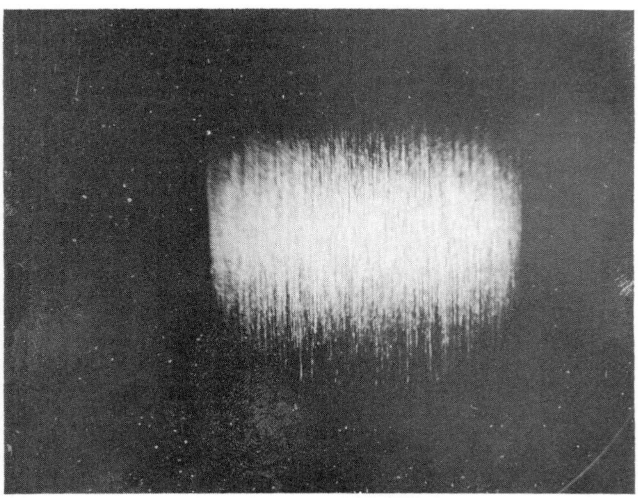

(b)

Fig. 9.1 (a) Equivalent circuit with a 'noiseless' load R_L at 0 K connected across a noisy resistance R at a temperature T; (b) oscillogram of resistive noise observed using an amplifier with a noise equivalent bandwidth of 100 kHz

connected across a noisy resistor R at a temperature T. Figure 9.1(b) shows an oscillogram of resistive noise observed using an amplifier with a noise equivalent bandwidth of 100 kHz. For the circuit of Fig. 9.1(a),

$$\overline{i_n^2} = \frac{\overline{e_n^2}}{(R+R_L)^2} \tag{9.2}$$

and the power dissipated in R_L will be $\overline{i_n^2}\, R_L$. Under matched conditions, $R = R_L$ and the maximum power drawn is known as the **available noise power** P_n, given by

$$P_n = \frac{\overline{e_n^2}\, R}{4R^2} = \frac{(4kTRB_n)\, R}{4R^2} = kTB_n \tag{9.3}$$

Thus P_n is independent of the value of R and of the frequency, i.e. it is constant over the entire frequency spectrum and hence the name **white noise** by analogy to white light.

The constancy of P_n with frequency begins to fail at about 1000 GHz at room temperature. The rigorously derived expression for thermal noise in a resistance R at a temperature T K[2,3] is

$$\overline{e_n^2} = \frac{4RB_n h\nu}{e^{h\nu/kT} - 1} \tag{9.4}$$

where h is Planck's constant and ν is the frequency. For small values of $(h\nu/kT)$ the exponential reduces to

$$e^{h\nu/kT} = 1 - \frac{h\nu}{kT} + \frac{(h\nu/kT)^2}{2!} \tag{9.5}$$

so that equation 9.4 reduces to equation 9.1, i.e.

$$\overline{e_n^2} = \frac{(4RB_n h\nu)kT}{h\nu} = 4kTRB_n$$

Brownian fluctuations in galvanometers and the theory of random fluctuations in physical measurements are extensively reviewed elsewhere.[4,5] In general, thermal noise, when regarded as a random stationary process, is characterised by the fact that its energy density function $S_x(\omega)$ or its energy per unit bandwidth is a constant up to frequencies of about 1000 GHz. Thermal noise can exist in any macroscopic parameter of a system above absolute zero. One method (the Langevin method[3]) of calculating the spectral intensities of the fluctuations is to write down its macroscopic differential equation and put on the right-hand side a *continuous* random source function of time, $n(t)$. For example, a second-order system may be described as

$$\frac{ad^2x}{dt^2} + \frac{bdx}{dt} + cx = n(t) \tag{9.6}$$

where ad^2x/dt^2 is the inertial term, bdx/dt is the damping term, cx is the restoring term and $n(t)$ is the forcing function.

If $X(f)$ is the Fourier transform of $x(t)$ and $N(f)$ is the Fourier transform of $n(t)$, then

$$X(f) = \bar{y}(f)N(f) \tag{9.7}$$

where

$$\bar{y}(f) = \text{frequency response function} = \frac{1}{c - 4\pi^2 f^2 a + j2\pi f b} \tag{9.8}$$

The energy density function is

$$G_x(f) = 2|X(f)|^2 \tag{9.9}$$

If $G_n(f)$ is the energy density function of the noise $n(t)$, then $G_n(f) = G_n$ = constant and

$$G_x(f) = |\bar{y}(f)|^2 G_n(f) \tag{9.10}$$

where $G_n(f) = 2|N(f)|^2 = G_n$ = constant. Hence

$$\overline{x^2(t)} = \int_0^\infty G_x(f)\,df$$

Therefore

$$\overline{x^2(t)} = G_n \int_0^\infty |\bar{y}(f)|^2 \mathrm{d}f \tag{9.11}$$

But

$$|\bar{y}(f)|^2 = \frac{1}{(c - 4\pi^2 f^2 a)^2 + 4\pi^2 f^2 b^2} \tag{9.12}$$

Therefore

$$\overline{x^2(t)} = G_n/4bc \tag{9.13}$$

Under thermal equilibrium conditions at a temperature T,

$$\tfrac{1}{2}c\,\overline{x^2(t)} = \tfrac{1}{2}kT \tag{9.14}$$

Therefore

$$G_n = 4bkT \tag{9.15}$$

In an electrical system b is the resistance so that equation 9.1 is a particular case of equation 9.15, which is the generalised **Nyquist theorem**.

It can be shown[3] that diffusion noise in semiconductors, i.e. noise caused by collisions of majority carriers with the lattice, can be described by the thermal noise equations 9.1 and 9.15 provided that Einstein's relationship $D_n = (kT/q)\mu_n$, equation (1.53), holds.

Shot noise

Just as thermal noise is characterised by noise due to the random movement of electrons in a resistor at a temperature T, shot noise is characterised by a saturated (temperature-limited) thermionic diode in which the spontaneous fluctuations in the number of electrons emitted per second follow a Poisson distribution. The electrons are emitted independently and at random, and their energy density function is given by

$$G_s(f) = 2q\bar{I} \tag{9.16}$$

where $\bar{I}$ is the average current through the diode and q is the electronic charge. Equation 9.16 is known as **Schottky's theorem** and applies for any mechanism consisting of a series of independent random events, such as carriers crossing potential barriers in p-n junctions and transistors. In the case of the saturated thermionic diode, the average current I_d in a small frequency band Δf can be represented by a noise current generator $\sqrt{2qI_d\Delta f}$ in parallel with the diode, Fig. 9.2. It is now common practice to express the rms value of the shot noise current i_{rms} in devices as

$$i_{rms} = (\overline{i^2})^{1/2} = (2qI_{eq}\Delta f)^{1/2} \tag{9.17}$$

where I_{eq} is the equivalent saturated diode current of the device. For example, if a device does not give 'pure' shot noise, then its noise current is expressed as

$$(\overline{i^2})^{1/2} = (2q\bar{I}K^2\Delta f)^{1/2} \tag{9.18}$$

where $I_{eq}/\bar{I}$ = average device current and K^2 is a constant. From equations 9.17 and 9.18, $K^2 = I_{eq}/I$, i.e. the noise suppression factor of the device.

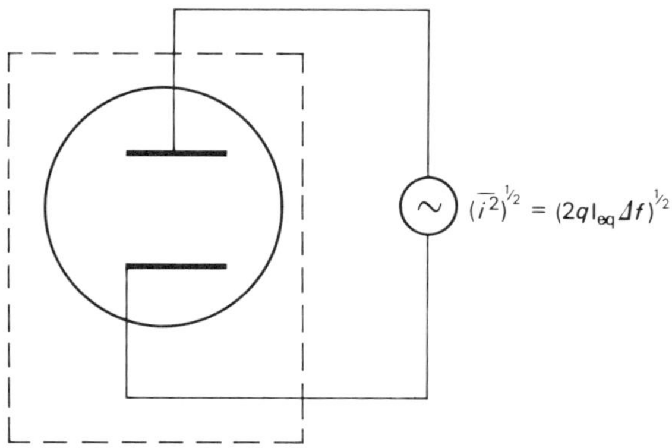

Fig. 9.2 Noise equivalent of a saturated thermionic diode

At low frequencies (below 10 kHz), the shot noise in semiconductors is masked by another kind of noise, often known as '$1/f$ noise' because its energy density function can be represented as

$$G_f = K/f^n \tag{9.19}$$

where K is a constant and n ranges from 0.8 to 1.5. When $n = 1$, the mean total energy in a band of frequencies from f_1 to f_2 will be

$$W = \int_{f_1}^{f_2} \frac{K\mathrm{d}f}{f} = K(\ln f_2 - \ln f_1) \tag{9.20}$$

Since the noise energy must remain finite, the $1/f$ law can only hold over a band of frequencies limited at the upper and lower end. Experimentally, the $1/f$ law is found to hold over frequencies ranging from a fraction of a cycle per second to several decades, and is also applicable to contact noise and vacuum tube flicker noise. The mechanism of generation of $1/f$ noise in junction diodes, transistors and MOSFETs is discussed by van der Ziel[3].

9.3 Environmental noise: sources and minimisation

The main sources of environmental noise are as follows:

a) interference from AC mains ('hum') and coupling between signal lines ('crosstalk');
b) transients and fluctuations in the AC mains;
c) noise of mechanical or acoustic origin and transmission line reflections;
d) noise radiated over a wide range of frequencies from sources such as circuit breakers, power switching circuits and sparking relay contacts, and large-scale natural phenomena such as thunderstorms. This interference is often picked up by radio and television receivers.

Environmental noise can be minimised by taking the following precautions:

a) shielding;
b) decoupling and avoiding ground loops;
c) using antivibration mountings, low noise differential amplifiers, transient suppressors and compensated stabilising transformers.[6, 7]

Shielding

Noise in the form of pulses, continous waveforms of known frequency such as 50 Hz, or repetitive complex waveforms, enter electronic systems by radiative coupling. The effect of electric fields, magnetic fields or both can be minimised by using shields or screens. Shielding materials range from aluminium, copper and brass to modern nickel–iron alloys having an extremely high permeability at low field strengths. The thickness of the shield is dictated by the **skin depth** δ of the electromagnetic wave in the material of which the shield is made. The skin depth is given by

$$\delta = (\pi f \sigma \mu \mu_0)^{-1/2} \tag{9.21}$$

where f is the frequency, σ is the conductivity of the shielding material, μ its permeability and $\mu_0 = 4\pi \times 10^{-7}\,\mathrm{H\,m^{-1}}$.

High permeability ferromagnetic foils such as 'Telshield'[8] provide a simple and economical answer to shielding components, reed relays, printed circuit modules, CR tubes, etc. from unwanted static and alternating fields. The effectiveness of a shield can be expressed in terms of a **shielding factor** which, for a cylindrical shield in a *static field*, is given by

$$\frac{H_0}{H_i} = \frac{\mu}{2}\frac{t}{d} \tag{9.22}$$

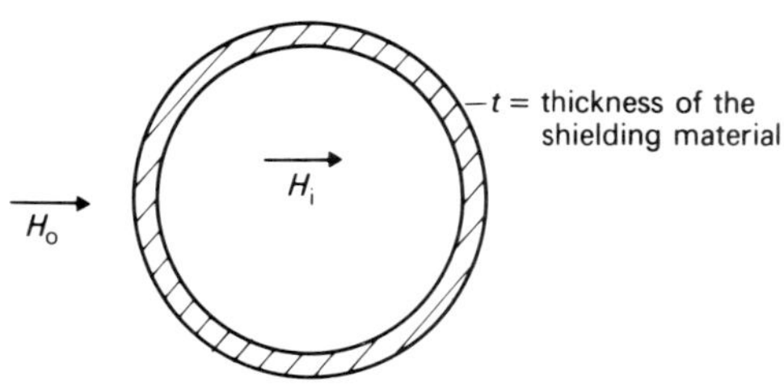

Fig. 9.3 A cylindrical shield in a static field

where t is the thickness of the material; H_0 is the field to be shielded; H_i is the field inside the shield; μ is the permeability of the material $\approx 10^4$; and d is the diameter of the cylinder.

For low-frequency AC fields, the shielding factor is increased because of the reflection and absorption of the electromagnetic waves at the shield. However, these shields are no better than those made of copper or aluminium at frequencies above 100 kHz because their effective permeability decreases rapidly with frequency.

Decoupling and grounding

'Ground' refers to the zero signal reference of an electronic system and is not necessarily the 'earth' (the third wire on a mains power cord). For example, the ground of the operational amplifier in Fig. 9.4(a) is the centre tap of the $\pm V$ power supply. To prevent pick-up from stray electric fields, mainly those at mains frequency, present in most experimental environments, electrostatic shields such as closed metal boxes S, Fig. 9.4(a), are used for amplifiers, and copper braiding is used for connecting cables. Screening of this type, to be effective, must be kept at ground potential as in Fig. 9.4(b), or else an undesirable feedback loop such as ABCD in Fig. 9.4(a) results. Earthing or grounding of a shield does not help if one side of the signal input is not earthed or grounded accordingly. Also, the conducting

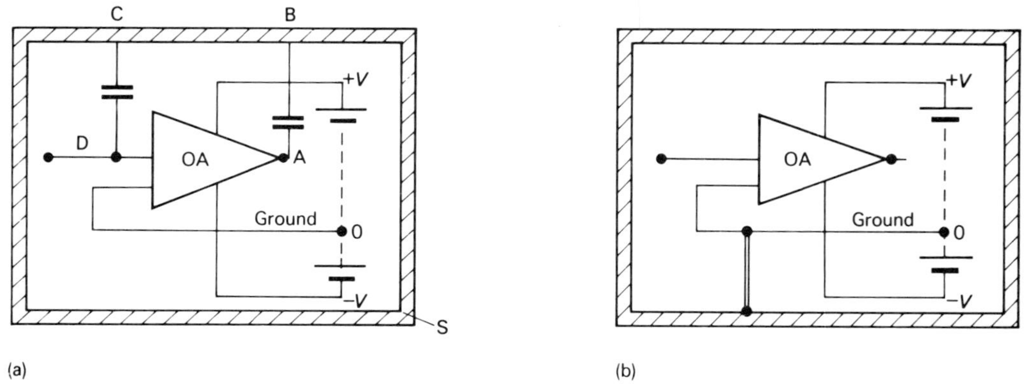

Fig. 9.4 (a) An electrostatic shield without ground; (b) a grounded electrostatic shield

shield should be connected at a single point to the zero signal reference potential at the signal earth end as in Fig. 9.5(a), so that parasitic currents will flow in the shield in loops such as 1, 2, 3, 4, 1 and not in the signal path. Incorrect grounding such as that shown in Fig. 9.5(b) will result in ground loops such as 1, 2, 3, 4, 1. The current flows in the signal path 1, 2 and results in unwanted pick-up.

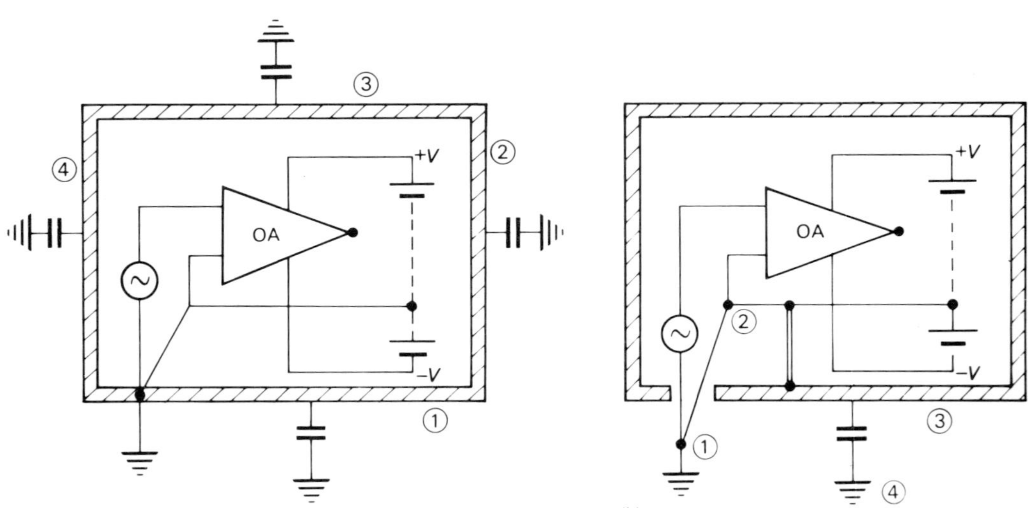

Fig. 9.5 (a) Correct grounding in which parasitic currents flow in the shield in loops such as 1, 2, 3, 4, 1; (b) incorrect grounding gives a 'ground loop' 1, 2, 3, 4, 1: current flows in the signal path 1, 2 and results in unwanted pick-up

Inductive or capacitive coupling between the input and output signal lines, or the power supply, often gives rise to instabilities in linear circuits because of undesirable feedback loops. Coupling between the input and output lines is usually cured by a suitable layout of the components and by using RC decoupling filters at appropriate points in the circuitry.

Coupling with the power supply is avoided by using voltage-regulated low internal impedance power supplies.

In digital circuits, noise spikes are often generated by poor grounding and reflections in transmission lines. In practical TTL circuits, the line termination must be high compared to the characteristic impedance of the transmission line. Also, the ground terminations, as well as the supply decoupling condensers, must always be situated close to the driving and receiving devices (see Section 9.5, p. 243).

Other precautions against environmental noise

Noise of mechanical or acoustic origin mainly arises from transducers with moving parts. It can be minimised by isolating the transducer by using antivibration mountings, or by using rejection filters if the disturbance has a fixed frequency[9].

In some experimental situations the use of differential amplifiers can minimise the effect of common mode signals, defined as the difference of potential V, Fig. 9.6, between the zero signal reference point and the signal input points A and B. Thus $V_A = (V_s + V)$ and $V_B = V$ so that the differential amplifier with gain K gives

$$V_{out} = K(V_A - V_B) = K(V_s + V - V) = KV_s \tag{9.23}$$

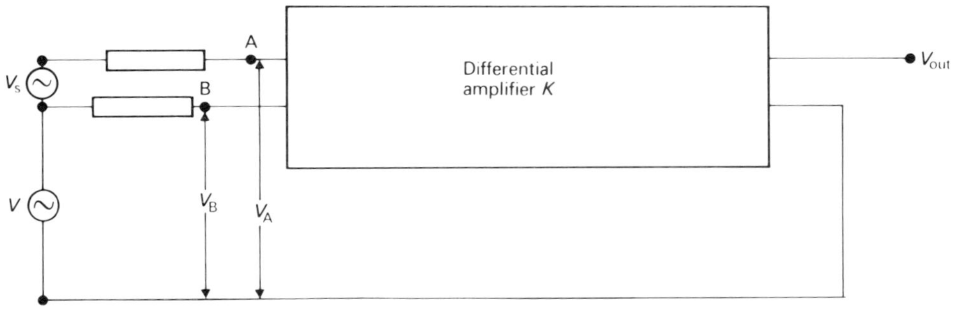

Fig. 9.6 A differential amplifier used to minimise the effect of common-mode signals

In practice, the effectiveness of the differential amplifier in eliminating common mode signals, which can be several orders of magnitude greater in level than the signal of interest, is assessed by the **common mode rejection ratio** (CMRR) given by

$$\text{CMRR} = \frac{\text{common mode signal}}{\text{unwanted pick-up at the input}}$$

The unwanted pick-up at the input is easily determined by measuring it at the output of the amplifier and dividing it by the gain of the amplifier. Thus, if a common mode voltage of 10 V causes a pick up of 10 μV at the input, then the rejection ratio is 10^6 or 120 dB. There are several types of 'flux coupled' differential DC amplifiers with chopper stabilisation.[6]

Voltage surges, due to switching of inductive loads and other electrical transients that cause component failure and interference in electronic circuitry, can be minimised at present by using varistors (see Chapter 1) which are commercially available in a wide variety of voltage and wattage ratings. When exposed to high-energy voltage transients, their resistance changes from a very high standby value to a very low conducting value, thereby clamping the transient voltage to a safe level. The energy of the incoming high voltage pulse is absorbed by the varistor and the voltage-sensitive component is adequately protected.

Fluctuations in the amplitude of the AC mains usually affect instruments that use unregulated DC or AC supplies. They are effectively combatted by using saturated core stabilising transformers[7] commonly called **constant voltage transformers**, which can reduce variations in the supply from, say, $\pm 15\%$ down to about 1%. The output waveform, however, is grossly distorted, and filters are often fitted in order to reduce its harmonic contents.

9.4 Noise measurements

The random noise generated by a device such as an amplifier is often described quantitatively in terms of the signal-to-noise power ratios at its input and output. Thus, when observations are made within a sufficiently narrow bandwidth, the **spot noise figure**, F, of the amplifier is defined as

$$F = \frac{S_i/N_i}{S_0/N_0} \tag{9.24}$$

where S_i is the signal power delivered into the input terminals; N_i is the noise power mixed with S_i; S_0 is the signal power dissipated in the load; and N_0 is the noise power mixed with S_0.

The spot noise figure F is generally a function of the signal source resistance as well as the frequency.

If the power gain of the amplifier is A, then

$$\frac{S_i}{S_0} = \frac{1}{A} \tag{9.25}$$

Therefore

$$F = \frac{N_0}{AN_i} \tag{9.26}$$

Let

$$N_0 = A(N_i + N_d) \tag{9.27}$$

where N_d is the noise power generated by the device referred to the input terminals. Then

$$F = \frac{A(N_i + N_d)}{AN_i} = 1 + \frac{N_d}{N_i} \tag{9.28}$$

Therefore

$$N_d = N_i(F - 1) \tag{9.29}$$

If N_i is standardised as the thermal noise from a matched source at 290 K, then

$$N_i = 290\, k\, \mathrm{d}f \tag{9.30}$$

and

$$N_d = T_d\, k\, \mathrm{d}f \tag{9.31}$$

where k is the Boltzmann constant and T_d is the **noise input temperature** which can be expressed as

$$T_d = 290(F - 1) \quad \text{degrees Kelvin}$$

Over a wide bandwidth, the **average noise figure** $\overline{F}$ may be interpreted as

$$\overline{F} = \frac{\text{Noise output from the amplifier}}{\text{Noise expected from a noiseless amplifier of the } \textit{same gain}}$$

For white noise at input under matched conditions from equations 9.3 and 9.30, N_i = (constant) df, so that from equation 9.26,

$$\overline{F} = \int_0^\infty FA\,\mathrm{d}f \Big/ \int_0^\infty A\,\mathrm{d}f \tag{9.32}$$

In order to permit calculations independent of bandwidth and temperature, the noise generated by a device is often represented by an **equivalent noise resistance**. All the noise is assumed to be generated in this equivalent resistance R_n with a noiseless amplifier following it. For example, a noisy transistor can be represented by two uncorrelated noise sources, one of which could be written in terms of an equivalent noise resistance R_n as

$$\overline{e_n^2} = (4kTB_n)R_n \tag{9.33}$$

or $\quad R_n = \overline{e_n^2}/4kTB$

This indicates that the noise source generates as much available noise power at the input as a resistance numerically equal to R_n (see Section 9.5).

Lastly, in the experimental measurement of noise figures of devices such as low noise amplifiers in order to evaluate their performance, the input and output circuit configurations and the characteristics of the measuring instruments are of utmost importance, as pointed out by Greene in a satirical article.[10]

9.5 Some commonly-encountered experimental situations

Amplification of low-level signals

Typical low-level signals of the order of a few microvolts are obtained from active or passive transducers (see Chapter 7). The limit to the amplification of such signals is obviously set by their signal-to-noise ratio, which must not deteriorate radically during the amplification process. The parameters of the amplifier must, therefore, be optimised in relation to the parameters of the given signal source.

In most experimental situations, the signal source can be represented as a voltage V_s in series with an impedance Z_s and shunted by a noise source I_{ns}, Fig. 9.7(a), or as a current source I_s shunted by an impedance Z_s and a noise source I_{ns}, Fig. 9.7(b). In either case, the input–output relations can be written as

$$V_{0a}(j\omega) = H_a(j\omega)V_s(j\omega) \tag{9.34}$$

$$V_{0b}(j\omega) = H_b(j\omega)I_s(j\omega) \tag{9.35}$$

where $V_{0a}(\omega)$ and $V_{0b}(j\omega)$ are the Fourier transforms of the outputs and $V_s(j\omega)$ and $I_s(j\omega)$ are

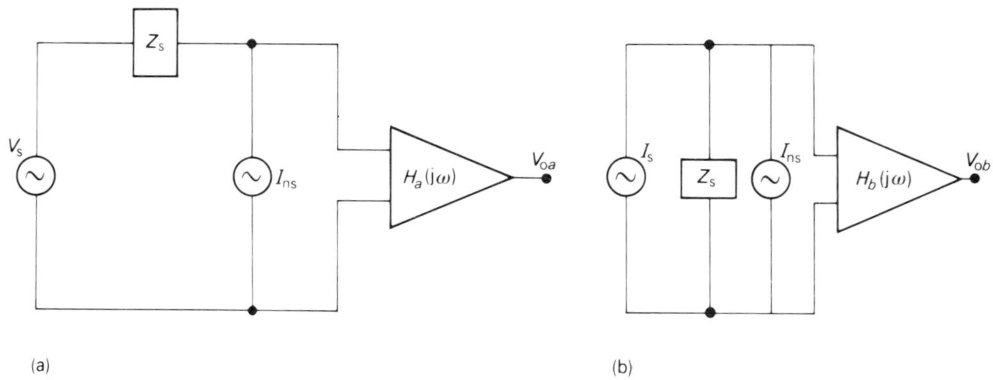

Fig. 9.7 (a) Voltage source V_s in series with an impedance Z_s; (b) current source I_s shunted by an impedance Z_s

the Fourier transforms of the inputs. The quantity $H_a(j\omega)$ is a dimensionless frequency response function of the amplifier in Fig. 9.7(a), and $H_b(j\omega)$ is the transfer impedance of the amplifier in Fig. 9.7(b).

The output voltages V_{0a} or V_{0b} can be made to have specific relationships with the inputs by a suitable choice of the functions $H_a(j\omega)$ and $H_b(j\omega)$ of the amplifiers. For example, if a system is required to respond to a signal voltage generator in a way that is independent of the source impedance Z_s, then the sufficient conditions are $|Z_{in}| \gg |Z_s|$ where $Z_{in}(j\omega)$ is the input impedance of the amplifier and the voltage gain frequency response function $A_v(j\omega)$ should be equal to the overall frequency response function $H_a(j\omega)$.

The input impedance of a signal amplifier must be chosen to suit the current–voltage characteristics of the source, and maximising the signal at the input of the amplifier does not necessarily give the best signal-to-noise ratio at the output. Apart from the spectral case where an impedance matched condition is required to prevent reflections at the end of a transmission line, the use of an input matching transformer seldom gives the best signal-to-noise ratio at the output, for the following reasons:

1) the value of the turns ratio n corresponding to the matched condition is not in general the one that gives the best signal-to-noise ratio at the output;
2) at frequencies below 100 kHz, transformers introduce unwanted resistance and reactance at the input and are subject to electrical and mechanical interference;
3) transformers leave the overall transfer function sensitive to variations in the source impedance Z_s.

The noise properties of an input device at low frequencies can be expressed in terms of two uncorrelated noise generations, i.e. a voltage generator V_n and a current generator I_n. The onset of correlation effects with increasing frequency can be represented by an additional current generator $(V_n)j\omega C_n$, where C_n is called the **noise capacitance** of the device. The energy density functions of these noise generators can be expressed in terms of a series noise resistance R_{nV} and a parallel noise resistance R_{nI}, both being functions of frequency. The dimensionless ratio R_{nI}/R_{nV} is a measure of the low noise qualities of the device and can be represented by a figure of merit given by

$$M = (R_{nI}/R_{nV})^{1/2} \tag{9.36}$$

In bipolar junction transistors the energy density function of I_n shows an increase at low frequencies, whereas in JFETs the energy density function of V_n shows a similar increase. The effect can be approximately described by a function $[1 + (\omega_f/\omega)]$, where ω_f is called the **flicker noise characteristic frequency** which, for modern transistors, is about 100 Hz. As a general-purpose device, the JFET performs better than the bipolar junction transistor in this respect, and is therefore widely used in the first stage of low-noise amplifiers.

Lastly, the definition of a signal-to-noise ratio depends very much on the nature of the signal and the system. The simplest case is one in which the signal is essentially monochromatic, i.e. occupies a small frequency range around a frequency ω_0. The other extreme is the case of a pulse signal which occupies a wide frequency range. The precise interpretation of the signal-to-noise ratio depends on how the information is to be extracted at the output of an amplifier or device. A detailed discussion of the above two cases is given elsewhere.[11] In short, the signal-to-noise ratio obtained in any experimental situation depends on the relation of R_{nV}, R_{nI} and C_n to the impedance of the signal source, and on **noise matching**, or conforming the noise characteristics of a device to those of the signal source.

Noise in photomultipliers

A photomultiplier is fundamentally a charge-transfer device, and in applications where the light intensity is high enough to produce an output of several hundred nanoamps at the collector, a DC measurement is made with the help of suitable low-noise amplifiers. For low intensities, the light beam is chopped and signal recovery equipment is used to measure the average value of the voltage across the load resistor (see Chapter 10). For still lower light intensities, **photon counting** or counting of the pulses at the collector via a wide-band amplifier and a pulse height discriminator yields better results.

Apart from the environmental noise caused by magnetic fields, electrostatic fields and radioactive sources, the three fundamental factors that set a limit to the lowest intensity of light that a photomultiplier is capable of detecting are as follows:

i) dark current;
ii) shot noise;
iii) thermal noise in the load resistor.

When a photomultiplier is in complete darkness, thermonic emission from the cathode still takes place, giving rise to a **dark current** i_t approximately obeying Richardson's equation

$$i_t = AT^2 e^{-(B/T)\phi} \quad \text{A cm}^{-2} \tag{9.37}$$

where $A = 1.2 \times 10^2$; $B = 1.16 \times 10^4$; ϕ is the work function of the cathode material; and T is the absolute temperature in degrees K.

The dark current electrons are emitted singly, and the statistics of the secondary emission process for single electron inputs into the first dynode are Poissonian to a good approximation. Taking into account the random variations in the secondary emission from all the dynodes, the most probable pulse height at the collector can be predicted.

Thus, in the photon counting mode, by cooling a photomultiplier with an antimony caesium cathode to 0 °C, the total background count can be reduced by a factor of ten, because of a large reduction in the thermal component of the dark current. For almost all photocathodes, this component of the dark current can be virtually eliminated by cooling to about −40 °C and any further cooling produces no significant improvement. Cooling does not affect the high amplitude pulses, which are not thermal in origin but are usually generated by cosmic rays and radioactive sources such as K^{40} which occur naturally in the window material.

Shot noise, a consequence of the discrete nature of the incident light quanta as well as that of the emitted electrons, occurs both in the signal and in the dark current. Fluctuations in the dark current are often referred to as **dark noise** and should not be confused with the dark current itself. It is the dark current shot noise that limits photomultiplier performance.

With the availability of modern low-noise current amplifiers, the use of large ($\approx 10^{10}\,\Omega$) collector resistors in order to obtain a large output voltage has the following disadvantages:

1) they give rise to a higher value of thermal noise;
2) they are susceptible to microphony and phase shifts in the case of chopped beam systems;
3) they give rise to excessively long time constants.

Noise in digital circuits

Noise in digital circuits commonly appears as sharp voltage or current pulses that find their way into logic systems and are often large enough to upset operation. For example, a JK flip-flop which under normal operation changes state at the trailing edge of a clock pulse (see Chapter 4) will ignore a small noise pulse n_t, Fig. 9.8(a), but will trigger if n_t is large enough, thus leading to faulty operation, Fig. 9.8(b).

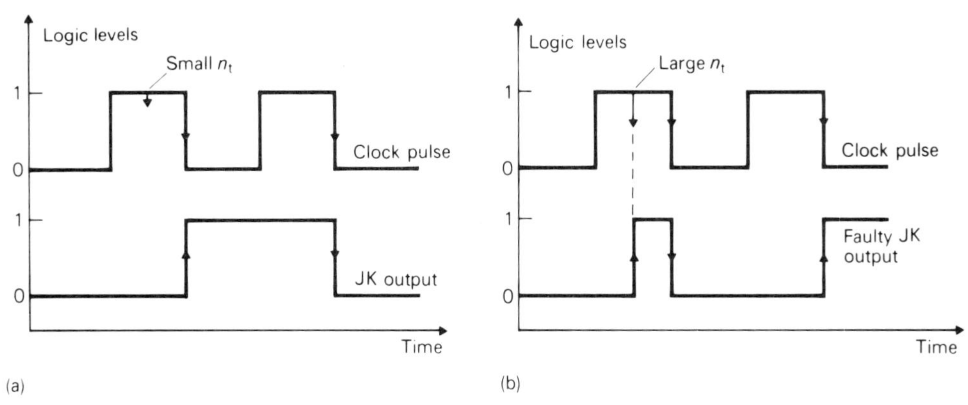

Fig. 9.8 (a) A small noise pulse n_t in the clock output is ignored; (b) a large noise pulse in the clock output leads to faulty operation of the JK flip-flop

The sources of such **noise pulses**, especially those with sharp (a few nanoseconds) leading or trailing edges can be listed as follows:

i) arcing relay contacts, circuit breakers, motor brushes, thyristors, etc., generate sharp pulses that interfere with digital circuitry via the power supplies or by radiative coupling;
ii) sharp coupling between signal lines, often known as **cross talk**;
iii) noise from unterminated transmission lines that cause **ringing** and **overshoots**;
iv) noise in the form of sharp current spikes caused by the switching of totem-pole output stages (see Chapter 4).

The problem of minimisation of environmental noise has been discussed in general terms in Section 9.3. However, it is worthwhile illustrating the following typical situations commonly encountered in digital circuitry.

Noise from ground loops
When gates such as G_1 and G_2 are used at the ends of a transmission line, Fig. 9.9(a), a poor ground return connection is equivalent to a high impedance Z_g in the return current path. A noise pulse n_{t_1} is thus generated at P and will interfere with any other circuitry such as gates G_3 and G_4, producing further unwanted pulses such as n_{t_2}. Two empirical rules are very effective in such situations. Firstly, a single ground connection (G) close to the driving and receiving devices, Fig. 9.9(b), must be used; secondly, the supply V_{cc} must be decoupled with condensers such as C_{d_1} and C_{d_2} ($\approx .1\ \mu$F) close to the driving and receiving devices.

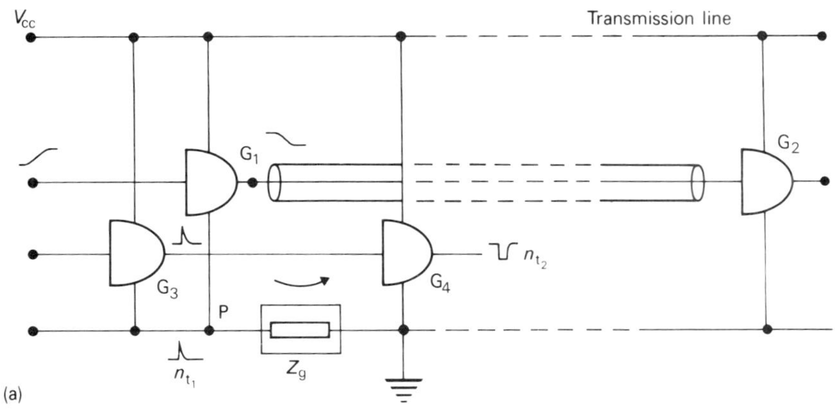

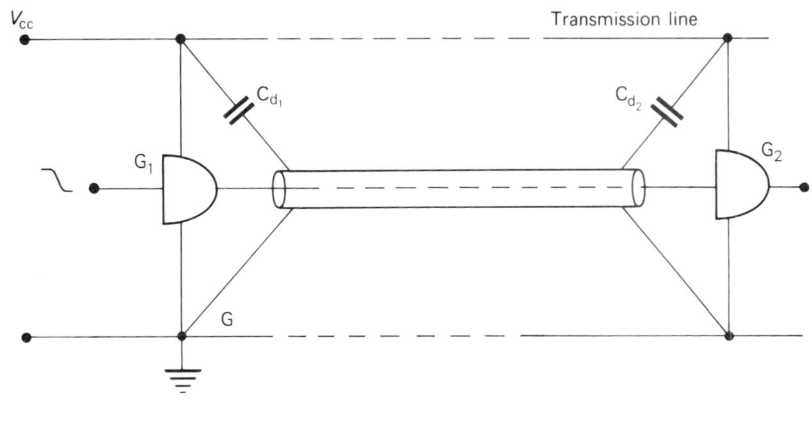

Fig. 9.9 (a) A poor ground return connection generating noise pulses; (b) a single ground connection G close to the driving and receiving devices and decoupling condensers C_{d_1} and C_{d_2} are very effective in noise reduction

Crosstalk

There are three types of transmission lines commonly used for sending and receiving digital signals:

a) straight wire lines;
b) twisted wire lines;
c) coaxial cable lines.

Direct wire leads (25–50 cm long) are the simplest and cheapest, provided that they are not cabled tightly together and are routed close to the ground. When the length of the transmission lines is long, i.e. about one hundred metres or more, the coupling impedance represented by Z_c and the characteristic impedance Z_o, Fig. 9.10, must be carefully considered. For example, in Fig. 9.10, if a signal pulse V_s from a source of impedance R_s is sent down the transmission line with a characteristic impedance Z_o to gate G_1, then due to coupling impedance Z_c a spurious voltage pulse V_n will appear at the input of G_2. The magnitude of the pulse will be given by

$$V_n \approx \left(\frac{Z_o}{Z_o + Z_c}\right) V_s$$

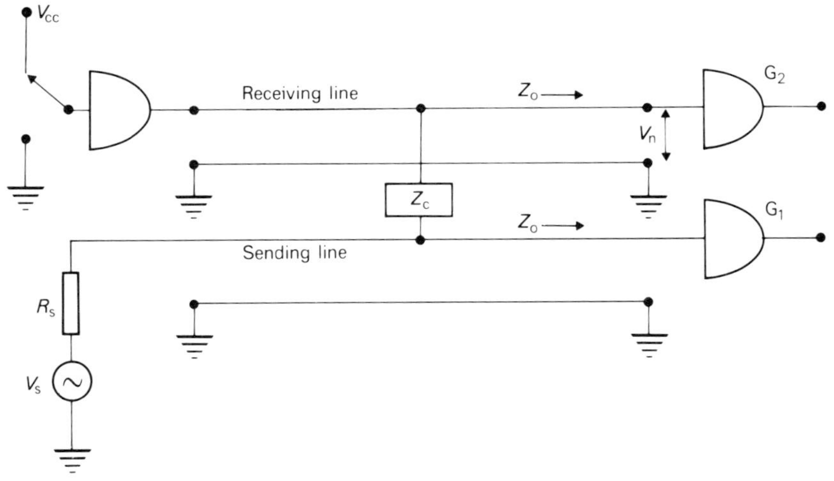

Fig. 9.10 Circuit configuration of a long transmission line

provided that $R_s \ll Z_o$. The worst case for signal crosstalk occurs when the sending and receiving lines are close together and widely separated from the ground return path. In this case Z_c is small ($\approx 80\,\Omega$) and Z_o is large ($\approx 200\,\Omega$), so that

$$V_n = \frac{1}{1 + (80/200)} V_s = 0.7 V_s$$

As the amplitude of V_s will be equal to the supply voltage, the amplitude of V_n will certainly interfere with the normal operation of the logic circuitry which, in most cases, has a noise margin of about $0.3\,V_s$, where V_s is the 'logic swing'. In such cases, close spacing of conductors giving rise to a small value of Z_c must be avoided.

When the sending and receiving lines are twisted pairs laid closely side by side, $Z_o \approx 80\,\Omega$ and $Z_c \approx 400\,\Omega$ so that

$$V_n = 0.17\ V_s$$

This noise level is safe and practical for most logic circuitry.

Coaxial cables combine a low characteristic impedance with a high value of Z_c and provide shielding as well. However, they are expensive and are generally used only in very noisy environments where twisted pairs are inadequate.

Reflections from transmission lines

When transmission lines are long enough to cause propagation delays equal to or greater than the pulse transition times, the effects of reflections in transmission lines due to termination mismatch must be considered. Reflections mainly give rise to overshoots and ringing, the magnitudes of which might well exceed the noise margins of the digital circuitry. Basic transmission line equations are applicable but unwieldy because the input and output impedances of the receiving and sending gates are often nonlinear functions. Hence transmission line characteristics of logic circuit interconnections are best analysed by graphical techniques.[12]

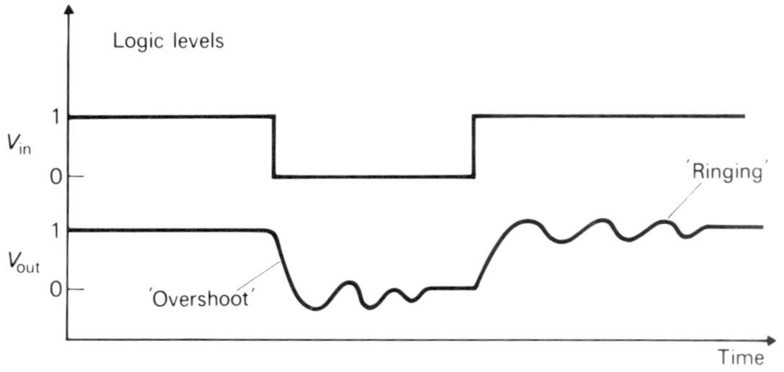

Fig. 9.11 Reflections in transmission lines giving rise to 'overshoots' and 'ringing'

References

(1) Johnson, J. B., Thermal agitation of electricity in conductors. *Phys. Rev.*, **32**, p. 97 (July 1928)
(2) Nyquist, H., Thermal agitation of electric charges in conductors. *Phys. Rev.* **32**, p. 110 (July 1928)
(3) Van der Ziel, A., *Noise: Sources, Characterisation, Measurement*. Prentice-Hall (1970)
(4) Jones, R. V., and McCombie, C. W., Brownian fluctuations in galvanometers and galvanometer amplifiers. *Phil. Trans. Roy. Soc.*, **244A**, p. 205 (1952)
(5) McCombie, C. W., Fluctuation theory in physical measurements. *Rep. Prog. Phys.*, **16**, pp. 266–320 (1953)
(6) Morrison, Ralph, *Grounding and Shielding Techniques in Instrumentation*. John Wiley and Sons (1967)
(7) Connelly, F. C., *Transformers*. Sir Isaac Pitman and Sons Ltd (1962)
(8) Telcon Metals Ltd, Publication Nos. 17–369, *'Telshield' wrap-around magnetic shielding*
(9) De Sa, A., et al., The signal-to-noise ratio of astatic magnetometers with negative feedback. *J. Phys. E.*, **7**, p. 1015 (1974)
(10) Greene, J. C., Noisemanship – the art of measuring noise figures nearly independent of device performance. *Proc. IEE*, **49**, p. 1223 (1961)

(11) Faulkner, E. A., The principles of impedance optimisation and noise matching. *J. Phys. E.*, **8**, pp. 533–40 (1975)
(12) Morris, L., and Miller, J. R., (eds), *Designing with TTL integrated circuits.* Texas Instruments Ltd and McGraw-Hill (1971)

Further reading

Bennett, W. R., *Electrical Noise.* McGraw-Hill (1960)

10

Principles of signal recovery

10.1 Introduction

In most experimental situations the signal of interest is a function of a single real variable, usually time. It may or may not be periodic. The time of observation is limited and it is desired to separate the signal from a combination of signal and noise as clearly as possible. In some cases it is the derivative or the time integral of the signal that is required, and in others it is the waveform that must be recovered.

All signal recovery devices basically rely on the fact that the signal and noise have radically different properties. In the following sections, basic mathematical concepts of cross-correlation and autocorrelation are outlined and their implementation in modern signal recovery instruments is described. Waveform recovery is discussed together with the improvement in the signal-to-noise ratio obtained by using several digital and analogue techniques.

10.2 Basic mathematical concepts and their physical implementation

Correlation functions

The cross-correlation function for two signals $x(t)$ and $y(t)$ is defined as

$$R_{xy}(\tau) = \lim_{T \to \infty} \frac{1}{T} \int_0^T x(t-\tau)y(t)\mathrm{d}t \tag{10.1}$$

This process is represented physically in Fig. 10.1.

In a time-invariant system, either $y(t)$ may be advanced or $x(t)$ may be delayed, so that

$$R_{xy}(\tau) = \lim_{T \to \infty} \frac{1}{T} \int_0^T x(t)y(t+\tau)\,\mathrm{d}t \tag{10.2}$$

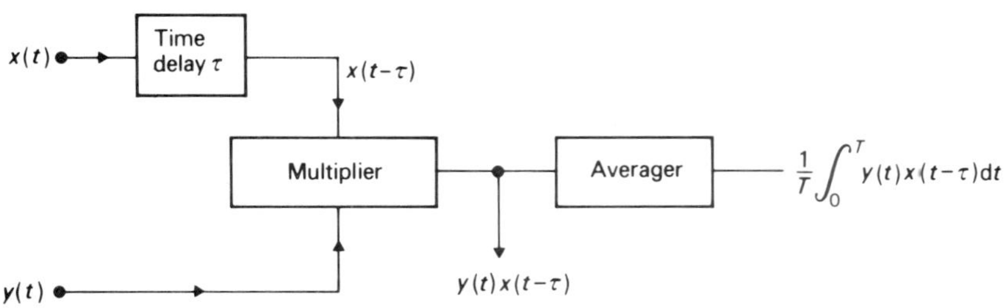

Fig. 10.1 Physical implementation of the cross-correlation process

However, time delays are physically realisable in instrumentation, and equation 10.1 is more often used. Also, as observation time is finite, a fairly large value of T is acceptable.

The plot of $R_{xy}(\tau)$ versus τ (i.e. a **correlogram**) indicates the degree of similarity between the two waveforms, since the time average of the product $y(t)x(t-\tau)$ will exist only if the two functions go positive or negative together.

If $x(t) = y(t)$ then the cross-correlation function becomes the autocorrelation function, defined as

$$R_{xx}(\tau) = \lim_{T \to \infty} \frac{1}{T} \int_0^T x(t-\tau)x(t)\,dt \tag{10.3}$$

The important properties of correlation functions may be listed as follows.

1) $R_{xx}(\tau) = R_{xx}(-\tau)$, i.e. it is an even function of τ.
2) $R_{xx}(0) \geqslant R_{xx}(\tau)$, i.e. it is a maximum at $\tau = 0$.
3) If $x(t)$ is periodic, then $R_{xx}(\tau)$ is also periodic.
4) If $x(t)$ is random, then $R_{xx}(\tau) \to 0$ as $\tau \to \infty$.
5) If $x(t)$ and $y(t)$ are uncorrelated and if $z(t) = x(t) + y(t)$ then

$$R_{zz}(\tau) = R_{xx}(\tau) + R_{yy}(\tau) \tag{10.4}$$

since $R_{xy}(\tau) = 0$.

The autocorrelation function gives an indication of the relationship between signal samples spaced τ seconds apart. Thus, a wide-band signal will have its autocorrelation function confined to small values of delay and vice versa. The relationship between the energy density function of a signal (Chapter 8) and its autocorrelation function is given by the Wiener–Kinchine theorem which may be stated as follows.

If $F_x(j\omega)$ is the Fourier transform of a stationary random signal $x(t)$, then its energy density function

$$S_x(\omega) = |F_x(j\omega)|^2 \tag{10.5}$$

may be expressed in terms of its autocorrelation function $R_{xx}(\tau)$ as

$$S_x(\omega) = \int_{-\infty}^{\infty} R_{xx}(\tau)e^{-j\omega\tau}\,d\tau \tag{10.6}$$

The inverse relation is

$$R_{xx}(\tau) = \frac{1}{2\pi} \int_{-\infty}^{\infty} S_x(\omega)e^{j\omega\tau}\,d\omega \tag{10.7}$$

The function $S_x(\omega)$ is a real, non-negative, even function defined for both positive and negative frequencies, but for practical situations it is often defined for positive frequences only as $G_x(f) = 2S_x(f)$ for $f > 0$, and $G_x(f) = 0$ otherwise.
Thus

$$G_x(f) = 4\int_0^{\infty} R_{xx}(\tau)\cos 2\pi f\tau\,d\tau \tag{10.8}$$

and the inverse relation is

$$R_{xx}(\tau) = \int_0^{\infty} G_x(f)\cos 2\pi f\tau\,df \tag{10.9}$$

where $\omega = 2\pi f$ and $G_x(f)$ is in $V^2\,Hz^{-1}$.

At low frequencies, it is easier to compute the energy density spectrum of a signal from its autocorrelation function than to measure it directly with a wave analyser having a mean square voltmeter.

The Wiener–Kinchine theorem leads to the following results.

1) If $x(t)$ is completely random, then $R_{xx}(\tau) = 0$ for $\tau > 0$ so that from equation 10.8, $G_x(f)$ is a constant independent of frequency, i.e. the energy spectrum is flat.
2) If $x(t)$ is periodic with a frequency f_0, then $R_{xx}(\tau)$ is periodic and from equation 10.8, the energy spectrum will be a δ-function at f_0.

Lastly, the difference between the autoconvolution function (Chapter 8) and the autocorrelation function is that the former defines the complex spectrum while the latter is related to the amplitude spectrum.

Spectral bandwidth and correlation time

If the energy spectrum of the signal is represented by Fig. 10.2 in terms of its energy density function, equation 10.5, then the spectral bandwidth $\Delta\omega$ can be considered as the range of frequencies which contain most of the energy of the signal. Quantitatively $\Delta\omega$ may be defined as

$$2\int_0^{\infty} S(\omega)\,d\omega = 2S(0)\Delta\omega \tag{10.10}$$

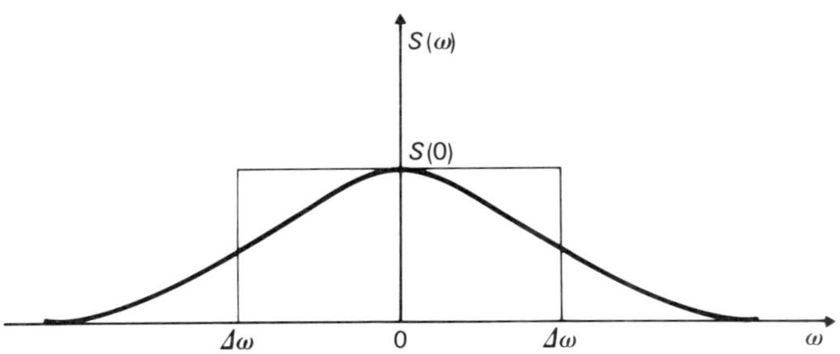

Fig. 10.2 Energy spectrum of a signal and its spectral bandwidth

which is the area of a rectangle of height $S(0)$ and width $2\Delta\omega$. This area is approximately equal to the area under the $S(\omega)$ curve.

Since the energy density function and the autocorrelation function are a Fourier transform pair, equations 10.6 and 10.7, the correlation time $\Delta\tau$ associated with a signal recovery device may be defined as

$$2\int_0^{\infty} R(\tau)\,d\tau = 2R(0)\Delta\tau \tag{10.11}$$

where

$$S(0) = \int_{-\infty}^{\infty} R(\tau)\,d\tau \tag{10.12}$$

$$R(0) = \frac{1}{2\pi} \int_{-\infty}^{\infty} S(\omega)\, d\omega \tag{10.13}$$

Hence, from equations 10.10 and 10.11,

$$\Delta\omega\Delta\tau = \pi/2 \tag{10.14}$$

The quantity $\Delta\tau$ thus defines the time interval during which the input signal affects the value of the output signal at a given time.

10.3 Filtering

Some general principles of optimising the performance of linear filters

When linear filters are used for separating signals with different spectral characteristics, the accuracy of the process can be assessed mathematically by considering the filter as a black box. Figure 10.3 has an input $f(t)$ given by

$$f(t) = s(t) + n(t) \tag{10.15}$$

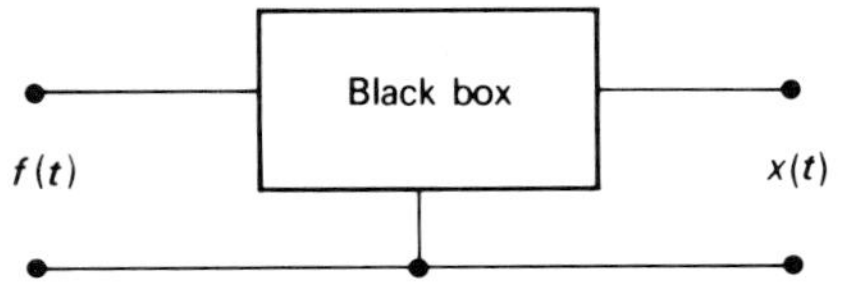

Fig. 10.3 An optimised filter as a black box

i.e. a mixture of two signals $s(t)$ and $n(t)$. Let the signal of interest $s(t)$ to be recovered from the mixture appear at the output as $x(t)$. If $h(t)$ is the true version of $s(t)$, after passing through the filter, the instantaneous error in the filtering process may be expressed as

$$\varepsilon(t) = x(t) - h(t) \tag{10.16}$$

In general, $s(t)$ and $n(t)$ could be random processes, homogeneous in time with distinct autocorrelation functions and mean values identically equal to zero.

The black box can be considered as a stationary linear filter performing a linear operation on the input so that the output is given by

$$x(t) = \int_{-\infty}^{\infty} k(t_1) f(t - t_1)\, dt_1 \tag{10.17}$$

where $k(t_1)$ is the impulse response of the filter.

The value of $\varepsilon(t)$ will fluctuate with time and its mean square value $\overline{\varepsilon^2(t)}$ will be a measure of the accuracy of the separation process.

From equations 10.16 and 10.17,

$$\varepsilon^2(t) = \left(\int_{-\infty}^{\infty} k(t_1) f(t - t_1)\, dt_1 \right)^2 - \left(2h(t) \int_{-\infty}^{\infty} k(t_1) f(t - t_1)\, dt_1 \right) + h^2(t) \tag{10.18}$$

Writing the first term as a double integral,

$$\overline{\varepsilon^2(t)} = \left(\iint_{-\infty}^{\infty} k(t_1)k(t_2)\overline{f(t-t_1)f(t-t_2)}\,\mathrm{d}t_1\mathrm{d}t_2\right)$$
$$-\left(2\int_{-\infty}^{\infty} k(t_1)\,\overline{h(t)f(t-t_1)}\,\mathrm{d}t_1\right)+\overline{h^2(t)} \tag{10.19}$$

If we consider $f(t)$ and $h(t)$ to be identical, then

$$\overline{h(t)f(t-t_1)} = R_{hf}(t_1) \tag{10.20}$$

which is the cross-correlation function of $h(t)$, and $f(t)$ becomes the autocorrelation function of $h(t)$, i.e.

$$\overline{h(t)h(t-t_1)} = R_{hh}(t_1) \tag{10.21}$$

For $t_1 = 0$,

$$R_h(0) = \overline{h^2(t)} \tag{10.22}$$

Similarly,

$$\overline{f(t-t_1)f(t-t_2)} = R_{ff}(t_1 - t_2) \tag{10.23}$$

which is the autocorrelation function of the stationary random process $f(t)$.

Substituting equations 10.20, 10.22 and 10.23 into 10.19 gives

$$\overline{\varepsilon^2(t)} = \iint_{-\infty}^{\infty} k(t_1)k(t_2)\,R_{ff}\,(t_1-t_2)\mathrm{d}t_1\mathrm{d}t_2$$
$$-\left(2\int_{-\infty}^{\infty} k(t_1)R_{hf}(t_1)\,\mathrm{d}t_1\right)+R_h(0) \tag{10.24}$$

Equation 10.24 shows that $\overline{\varepsilon^2(t)}$ does not depend on $h(t)$ and $f(t)$ but on their correlation functions.

In order to optimise the performance of the filter, the value of $k(t_1)$ must be found for which $\overline{\varepsilon^2(t)}$ is a minimum.

Let the mean square error $\overline{\varepsilon^2(t)}$ corresponding to the optimum value of $k(t_1)$ be E. If $k(t_1)$ deviates from its optimum value by $\delta k(t_1)$ and if the mean square error corresponding to $[k(t_1)+\delta k(t_1)]$ is E' then the condition for $(E'-E)=0$ can be found as follows. Replacing $k(t_1)$ by $[k(t_1)+\delta k(t_1)]$ and $k(t_2)$ by $[k(t_2)+\delta k(t_2)]$ in equation 10.24 gives

$$E' = \iint_{-\infty}^{\infty} [k(t_1)+\delta k(t_1)]\,[k(t_2)+\delta k(t_2)]\,R_{ff}(t_1-t_2)\mathrm{d}t_1\mathrm{d}t_2$$
$$-\left(2\int_{-\infty}^{\infty} [k(t_1)+\delta k(t_1)]\,R_{hf}(t_1)\,\mathrm{d}t_1\right)+R_h(0) \tag{10.25}$$

Expanding and simplifying,

$$E' = E + 2\int_{-\infty}^{\infty} \delta k(t_1)\,\mathrm{d}t_1\left(\int_{-\infty}^{\infty} k(t_2)R_{ff}(t_1-t_2)\,\mathrm{d}t_2 - R_{hf}(t_1)\right)+J$$

where

$$J = \left(\int_{-\infty}^{\infty} \overline{\delta k(t_1) f(t - t_1)}\, \mathrm{d}t_1 \right)^2$$

Since $J \geqslant 0$ the condition for $(E' - E) = 0$ is

$$\left(\int_{-\infty}^{\infty} k(t_2) R_{ff}(t_1 - t_2) \mathrm{d}t_2 - R_{hf}(t_1) \right) = 0$$

Therefore

$$R_{hf}(t_1) = \int_{-\infty}^{\infty} k(t_2) R_{ff}(t_1 - t_2) \mathrm{d}t_2 \tag{10.26}$$

Multiplying both sides of equation 10.26 by $\mathrm{e}^{-\mathrm{j}\omega t_1}$ and integrating with respect to t_1 from $-\infty$ to ∞ gives

$$\int_{-\infty}^{\infty} R_{hf}(t_1) \mathrm{e}^{-\mathrm{j}\omega t_1} \mathrm{d}t_1 = \left(\int_{-\infty}^{\infty} \mathrm{e}^{-\mathrm{j}\omega t_1} \mathrm{d}t_1 \right) \left(\int_{-\infty}^{\infty} k(t_2) R_{ff}(t_1 - t_2) \mathrm{d}t_2 \right) \tag{10.27}$$

Changing the variable $(t_1 - t_2) = t$ and rearranging the right-hand side of equation 10.27 gives

$$\int_{-\infty}^{\infty} R_{hf}(t_1) \mathrm{e}^{-\mathrm{j}\omega t_1} \mathrm{d}t_1 = \left(\int_{-\infty}^{\infty} k(t_2) \mathrm{e}^{-\mathrm{j}\omega t_2} \mathrm{d}t_2 \right) \left(\int_{-\infty}^{\infty} R_{ff}(t) \mathrm{e}^{-\mathrm{j}\omega t} \mathrm{d}t \right) \tag{10.28}$$

Thus

$$F[R_{hf}(t_1)] = F[k(t_2)]\, F[R_{ff}(t)] \tag{10.29}$$

As shown in Chapter 8, the Fourier transforms of the correlation functions are their energy density functions and give the amplitude spectra. Hence from equations 10.28 and 10.29,

$$F[R_{hf}(t_1)] = S_{hf}(\omega) = \int_{-\infty}^{\infty} R_{hf}(t_1) \mathrm{e}^{\mathrm{j}\omega t_1} \mathrm{d}t_1 \tag{10.30}$$

and the inverse transform will be

$$R_{hf}(t_1) = \frac{1}{2\pi} \int_{-\infty}^{\infty} \mathrm{e}^{\mathrm{j}\omega t} S_{hf}(\omega)\, \mathrm{d}\omega \tag{10.31}$$

For $$F[R_{ff}(t)] = S_f(\omega) = \int_{-\infty}^{\infty} R_{ff}(t) \mathrm{e}^{-\mathrm{j}\omega t} \mathrm{d}t \tag{10.32}$$

the inverse transform will be

$$R_{ff}(t) = \frac{1}{2\pi} \int_{-\infty}^{\infty} \mathrm{e}^{\mathrm{j}\omega t} S_f(\omega)\, \mathrm{d}\omega \tag{10.33}$$

For $$F[k(t_2)] = k(\omega) = \int_{-\infty}^{\infty} k(t_2) \mathrm{e}^{-\mathrm{j}\omega t_2} \mathrm{d}t_2 \tag{10.34}$$

the inverse transform will be

$$k(t_2) = \frac{1}{2\pi} \int_{-\infty}^{\infty} \mathrm{e}^{\mathrm{j}\omega t_2} k(\omega)\, \mathrm{d}\omega \tag{10.35}$$

Equation 10.29 can be written as

$$S_{hf}(\omega) = k(\omega)S_f(\omega)$$

i.e. the transfer function of the optimised filter must be

$$k(\omega) = S_{hf}(\omega)/S_f(\omega) \tag{10.36}$$

If $s(t)$ and $n(t)$ of equation 10.15 are uncorrelated, then in the case of the filter where the signal of interest $s(t)$ is not modified, i.e. $h(t) = s(t)$, the cross-correlation function is given by $R_{hf}(t_1) = R_{sf}(t_1) = R_{ss}(t_1) + R_{sn}(t_1)$.

But $R_{sn}(t_1) = 0$ so that $S_{hf}(\omega) = S_s(\omega)$ and $S_f(\omega) = S_s(\omega) + S_n(\omega)$. Therefore equation 10.36 may be written as

$$k(\omega) = \frac{S_s(\omega)}{S_s(\omega) + S_n(\omega)} \tag{10.37}$$

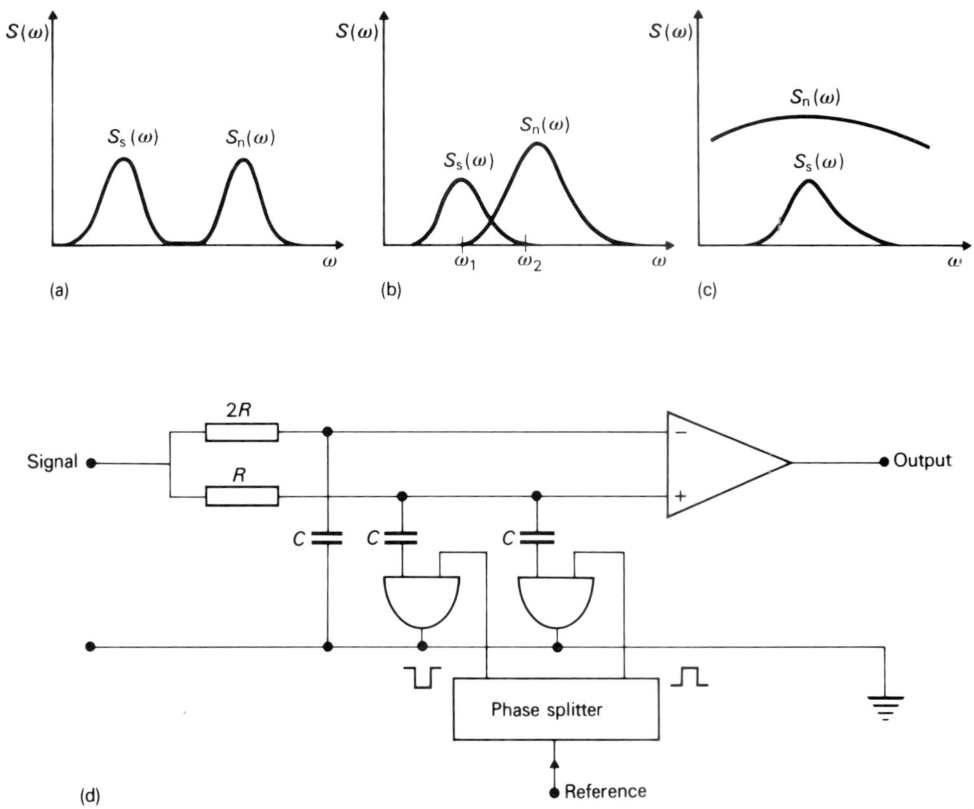

Fig. 10.4 (a) $S_s(\omega)$ and $S_n(\omega)$ do not overlap; (b) $S_s(\omega)$ and $S_n(\omega)$ overlap in the frequency range; (c) $S_n(\omega) > S_s(\omega)$ and occupies the same frequency range $\omega_1 < \omega < \omega_2$; (d) principles of a coherent filter

If $S_s(\omega)$ and $S_n(\omega)$ do not overlap, Fig. 10.4(a), then $k(\omega) = 1$ and the recovery is without error. If $S_s(\omega)$ and $S_n(\omega)$ overlap then the error is not zero because part of the $S_n(\omega)$ spectrum in the range $\omega_1 < \omega < \omega_2$, Fig. 10.4(b), passes through the filter and will distort the signal of interest. Lastly, if $S_n(\omega) > S_s(\omega)$ and occupies the same frequency range, Fig. 10.4(c), then

$$k(\omega) \approx S_s(\omega)/S_n(\omega) \tag{10.38}$$

which is the ratio of the energy density functions.

Substituting equation 10.26 in 10.24 gives

$$\overline{\varepsilon^2(t)} = R_h(0) - \iint_{-\infty}^{\infty} k(t_1)k(t_2)R_{ff}(t_1 - t_2)\mathrm{d}t_1\mathrm{d}t_2 \tag{10.39}$$

Writing

$$R_h(0) = \frac{1}{2\pi}\int_{-\infty}^{\infty} S_h(\omega)\mathrm{d}\omega \tag{10.40}$$

and from equation 10.33,

$$\begin{aligned}
&\iint_{-\infty}^{\infty} k(t_1)k(t_2)R_{ff}(t_1 - t_2)\mathrm{d}t_1\mathrm{d}t_2 \\
&= \frac{1}{2\pi}\iint_{-\infty}^{\infty} (k(t_1)k(t_2)\mathrm{d}t_1\mathrm{d}t_2)\left(\int_{-\infty}^{\infty} \mathrm{e}^{\mathrm{j}\omega(t_1 - t_2)} S_f(\omega)\,\mathrm{d}\omega\right) \\
&= \frac{1}{2\pi}\int_{-\infty}^{\infty} S_f(\omega)\,\mathrm{d}\omega \int_{-\infty}^{\infty} \mathrm{e}^{\mathrm{j}\omega t_1}k(t_1)\mathrm{d}t_1 \int_{-\infty}^{\infty} \mathrm{e}^{-\mathrm{j}\omega t_2}k(t_2)\mathrm{d}t_2 \\
&= \frac{1}{2\pi}\int_{-\infty}^{\infty} S_f(\omega)\mathrm{d}\omega\, k(-\omega)\,k(\omega)\,\mathrm{d}\omega \\
&= \frac{1}{2\pi}\int_{-\infty}^{\infty} S_f(\omega)|k(\omega)|^2\mathrm{d}\omega
\end{aligned}$$

Therefore

$$\overline{\varepsilon^2(t)} = \frac{1}{2\pi}\int_{-\infty}^{\infty} S_h(\omega) - (S_f(\omega)\,|\,k(\omega)\,|^2\,\mathrm{d}\omega)$$

From equations 10.36 and 10.37,

$$\overline{\varepsilon^2(t)} = \frac{1}{2\pi}\int_{-\infty}^{\infty}\left(\frac{S_s(\omega)\,S_n(\omega)}{S_s(\omega) + S_n(\omega)}\right)\mathrm{d}\omega \tag{10.41}$$

For $S_n(\omega) > S_s(\omega)$,

$$\overline{\varepsilon^2(t)} = \frac{1}{2\pi}\int_{-\infty}^{\infty} S_s(\omega)\,\mathrm{d}\omega = R_s(0) \tag{10.42}$$

The average power of the output of the filter, Fig. 10.3, can be calculated using the procedure outlined earlier as

$$\begin{aligned}
\overline{x^2(t)} &= \iint_{-\infty}^{\infty} k(t')\,k(t'')\,\overline{f(t - t')\,f(t - t'')}\,\mathrm{d}t'\,\mathrm{d}t'' \\
&= \iint_{-\infty}^{\infty} k(t')\,k(t'')\,R_f(t' - t'')\,\mathrm{d}t'\,\mathrm{d}t'' \\
&= \frac{1}{2\pi}\int_{-\infty}^{\infty} k(\omega)\,k(-\omega)\,S_f(\omega)\,\mathrm{d}\omega
\end{aligned}$$

Therefore

$$\overline{x^2(t)} = \frac{1}{2\pi}\int_{-\infty}^{\infty} |k(\omega)|^2 S_f(\omega)\,d\omega \tag{10.43}$$

In the particular case where the useful signal $s(t)$ is not a random process but is instead a given function of time, with parameters such as amplitude and phase, the purpose of filtering is to measure these parameters in the presence of the random 'noise' $n(t)$. In such cases, the index of performance of the filter is the 'signal'-to-'noise' ratio at the output. Thus in equation 10.15, if $s(t)$ represents the signal of known form and $n(t)$ the noise, regarded as a stationary random process, then let the output of a filter with transfer function $k(\omega)$ be

$$\phi(t) = \mu(t) + v(t) \tag{10.44}$$

where $\mu(t)$ is the result of passing $s(t)$ through the filter and $v(t)$ is the result of passing $n(t)$ through the filter.

The useful signal power at the output of the filter can be expressed in terms of its energy density function $S(\omega)$ at some time t_0 as

$$\mu^2(t_0) = \frac{1}{4\pi^2}\left|\int_{-\infty}^{\infty} e^{j\omega t_0} k(\omega)\,S(\omega)\,d\omega\right|^2 \tag{10.45}$$

whereas the average noise power as shown in equation 10.43 will be

$$\overline{v^2(t)} = \frac{1}{2\pi}\int_{-\infty}^{\infty} |k(\omega)|^2 S_n(\omega)\,d\omega \tag{10.46}$$

The signal-to-noise power ratio at the output will be

$$\rho = \frac{\mu^2(t_0)}{\overline{v^2(t)}} = \frac{1}{2\pi}\,\frac{\left|\int_{-\infty}^{\infty} e^{j\omega t_0} k(\omega)\,S(\omega)\,d\omega\right|^2}{\int_{-\infty}^{\infty} |k(\omega)|^2 S_n(\omega)\,d\omega} \tag{10.47}$$

Using the Schwartz inequality which states that

$$\left|\int_{-\infty}^{\infty} A(\omega)\,B(\omega)\,d\omega\right|^2 \leqslant \int_{-\infty}^{\infty} \left|A(\omega)\right|^2 d\omega \int_{-\infty}^{\infty} \left|B(\omega)\right|^2 d\omega$$

and comparing with the numerator of equation 10.47 gives

$$A(\omega) = e^{j\omega t_0} k(\omega)\sqrt{S_n(\omega)}$$

and $\quad B(\omega) = S(\omega)/\sqrt{S_n(\omega)}$

Hence,

$$\left|\int_{-\infty}^{\infty} e^{j\omega t_0} k(\omega)\,S(\omega)\,d\omega\right|^2 \leqslant \int_{-\infty}^{\infty} |k(\omega)|^2 S_n(\omega)\,d\omega \int_{-\infty}^{\infty} \frac{|S(\omega)|^2}{S_n(\omega)}\,d\omega \tag{10.48}$$

Therefore

$$\rho \leqslant \frac{1}{2\pi} \int_{-\infty}^{\infty} \frac{|S(\omega)|^2}{S_n(\omega)} \, d\omega$$

and the maximum value of ρ is given by

$$\rho = \frac{1}{2\pi} \int_{-\infty}^{\infty} \frac{|S(\omega)|^2}{S_n(\omega)} \, d\omega \tag{10.49}$$

In the case of white noise, the frequency spectrum of the noise is uniformly distributed over the frequency band occupied by the signal so that $S_n(\omega) = \text{constant}$, and

$$\rho = \frac{1}{2\pi S_n} \int_{-\infty}^{\infty} |S(\omega)|^2 \, d\omega \tag{10.50}$$

The signal-to-noise power ratio at the output of a matched filter is defined by the total energy of the signal and the energy density function of the noise.

Lastly, it is convenient to define a **noise equivalent bandwidth**, B_n, as follows. Consider a filter with a rectangular passband B_n. Let the maximum amplitude of its transfer function be $k(\omega_{max})$. If a device is to pass the same amount of noise power as the ideal filter with a rectangular passband, then from equation 10.46,

$$B_n |k(\omega_{max})|^2 S_n = S_n \int_{-\infty}^{\infty} |k(\omega)|^2 \, d\omega$$

Therefore

$$B_n = \frac{\int_{-\infty}^{\infty} |k(\omega)|^2 \, d\omega}{|k(\omega_{max})|^2} \tag{10.51}$$

Practical applications of filters

In situations where the input noise bandwidth is large, considerable improvement in the signal-to-noise ratio can be obtained by using selective filters. For band-limited white noise $n(t)$ having an energy density function $S_n(\omega)$ up to a maximum frequency of ω_{max} and zero above this value, the improvement in signal-to-noise ratio can easily be related to the noise equivalent bandwidth of the filter as follows.

For simplicity, consider a noisy signal of the form

$$E_s \cos(\omega t) + n(t) \tag{10.52}$$

Before filtering, the signal-to-noise ratio is given by

$$\left(\frac{S}{N}\right)_{in} = 10 \log_{10} \left(\frac{E_s^2}{2S_n(\omega)\omega_{max}}\right) \tag{10.53}$$

If the noise equivalent bandwidth of the filter is expressed by equation 10.51, where $k(\omega)$ is the transfer function of the filter, then the signal-to-noise ratio at the output will be

$$\left(\frac{S}{N}\right)_{out} = 10 \log_{10} \left(\frac{E_s^2}{2S_n(\omega)B_n}\right) \tag{10.54}$$

The improvement in the signal-to-noise ratio is

$$\left(\frac{S}{N}\right)_{\text{out}} - \left(\frac{S}{N}\right)_{\text{in}} = 10\log_{10}\left(\frac{\omega_{\text{max}}}{B_{\text{n}}}\right) \tag{10.55}$$

Thus when ω_{max} is small, high selectivity filters or piezoelectric resonators must be used in order to obtain a reasonable improvement in the signal-to-noise ratio. However, in cases where the signal has a complex waveform, simple narrow band filters cannot obviously be used efficiently and other signal recovery techniques must be chosen.

In situations where the frequency of a signal is unstable, a commercial device known as a **coherent filter** (e.g. Brookdeal type 467 may be used to overcome effectively the problems of conventional tuned circuits. It is essentially a combination of a two-path switching integrator actuated by a reference signal and a low-pass filter, Fig. 10.4(d). The differential output is a square wave at the reference frequency. The device has a large dynamic range ($\approx 100\,\text{dB}$) and is conveniently used in the signal channel of a phase-sensitive detector, since both the devices can utilise the same reference (see Section 10.6).

10.4 Signal recovery by cross-correlation

The amplitude and phase of a signal buried in noise can be recovered by cross-correlating it with a clean reference signal. For simplicity, consider a sinusoidal signal $E_s \cos\omega t$ of known frequency buried in white noise, $n(t)$. Let the reference of the same frequency be $E_r \cos(\omega t + \phi)$. The cross-correlation function, equation 10.1 and Fig. 10.1, can be written as

$$R_{xr}(\tau) = \frac{1}{2T}\int_{-T}^{T} E_s \cos\omega(t-\tau) + n(t-\tau)E_r\cos(\omega t + \phi)\mathrm{d}t \tag{10.56}$$

where $x(t) = E_s\cos(\omega t) + n(t)$ (10.57)

and $y(t) = E_r\cos(\omega t + \phi)$ (10.58)

The limits of T, the integration time, comprise a 'time window' through which one can observe the two waveforms

$$R_{xr}(\tau) = \frac{1}{2T}\int_{-T}^{T} E_s\cos\omega(t-\tau)E_r\cos(\omega t+\phi)\mathrm{d}t + \frac{1}{2T}\int_{-T}^{T} n(t-\tau)E_r\cos(\omega t+\phi)\mathrm{d}t$$

Therefore

$$R_{xr}(\tau) = R_{sr}(\tau) + R_{nr}(\tau) \tag{10.59}$$

The cross-correlation of the noise with the reference is zero for all τ, provided that the noise is truly random and the averaging time T is long.

$$R_{xr}(\tau) = R_{sr}(\tau) = \frac{E_rE_s}{2}\cos(\omega\tau+\phi) + \frac{E_rE_s}{4T}\int_{-T}^{T}\cos(2\omega t - \omega\tau + \phi)\mathrm{d}t \tag{10.60}$$

The first term of equation 10.60 gives rise to a DC component. The ripple component at a frequency 2ω, as well as the cross-modulation products, should in principle have zero mean value and be eliminated by the averager. In practice, these terms do not average to zero and give rise to statistical errors in the measurement of the DC.

10.5 Phase-sensitive detection

In addition to its traditional use in AC bridges and feedback systems, a **phase-sensitive detector** (PSD) can be used as a cross-correlator for the enhancement of poor signal-to-noise ratios in experiments where a clean reference voltage is available at the same frequency as the signal. For instance, in Fig. 10.5 a photomultiplier (PM) is used in a fluorometer for detecting low light intensities. The incident light beam is chopped at about 600 Hz with a rotating slotted disc, and the reference signal is derived by using a photoelectric module consisting of a light-emitting diode (LED) and a phototransistor. The output of the phototransistor is converted to a square wave and fed to the reference channel of the PSD via a phase shifter. The output from the photomultiplier is amplified and band-limited to about 10 kHz by the AC amplifier, and is connected to the signal channel of the PSD. Figure 10.6 shows the output of the PSD when (a) the phase difference ϕ between the signal and reference is 0; (b) when $\phi = 180°$; (c) when $\phi = 90°$; and (d) when $\phi = 270°$.

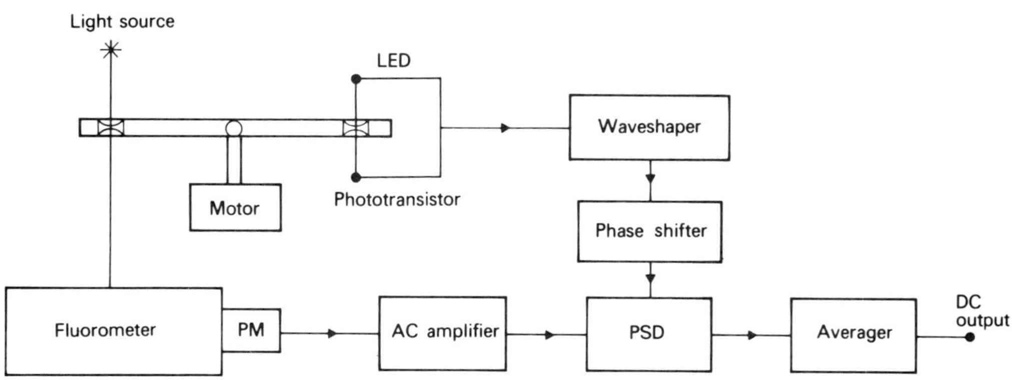

Fig. 10.5 A typical signal recovery device using phase-sensitive detection technique

In (a) and (b) the outputs of the averager will be equal in magnitude but opposite in sign. In (c) and (d) the outputs will be zero. An expression for the cross-correlation function similar to equation 10.59 may be derived by replacing the cosine waveforms by square wave ones in equations 10.57 and 10.58.

If the averager consists of a single-stage low-pass RC filter with a transfer function $1/(1+j\omega RC)$, its noise equivalent bandwidth can be calculated from equation 10.51 as

$$B_n = \int_0^\infty \frac{d\omega}{1+\omega^2 C^2 R^2} = \frac{\pi}{2RC} = \frac{1}{4RC} \quad \text{Hz} \tag{10.61}$$

As in equation 10.53, if the signal-to-noise ratio at the input is

$$\left(\frac{S}{N}\right)_{in} = 10\log_{10}\left(\frac{E_s{}^2}{2S_n(\omega)\omega_{max}}\right)$$

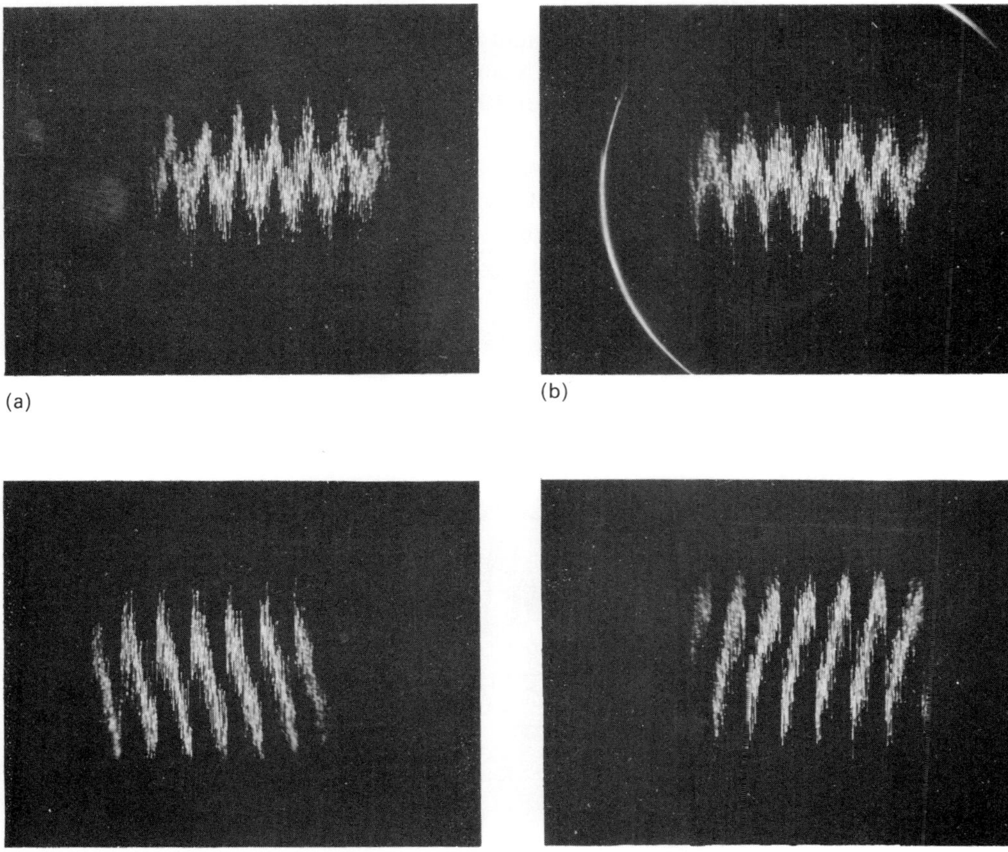

(a) (b) (c) (d)

Fig. 10.6 Output waveforms of the PSD with a square wave reference signal: (a) phase difference $\phi = 0$; (b) $\phi = 180°$; (c) $\phi = 90°$; (d) 270°

where $S_n(\omega)$ is the energy density function of the band-limited white noise $n(t)$, up to a maximum frequency ω_{max} and zero above this value; E_s is the signal amplitude.

At the output, since both the sum and difference frequencies will be present in the bandwidth B_n, the energy density of the noise will be $2S_n(\omega)$. The signal-to-noise power ratio at the output, assuming optimum phase relations between the signal and reference, will be

$$\left(\frac{S}{N}\right)_{out} = 10\log_{10}\left(\frac{E_s^{\,2}}{4S_n(\omega)B_n}\right)$$

$$= 10\log_{10}\left(\frac{E_s^{\,2}RC}{2\pi S_n(\omega)}\right) \tag{10.62}$$

The improvement in the signal-to-noise ratio is

$$\left(\frac{S}{N}\right)_{out} - \left(\frac{S}{N}\right)_{in} = 10\log_{10}\left(\frac{RC}{\pi}\right)\omega_{max} \tag{10.63}$$

In the above example, with $\omega_{max} = 2\pi \times 10^4$ rad s^{-1} for $RC = 0.5$ s, an improvement of 40 dB will be obtained. However, if $S_n(\omega)$ is constant over a wide frequency range, this method is disadvantageous because if the noisy signal is connected directly to a low-pass filter, the noise power at the output will be $S_n(\omega)B_n$ (equation 10.54), whereas in the case of the PSD the noise power at the output is $2S_n(\omega)B_n$ (equation 10.62). Thus the modulation/demodulation scheme yields better results in situations where the noise arises at low frequencies, e.g. in mechanical vibration, thermal drift, transistor amplifiers, etc. Figure 10.7 shows a recording of low-frequency noise from a fluxgate magnetometer exposed to fluctuations in the ambient magnetic field.

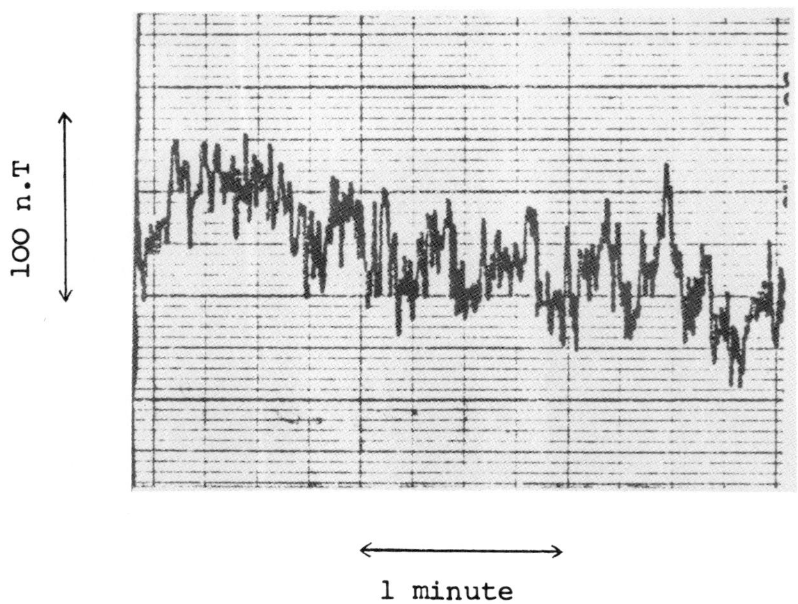

Fig. 10.7 A recording of low-frequency noise from a flux gate magnetometer exposed to fluctuations in the ambient magnetic field

Also, when a square wave instead of a sine wave is used as a reference, noise centred around frequencies equal to the odd harmonics of the reference will appear at the output. The noisy signal must therefore be modulated or chopped at such a frequency that its odd harmonics do not coincide with the frequency of the mains or that of any other interference. In experiments where the modulation frequency cannot be readily altered, considerable improvement in the signal-to-noise ratio can be achieved by including a 'coherent filter' in the signal input channel of the PSD.

Commercial instruments for measuring the amplitude of periodic signals buried in noise, incorporating a phase-sensitive detector, are known as **lock-in amplifiers**. The signal channel of the phase-sensitive detector contains a wide-band preamplifier, a band-pass filter that can be centred around the reference frequency, and a main amplifier. The reference channel contains waveshaping circuitry so that reference signals of any shape are acceptable. It also contains a phase shifter, so that components of the signal at the reference frequency, but at a different phase, can be found. A low-pass filter with a variable time constant is included in the output circuitry so that a suitable integrating time can be selected.

10.6 Signal recovery by autocorrelation

The autocorrelation function is a unique characteristic of a waveform, because it ignores all phase information and deals only with the power spectrum. In cases where the frequency of the signal to be recovered is not known, or when no reference signal is available, autocorrelation provides a means of detecting the periodicity of the signal. Figure 10.8 shows a scheme for obtaining an autocorrelogram of a noisy signal. The delay is generated by a magnetic tape loop and varied by changing the distance between the heads. If the noisy signal is represented as $E_s \cos(\omega t) + n(t)$ then the output of the autocorrelator may be written as

$$R_o(\tau) = \frac{1}{2T}\int_{-T}^{T} [E_s \cos\omega(t-\tau) + n(t-\tau)] \times [E_s \cos(\omega t) + n(t)]\,dt \tag{10.64}$$

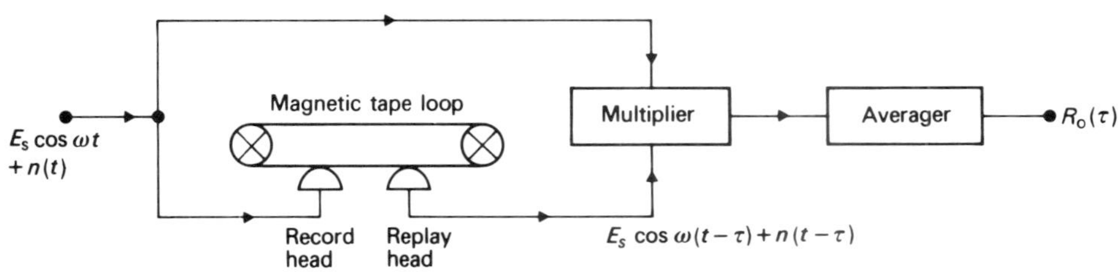

Fig. 10.8 A scheme for obtaining the autocorrelogram of a noisy signal

Expanding and rearranging,

$$R_o(\tau) = R_{ss}(\tau) + 2R_{sn}(\tau) + R_{nn}(\tau) \tag{10.65}$$

The cross-correlation function $R_{sn}(\tau)$ of the periodic component of the signal with noise component is zero for all τ. Hence,

$$R_o(\tau) = R_{ss}(\tau) + R_{nn}(\tau) \tag{10.66}$$

Figure 10.9 shows the autocorrelation function of the signal superimposed on the

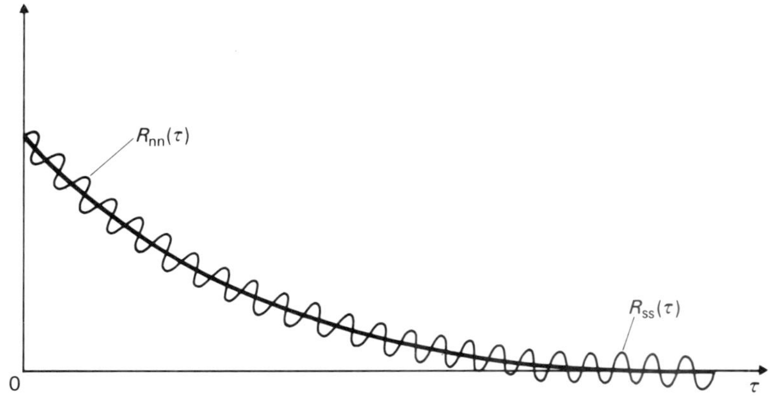

Fig. 10.9 Autocorrelation function $R_{ss}(\tau)$ of a signal superimposed on the autocorrelation function of noise, $R_{nn}(\tau)$

autocorrelation function of the noise. The decay time of the exponential depends on the bandwidth of the noise.

In practice, once the frequency of the periodic component is found, a local oscillator is phase-locked to it. The output of this oscillator is then used as reference for cross-correlation. This technique eliminates the noise correlation function $R_{nn}(\tau)$ from the output, and a shorter integration time can be used.

10.7 Multipoint correlation by digital techniques

The analogue correlator of Fig. 10.7 provides a simple means of constructing a correlogram. However, it has two main disadvantages:

1) it is difficult to generate long delays;
2) it takes a long time to build up a correlogram because the integrals of each value of the delay are computed serially.

For example, if discrete delay intervals are generated by stepping along the playback head, and if the averaging time is 1 s, then it would take 100 s to build up a 100-point correlogram. The digital multipoint correlator overcomes these difficulties by computing simultaneously the correlation functions of a large number of discrete delay points, each held in a separate 'channel' or digital memory location. A digital shift register driven by a crystal clock is often used for producing accurate delays. Each position of the shift register controls a channel in which the product of the instantaneous value of the reference and the delayed value of the noisy signal is held. The products are summed once every sampling cycle or 'sweep'. The continuous integration process is replaced by a summation over a large number of successive samples which are statistically independent.

Let $f(t, \tau)$ denote the product of one sample of the signal delayed by τ and the reference. As before, if the signal is represented by $E_s \cos(\omega t) + n(t)$ and the reference by $E_r \cos(\omega t + \phi)$ then

$$f(t, \tau) = [E_s \cos \omega(t - \tau) + n(t - \tau)] \times [E_r \cos(\omega t + \phi)] \tag{10.67}$$

If the variance of the output is σ_0^2 then

$$\sigma_0^2 = \sigma_s^2/N \quad \text{where } \sigma_s^2 = \overline{f^2(t, \tau)} - (\overline{f(t, \tau)})^2 \tag{10.68}$$

where N is the number of samples averaged. As in equation 10.23,

$$\overline{f(t, \tau)} = R_{sr}(\tau) + R_{nr}(\tau)$$

and since

$$R_{nr}(\tau) = 0 \quad \text{and} \quad R_{sr}(\tau) = \frac{E_r E_s}{2} \cos(\omega\tau + \phi)$$

$$\overline{f(t, \tau)} = \frac{E_r E_s}{2} \cos(\omega\tau + \phi) \tag{10.69}$$

Therefore

$$(\overline{f(t, \tau)})^2 = \frac{E_r^2 E_s^2}{8} [1 + \cos 2(\omega\tau + \phi)] \tag{10.70}$$

Let the mean squared value of the noise be $\overline{n^2(t)} = \sigma_n^2$ and its mean value $\overline{n(t)} = 0$.

$$\overline{f^2(t,\tau)} = \frac{E_r^2 E_s^2}{4}\left[\tfrac{1}{2}\{\cos 2(\omega\tau + \phi)\} + 1\right] + \frac{\sigma_n^2 E_r^2}{2} \tag{10.71}$$

From equations 10.68, 10.70 and 10.71,

$$\sigma_s^2 = \frac{E_r^2}{2}\left(\frac{E_s^2}{4} + \sigma_n^2\right) \tag{10.72}$$

From equation 10.70, the rms value of the output in the absence of noise is given by $E_r E_s / 2\sqrt{2}$. Hence the signal-to-noise ratio at the output is, from equations 10.68, 10.70 and 10.72,

$$\left(\frac{S}{N}\right)_{\text{out}} = 10\log_{10}\left(\frac{N E_s^2}{E_s^2 + 4\sigma_n^2}\right) \tag{10.73}$$

where N is the number of samples averaged.

The signal-to-noise ratio at the input is

$$\left(\frac{S}{N}\right)_{\text{in}} = 10\log_{10}\left(\frac{E_s^2}{2\sigma_n^2}\right) = 10\log_{10}\left(\frac{1}{\rho^2}\right) \tag{10.74}$$

where ρ, the noise-to-signal ratio, is given by $(\sqrt{2}/E_s)\sigma_n$.

The improvement in the signal-to-noise ratio is

$$\left(\frac{S}{N}\right)_{\text{out}} - \left(\frac{S}{N}\right)_{\text{in}} = 10\log_{10}\left(\frac{N\rho^2}{1 + 2\rho^2}\right) \tag{10.75}$$

When ρ is large, the improvement in the signal-to-noise ratio tends to $10\log_{10}(N/2)$.

10.8 Waveshape recovery

Repetitive signals of complex waveshape can be recovered from noise provided that the correlator responds to each of its harmonics without phase or amplitude distortion. This may be achieved by using the following **delta-function correlation technique**.

The δ-functions are represented by a train of unit impulses which repeat with a basic frequency of $1/T$. The correlation function may be written as

$$R_{s\delta}(\tau) = \frac{1}{T}\int_{-T/2}^{T/2} \delta(t)\, s(t+\tau)\, dt \tag{10.76}$$

$$= s(\tau)/T$$

because

$$\int_{-T/2}^{T/2} \delta(t)\, dt = 1 \qquad \text{when } t = 0$$

$$= 0 \qquad \text{when } t \neq 0$$

The correlation function thus reproduces the waveform $s(t)$. Physically, since $s(t+\tau)$, i.e. the value of $s(t)$ at a time τ later, is not accessible, the δ-function train is delayed by a time τ relative to the start. The δ-functions sample the waveform after this delay, yielding the value $s(\tau)$ at time τ.

If the signal contains random noise $n(t)$, then

$$R_{s\delta}(\tau) = \frac{1}{T}\int_{-T/2}^{T/2} \delta(t)[s(t+\tau)+n(t+\tau)]\,dt \tag{10.77}$$

Since $s(t)$ and $\delta(t)$ have the same period,

$$R_{s\delta}(\tau) = \frac{1}{2T}\int_{-T}^{T} \delta(t)\,s(t+\tau)\,dt + \frac{1}{2T}\int_{-T}^{T} \delta(t)\,n(t+\tau)\,dt$$

Therefore

$$R_{s\delta}(\tau) = \frac{s(\tau)}{T} + \frac{\overline{n(t)}}{T} \tag{10.78}$$

But $\overline{n(t)} = 0$, and if T is large enough for the successive samples to be uncorrelated then the variance of the output will depend on the mean square noise fluctuation $\overline{n^2(t)} = \sigma_n^{\,2}$ provided that $s(t)$ does not vary. Thus

$$\sigma_o^{\,2} = \sigma_n^{\,2}/N$$

where N is the number of samples averaged. If E_s is the rms amplitude of the input signal, then the output signal-to-noise ratio is given by

$$\left(\frac{S}{N}\right)_{\text{out}} = 10\log_{10}\left(\frac{E_s^{\,2}N}{\sigma_n^{\,2}}\right) \tag{10.79}$$

The input signal-to-noise ratio is given by

$$\left(\frac{S}{N}\right)_{\text{in}} = 10\log_{10}\left(\frac{E_s^{\,2}}{\sigma_n^{\,2}}\right) \tag{10.80}$$

The improvement in the signal-to-noise ratio is given by

$$\left(\frac{S}{N}\right)_{\text{out}} - \left(\frac{S}{N}\right)_{\text{in}} = 10\log_{10} N \tag{10.81}$$

which is independent of the input signal-to-noise ratio under all conditions.

The δ correlation process for the recovery of the waveshape of repetitive signals buried in noise is used in several commercially available instruments such as the multichannel averager and the 'box-car' detector.

10.9 Multichannel averager

The physical principle of the multichannel averager with capacitors as storage elements is illustrated in Fig. 10.10. The storage capacitors are connected one at a time to the signal line via linear gates driven by a shift register. The clock of the shift register is triggered by the reference which is synchronous with the signal. During a sequence, each capacitor charges to the mean value of a segment of the signal waveform. At the end of the sequence, the shift register is reset and the process repeated at every trigger from the reference. On account of the charge leakage from the capacitors, this storage technique is not suitable for situations where the observation time is long. In digital summation averages, each sample is digitised and

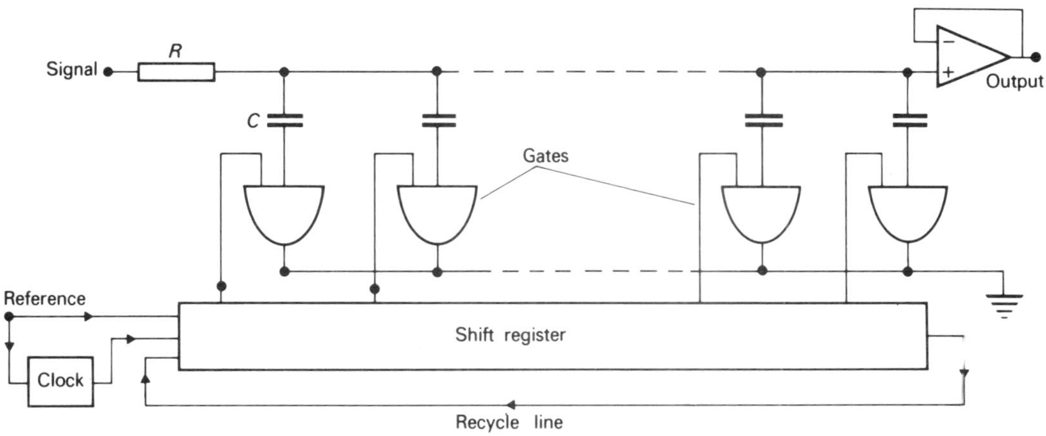

Fig. 10.10 Physical principle of a multichannel averager

corresponding samples are added and stored in a digital memory location. This technique tends to be expensive as it requires large memories for accumulation. A more recent technique utilises the mathematical definition of an average written as

$$\overline{X}_n = \overline{X}_{n-1} + \frac{X_n - \overline{X}_{n-1}}{n} \tag{10.82}$$

where n is the sweep number; X_n is the value for the nth sweep; and $\overline{X}_n$ is the average value after the nth sweep. Thus the value of the present average $\overline{X}_n$ is obtained from the differences of the present value X_n and the average value of the previous sweep. The shape and the true amplitude of the signal is therefore known at all times.

In general, the time resolution of the waveform is determined by the number of channels. Thus, if 1000 channels are used, 1000 samples of the waveform are taken at each repetition or **sweep**; for summation averaging the improvement in the signal-to-noise power ratio increases with the number of sweeps (equation 10.81). Figure 10.11(a) shows a noisy signal at the input of an averager and Fig. 10.11(b) the output after 256 sweeps.

According to the sampling theorem, the maximum frequency that a multipoint averager

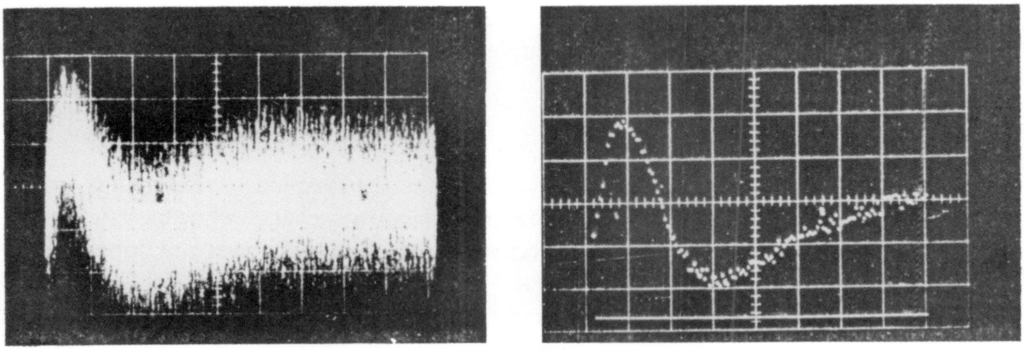

(a) (b)

Fig. 10.11 (a) A noisy signal at the input of an averager; (b) the output of the averager after 256 sweeps can display without ambiguity is $1/2T$, where T is the time between two successive samples in a single sweep. However, t_g, the time during which the signal is sampled, is very much less than T and so the averager will respond to frequencies higher than those allowed by the sampling theorem. Prefiltering must, therefore, be used to band-limit the noisy input.

10.10 The 'box-car' detector

Unlike the multichannel averager, the box-car detector samples and averages one point on the waveform at a time and then moves on to the next, so that a single storage capacitor is needed, Fig. 10.12.

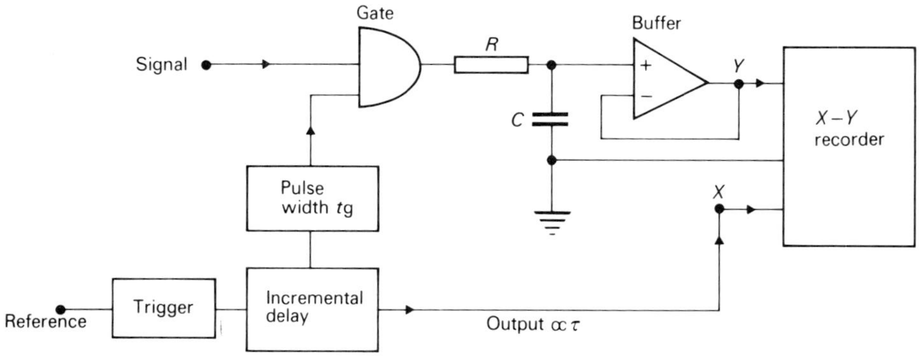

Fig. 10.12 Block diagram of a 'box-car' detector

The delay is changed at a predetermined rate by the incremental delay unit which simultaneously drives the X channel of an X– Y recorder proportional to the value of the delay τ.

Compared to the multichannel averager with n capacitor-gate units, the box-car will take n times as long to plot a waveform with equal noise reduction. Also, the instrument cannot resolve changes in the input signal which take place in a time interval shorter than t_g, the interval for which the gate is open. Hence, its upper bandwidth limit is $\omega_g = 2\pi/t_g$.

The box-car detector uses exponential averaging by an RC integrator to which a succession of sampled voltage levels are applied. If the averaging time is long enough to allow the output voltage to stabilise to the mean value of the input, then it can be shown using equation 10.61 and 10.68, that the effective number of samples after integration is given by

$$n = \frac{2RC}{\pi}\omega_s \tag{10.83}$$

where $\omega_s = 2\pi f_s = 2\pi/T$ is the equivalent bandwidth of the sampled waveform and f_s is the sampling frequency. Since the voltage on the capacitor can only change while the gate is open, for t_g seconds out of every T seconds, the effective value of RC is obtained by multiplying it by T/t_g. Hence,

$$n = \frac{2RC}{\pi}\frac{T}{t_g}\omega_s \tag{10.84}$$

Let the input noise bandwidth be greater than $\omega_g = 2\pi/t_g$ and let the energy density function of the noise $S_n(\omega)$ be zero above ω_{max}. The input signal-to-noise power ratio, as in equation 10.53, is

$$\left(\frac{S}{N}\right)_{in} = 10\log_{10}\left(\frac{E_s^2}{2S_n(\omega)\omega_{max}}\right)$$

and the output signal-to-noise ratio is given by equation 10.54,

$$\left(\frac{S}{N}\right)_{out} = 10\log_{10}\left(\frac{E_s^2 n}{2S_n(\omega)\omega_g}\right) \tag{10.85}$$

The improvement in the signal-to-noise ratio is thus

$$\left(\frac{S}{N}\right)_{out} - \left(\frac{S}{N}\right)_{in} = 10\log_{10}\left(\frac{n\omega_{max}}{\omega_g}\right) \tag{10.86}$$

Substituting for n from equation 10.84 and $\omega_g = 2\pi/t_g$,

$$\left(\frac{S}{N}\right)_{out} - \left(\frac{S}{N}\right)_{in} = 10\log_{10}\left(\frac{2RC}{\pi}\omega_{max}\right) \tag{10.87}$$

A similar expression is obtained for the phase-sensitive detector equation 10.63. As t_g is increased, the reduction in the value of n, equation 10.84, is compensated by the reduction in ω_g, equation 10.86, the upper bandwidth limit of the instrument.

When multiple samples per sweep are taken, as in the multichannel averager, an improvement in the signal-to-noise ratio is obtained in a wide variety of noise spectra, including the $1/f$ noise. Whereas in the case of the box-car detector where a single waveform sample is taken per sweep, averaging becomes less effective as the low-frequency noise components increase and gives no improvement for the $1/f$ noise. However, the time resolution of the box-car detector can be increased by increasing the sweep time up to a limit set by the width of the sampling pulse, in situations where the observation time is long.

10.11 Computer-aided signal recovery

It has been pointed out on several occasions that a waveform which is a function of time has its frequency spectrum represented by a Fourier transform pair. The primary characteristic of the transform that makes it so useful is related to the inverse nature of the transform variables. For example, a fine structure in the time domain becomes a gross structure in the frequency domain, so that features which might otherwise be hidden are disclosed. However, the conversion from the time to the frequency domain and vice versa involves a lot of tedious arithmetic, for which the modern digital computers are ideal. Hence, wherever possible, digital computer methods are preferred to analogue ones because of their inherent flexibility and the precision achieved by suitable programming. In most cases, the analogue signal is sampled at equally-spaced time intervals that are sufficiently small to present to the computer an adequate digital version of the signal (see Chapter 8).

Each sample may be looked upon as a weighted δ-function, so that the time function $f(t)$, represented in Fig. 10.13, can be described by the expression

$$f(t) = x_0\delta(t) + x_1\delta(t-T) + x_2\delta(t-2T) + x_3\delta(t-3T)\ldots \tag{10.88}$$

where T is the sampling interval.

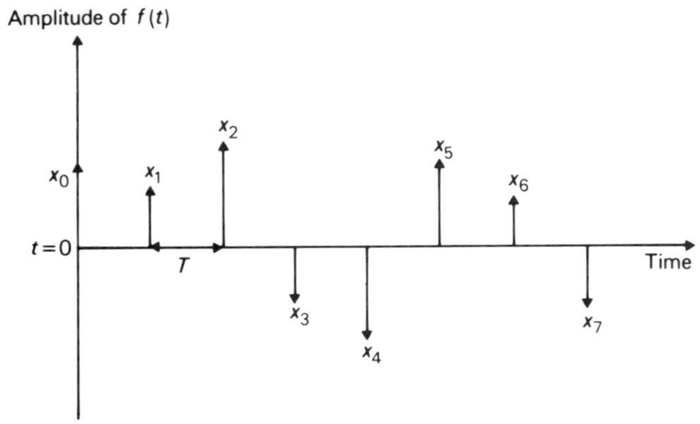

Fig. 10.13 A time function $f(t)$ represented by weighted δ-function samples

The Fourier transform of this series can be written as

$$F(\mathrm{j}\omega) = x_0 + x_1 \mathrm{e}^{-\mathrm{j}\omega T} + x_2 \mathrm{e}^{-\mathrm{j}\omega 2T} + x_3 \mathrm{e}^{-\mathrm{j}\omega 3T} + \ldots \tag{10.89}$$

$$= x_0 + (x_1 \cos \omega T + x_2 \cos 2\omega T + x_3 \cos 3\omega T + \ldots)$$
$$- \mathrm{j}(x_1 \sin \omega T + x_2 \sin 2\omega T + x_3 \sin 3\omega T + \ldots) \tag{10.90}$$

This series is known as the **Discrete Fourier Transform** (DFT), mainly because a digital computer estimates the Fourier transform for a set of discrete values of ω. For example, if the sampled signal is an even function of time, then it will have a purely real spectrum, and it is only necessary to compute the cosine terms of equation 10.90.

The value of $F(\mathrm{j}\omega)$ corresponding to the frequency of the kth harmonic may be denoted as

$$F_k(\mathrm{j}\omega) = \sum_{n=0}^{N-1} x_n \mathrm{e}^{-\mathrm{j}(2\pi kn)/N} \tag{10.91}$$

If the entire set of samples x_0 to x_{N-1} is used to calculate each harmonic term $F_k(\mathrm{j}\omega)$, the product $x_k \mathrm{e}^{-\mathrm{j}(2\pi kn)/N}$ is formed many times for different values of n and k. The **Fast Fourier Transform** (FFT) minimises the repeated calculation of products (see ref. (4), p. 233) and thereby achieves a reduction in computing time. Roughly speaking, the computation time for the evaluation of an N-sample transform using the FFT is proportional to $N \log_2 N$ as opposed to N^2 using the DFT of equation 10.90.

The Fast Fourier algorithm is most suitable when the number of signal samples is very large; the tedious correlation integrals involved in most signal recovery mechanisms are far simpler to evaluate using this technique. For example, to cross-correlate two functions of time, $x(t)$ and $y(t)$, the FFT algorithm is used to transform them to the complex frequency domain. The frequency-domain signals are conjugated, and the complex product $X^*(\omega)Y^*(\omega) = R(\omega)$ is formed. The cross-correlation function $R(\tau)$ is obtained by an inverse transformation of $R(\omega)$ to the time domain using the same FFT algorithm. This simple approach is not without pitfalls[1] and without adequate precautions, a periodicity that bears no relationship to the period of the signal is introduced in the cross-correlation function. Briefly, the effect can be attributed to the fact that the FFT assumes a period T, the 'sampling window' that contains chunks of the waveform to be transformed. For example, when sampling a periodic function with a narrow window T, if T contains a non-integral number of

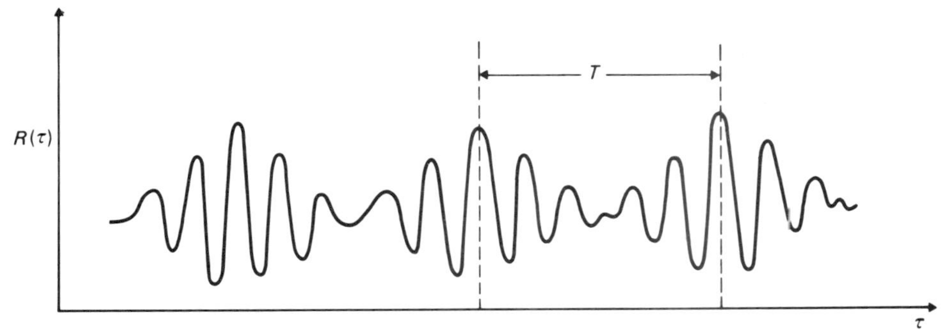

Fig. 10.14 Periodicity in a computed autocorrelation function when a periodic function is sampled with a narrow 'window' T containing a non-integral number of cycles of the waveform

cycles of the waveform then the computed autocorrelation function $R(\tau)$ will show a periodicity T, Fig. 10.14. A maximum is produced when the lag is equal to $T/2$. Details of digital processing of signals may be found in other texts.[2]

References

(1) Ramirez, Robert, Fast Fourier Transform makes correlation simpler. *Electronics* pp. 98–103 (June 26 1975)
(2) Gold, B., and Rader, C., *Digital Processing of Signals.* McGraw-Hill (1969)

Further reading

Blackman, R. B., and Tukey, J. W., *The Measurement of Power Spectra.* Dover (1958)
Oliver, B. M., and Cage, J. M., *Electronic Measurements and Instrumentation.* McGraw-Hill (1971)
Ortec/Brookdeal Electronics, *Your Guide to Lock-In Amplifiers.* (1974)
Whalen, A. D., *Detection of Signals in Noise.* Academic Press (1971)

Index